CW00631871

Peritoneal Dialysis – From Basic Concepts to Clinical Excellence

Contributions to Nephrology

Vol. 163

Series Editor

Claudio Ronco *Vicenza*

Peritoneal Dialysis – From Basic Concepts to Clinical Excellence

Volume Editors

Claudio Ronco *Vicenza*
Carlo Crepaldi *Vicenza*
Dinna N. Cruz *Vicenza*

22 figures and 32 tables, 2009

Basel · Freiburg · Paris · London · New York · Bangalore · Bangkok · Shanghai · Singapore · Tokyo · Sydney

Contributions to Nephrology
(Founded 1975 by Geoffrey M. Berlyne)

Claudio Ronco
Department of Nephrology
Viale Rodolfi 37
IT-36100 Vicenza (Italy)

Carlo Crepaldi
Department of Nephrology
Viale Rodolfi 37
IT-36100 Vicenza (Italy)

Dinna N. Cruz
Department of Nephrology
Viale Rodolfi 37
IT-36100 Vicenza (Italy)

Library of Congress Cataloging-in-Publication Data

Peritoneal dialysis : from basic concepts to clinical excellence / volume editors, Claudio Ronco, Carlo Crepaldi, Dinna N. Cruz.
p. ; cm. – (Contributions to nephrology ; v. 163)
Includes bibliographical references and indexes.
ISBN 978–3–8055–9202–4 (hard cover : alk. paper)
1. Peritoneal dialysis. I. Ronco, C. (Claudio), 1951– II. Crepaldi, Carlo. III. Cruz, Dinna N.
[DNLM: 1. Peritoneal Dialysis. W1 CO778UN v. 163 2009 / WJ 378 P4464 2009]
RC901.7.P48P4745 2009
617.4' 61059–dc22 2009018347

Bibliographic Indices. This publication is listed in bibliographic services, including Current Contents® and Index Medicus.

Drug Dosage. The authors and the publisher have exerted every effort to ensure that drug selection and dosage set forth in this text are in accord with current recommendations and practice at the time of publication. However, in view of ongoing research, changes in government regulations, and the constant flow of information relating to drug therapy and drug reactions, the reader is urged to check the package insert for each drug for any change in indications and dosage and for added warnings and precautions. This is particularly important when the recommended agent is a new and/or infrequently employed drug.

www.karger.com
Printed in Switzerland on acid-free and non-aging paper (ISO 9706) by Reinhardt Druck, Basel
ISSN 0302–5144
ISBN 978–3–8055–9202–4
e-ISBN 978–3–8055–9203–1

Contents

Preface

After more than 25 years the International Course on Peritoneal Dialysis of Vicenza has become a classic appointment for beginners and experts in the field of peritoneal dialysis. In 2009, the 18th edition is to be held once more in Vicenza with the classic format including basic principles as well as advanced studies and cutting-edge experimental reports. All these pieces of knowledge are included in this volume that, according to our tradition, is published in time for the conference.

Peritoneal dialysis represents more than ever an important approach to the therapy of chronic kidney disease. Its integration with other techniques in the therapy of uremia represents an important step in the optimization of the whole program of renal replacement therapy. Today's clinical practice and basic research in the field of peritoneal dialysis seem to proceed in parallel, leading to a true translational research program for peritoneal dialysis. Clinical results from multicenter studies and physiological findings from experimental studies both contribute to a significant progress in the knowledge and to the safe and adequate application of peritoneal dialysis. All these aspects are discussed by leading experts in the field and their contributions in this volume will represent an important tool for the reader to become familiar with today's art of peritoneal dialysis. Our thanks go to the authors for their efforts to convey their knowledge in a concise but complete text, and to Karger publishers for the efficient and competent publication of the present volume. This book, in conjunction with the previous volumes published in the series *Contribution to Nephrology*, represents an important source of information for beginners and experts, basic scientists and clinical physicians, and students and investigators who want to have a true update on the current research and clinical practice in peritoneal dialysis.

Claudio Ronco, Vicenza
Carlo Crepaldi, Vicenza
Dinna N. Cruz, Vicenza

Ronco C, Crepaldi C, Cruz DN (eds): Peritoneal Dialysis – From Basic Concepts to Clinical Excellence. Contrib Nephrol. Basel, Karger, 2009, vol 163, pp 1–6

Application of Body Composition Monitoring to Peritoneal Dialysis Patients

Carlo Crepaldi[a], *Sachin Soni*[a], *Chang Yin Chionh*[a], *Peter Wabel*[b], *Dinna N. Cruz*[a], *Claudio Ronco*[a]

[a]Department of Nephrology, St. Bortolo Hospital, Vicenza, Italy; [b]Fresenius Medical Care, Bad Homburg, Germany

Abstract

Assessment of body fluids in peritoneal dialysis is an important issue in the treatment of renal failure. Overhydration is related with hypertension and left ventricular hypertrophy and dehydration leads to hypotension and reduction of residual renal function. Bioimpedance analysis (BIA) provides objective information in assessment of hydration status of the patients. In the past BIA was not widely used in patients undergoing peritoneal dialysis. Our aim was to estimate the status of hydration in our peritoneal dialysis population by body composition monitoring (BCM) device to modify our pharmacological and dialysis policy. We used a Fresenius Body Composition Monitor, a whole-body bioimpedance spectroscopy (50 frequencies, 5–1,000 kHz), to assess the body composition of 97 patients on peritoneal dialysis in our center. The patients were subjected to a physical examination every three months: We measured body weight, 24 h diuresis and performed a BIA session. BIA measurements were repeated according to different clinical situations. Every patient underwent BIA at least on two different occasions. Our preliminary results have found a strict correlation between weight increase or decrease and the results (total body water, extracellular water, lean mass index) shown by BCM. Modifications of therapy in patients dehydrated restored a satisfying amount of diuresis. Hypertensive overhydrated patients changed their scheduled treatment improving their blood pressure and achieving a lower body weight. Bio impedance analysis is the most reliable, repetitive, not invasive, simple, portable and relatively inexpensive technique to assess the fluid status of a dialysis patient is bioimpedance.

Bioimpedance and Dialysis

Fluid overload is not uncommon in patients on chronic peritoneal dialysis (PD). Previous studies suggested that residual renal function (RRF) plays an important role in maintaining fluid balance. However, good fluid status should be a balance between fluid intake and removal [1].

Residual renal function (RRF) is of paramount importance in patients with end-stage renal disease, with benefits that go beyond contributing to achievement of adequacy targets. Several studies have found that RRF rather than overall adequacy (as estimated from total small solute removal rates) is an essential marker of patient and, to a lesser extent, technique survival during chronic PD therapy.

In addition, RRF is associated with a reduction in blood pressure and left ventricular hypertrophy, increased sodium removal and improved fluid status, lower serum beta-2-microglobulin, phosphate and uric acid levels, higher serum hemoglobin and bicarbonate levels, better nutritional status, a better lipid profile, decreased circulating inflammatory markers, and lower risk for peritonitis in PD.

As compared with conventional hemodialysis, PD is associated with a slower decrease in RRF. This highlights the usefulness of strategies oriented to preserve both RRF and the long-term viability of the peritoneal membrane [2].

Fluid status is an important adequacy issue in the treatment of renal failure, but it remains a difficult task to establish reliable end points indicative of normovolemia during routine clinical practice. There is little doubt in the community regarding the clinical benefit of optimal fluid status or normovolemia. States of hypervolemia are regarded to be the most important factor predisposing to hypertension in dialysis patients. Furthermore the long-term consequences of overhydration are well recognized [3].

On the other hand dehydration exposes the dialysis patients to hypotension and reduction of urine output. Dehydration may come from intercurrent illnesses like diarrhea, fever or large use of furosemide and of highly hypertonic solutions for PD. These considerations suggest the need of a reliable method for measure fluid status in dialysis patients.

Several methods have been proposed for non clinical dry weight assessment, methods like evaluation of serum natriuretic peptides or measurement of inferior vena cava diameter. Unfortunately, these methods suffer from several shortcomings, such as poor specificity (natriuretic peptides) operator dependence (inferior vena cava diameter measurements) and poor correlation with extracellular volume (continuous blood volume measurement).

After years of bioimpedence analysis (BIA) research in dialysis patients, this technique is now increasingly being used clinically [4]. Whole-body bioimpedance spectroscopy (BIS) (50 frequencies, 5–1,000 kHz) in combination with a physiologic tissue model can provide an objective target for normohydration [5].

The potential of BIA as a technique of body composition analysis is even greater when one considers that body water can be used as a surrogate measure of lean body mass [6].

Single Frequency Bioimpedance Analysis and Multifrequency Bioimpedance Spectroscopy

The majority of bioimpedance devices on the market are based on single-frequency measurements. These devices are useful for healthy subjects with a normal hydration status. The oversimplified measurement technique limits the use in patients with pathological conditions. With the single frequency measurement a separation between extra- and intracellular space is not possible and thus these devices have to rely on empirical regression equations. Single frequency approaches are usually called BIA-methods (bioimpedance analysis).

Another method, the bioimpedance vector analysis (BIVA), has been proposed by Piccoli [Kidney Int 1994;46]. It is a qualitative single-frequency method which is useful for a first impression of the hydration status but does not provide a quantitative value for fluid overload or hydration status.

BIS devices measure the impedance, i.e. the electrical AC resistance of the body, with a frequency sweep from the low kilohertz to the megahertz range. Impedance consists of two components: the resistance responsible for dissipative losses, and reactance which accounts for capacitive losses. The well accepted rationale behind BIS measurements is that low frequency current cannot penetrate cell membranes and thus flows exclusively through the ECW space, while high-frequency current flows through both ECW and ICW. Therefore, only BIS-based devices have the potential to independently determine ECW and TBW by separating these two compartments [7].

Body Composition Monitor

The body composition monitor (BCM) uses whole body BIS and measures at 50 different frequencies between 5 kHz and 1 MHz. The raw impedance measurements are extrapolated to obtain the conditions expected at zero and infinite frequency by means of Cole modeling [8]. Additionally, Hanai mixture theory equations [9] are used for ECW and ICW determination which takes into account the changes in resistivity due to the suspension of cells in ECW.

The hydration state of the patient is assessed in a two phase physiologic model approach. The fluid model transfers the resistance at zero and infinite frequency into fluid volumes [10]. The body composition model [11] is based on well documented properties of normohydrated tissue. The hydration status can be calculated using the total water content of normohydrated lean tissue and normohydrated adipose tissue and the ratio between intracellular and extracellular fluid in the respective normohydrated tissues.

Validation

The BCM was validated extensively against all available gold-standard reference methods in more than 1,000 patients and healthy volunteers. These methods included indicator dilution methods (sodium bromide or deuterium) for the fluid volumes, X-ray methods (DEXA) for the body composition but also clinical assessment for fluid overload.

Target of Normohydration

The aim of the BCM is to provide the long-term target of normohydration. Fluid overload and possible dehydration are calculated by comparing the actual fluid status of the patient to the normohydration state. Using the target the clinical personnel has the chance to set up the most appropriate therapy for a long term achievement of a state of normohydration. Figure 1 shows two exemplary patients illustrating the concept of normohydration range. First results in hemodialysis patients suggest that it is possible to achieve this target and that this leads to an improvement of the patient condition. But it is yet not known if this target of normohydration is the most appropriate target for all chronic kidney disease patients taking into account different co-morbidities and treatment modalities.

Vicenza Experience with Body Composition Monitoring on a Large Population of Peritoneal Dialysis Patients

Our PD unit in Vicenza is one of the biggest in Italy. We have more than 100 patients on regular follow up and on an average four to five new cases are initiated on PD each month. The catheter insertion is performed by a nephrologist by the surgical method in an independent operation theater of the department. The unit has 4 qualified and experienced nephrologists and 6 trained

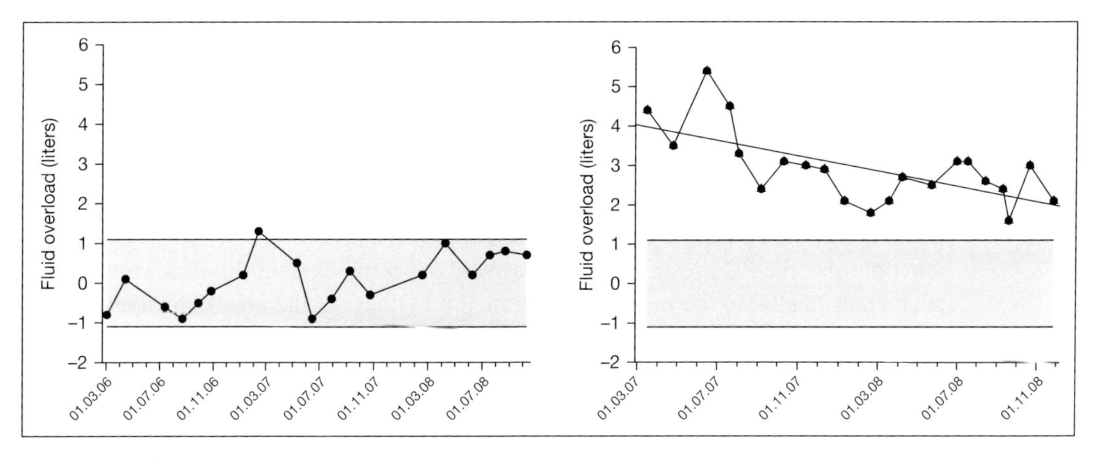

Fig. 1. Use of the normohydration range (gray area) as a target: two exemplary PD patients. Left: the patient was kept in the normohydration range (±1.1 liters) for the whole observation period. Right: this patient was slowly guided towards the normohydration range [unpubl. data provided by Dr. Helga Petrov, Berlin].

nurses. The unit performs initial assessment and counseling for the candidate patients for PD, placement of PD catheters, training of patients and/or families for PD and regular follow-up of the patients.

Over the period of last 6 months, we have assessed 97 patients by the Fresenius BCM device. The patients were subjected to a physical examination every 3 months: we measured body weight, 24-hour urine output and performed a BIS session in addition to the routine assessment of adequacy and nutritional parameters. BIS measurements were repeated if necessary. Every patient underwent BIS at least on two different occasions during the last 6 months.

The Fresenius BCM monitor has been specially designed for the patients with chronic kidney disease. It is a small, portable, battery operated, point of care device needing only couple of minutes for a single measurement.

The device measurement is stored in a chipcard which can be subsequently connected to a computer through a user interface. Each measurement gives detailed information about the body composition, status of hydration, amount of overhydration in litres, urea distribution volume, total, intracellular and extracellular water. The measurement also gives information about the nutritional status of the patient.

After the initial encouraging results with BCM device, we have started a prospective study on incident PD patients called IPOD-PD (Initiative for Preservation Of Diuresis in Peritoneal Dialysis). The aim of this study is to preserve residual renal function for longer period in PD patients, without overloading the patients with fluid.

The study protocol includes assessment of patient's fluid status using the BCM device. The measurements will be made at regular intervals – at the time of initiation of PD and at 1, 3, 6 and 12 months. Additional measurements will be made if required. The PD prescription will be tailored to get correct hydration status as assessed by BCM. We postulate that this objective assessment will help optimize the PD prescription and to avoid overzealous ultrafiltration which can be detrimental to the residual renal function. Other regular checks to assess adequacy of the therapy will also be carried out regularly.

References

1 Cheng LT, Chen W, Tang W, Wang T: Residual renal function and volume control in peritoneal dialysis patients. Nephron Clin Pract 2006;104:c47–c54.
2 Marrón B, Remón C, Pérez-Fontán M, Quirós P, Ortíz A: Benefits of preserving residual renal function in peritoneal dialysis. Kidney Int Suppl 2008;108:S42–S51.
3 Chamney PW, Krämer M, Rode C, Kleinekofort W, Wizemann V: A new technique for establishing dry weight in hemodialysis patients via whole body bioimpedance. Kidney Int 2002;61:2250–2258.
4 Kotanko P, Levin NW, Zhu F.: Current state of bioimpedance technologies in dialysis. Nephrol Dial Transplant 2008;23:808–12.
5 Wabel P, Chamney P, Moissl U, Jirka T: Importance of whole-body bioimpedance spectroscopy for the management of fluid balance. Blood Purif 2009;27:75–80.
6 Thomas BJ, Ward LC, Cornish BH: Bioimpedance spectrometry in the determination of body water compartments: accuracy and clinical significance. Appl Radiat Isot 1998;49:447–455.
7 Matthie JR: Bioimpedance measurements of human body composition: critical analysis and outlook. Expert Rev Med Devices 2008;5:239–261.
8 Cole KS, Cole RH: Dispersion and adsorption in dielectrics. J Chem Phys 1941;9:341–351.
9 Hanai T: Electrical Properties of Emulsions: Emulsion Science. London, Academic Press, 1968, pp 354–477.
10 Moissl UM, Wabel P, Chamney PW, et al: Body fluid volume determination via body composition spectroscopy in health and disease. Physiol Meas 2006;27:921–933.
11 Chamney PW, Wabel P, Moissl UM, et al: A whole-body model to distinguish excess fluid from the hydration of major body tissues. Am J Clin Nutr 2007;85:80–89.

Carlo Crepaldi, MD
Department of Nephrology, St. Bortolo Hospital
Viale Rodolfi
IT–36100 Vicenza (Italy)
E-Mail carlo.crepaldi@ulssvicenza.it

Ronco C, Crepaldi C, Cruz DN (eds): Peritoneal Dialysis – From Basic Concepts to Clinical Excellence. Contrib Nephrol. Basel, Karger, 2009, vol 163, pp 7–14

Peritoneal Ultrafiltration: Physiology and Failure

Michael F. Flessner

Department of Medicine, Division of Nephrology, University of Mississippi Medical Center, Jackson, Miss., USA

Abstract

Net ultrafiltration in peritoneal dialysis results from a complex set of forces within the tissue space surrounding the peritoneal cavity. Hydrostatic pressure due to the large volume of fluid drives water and solute into the surrounding tissue, and therefore a high osmotic pressure must be maintained in the cavity to draw fluid from blood capillaries distributed in the tissue adjacent to the peritoneum. The osmotic pressure in the interstitium decreases from that of the cavity to equilibration with the plasma in the first millimeter of tissue below the peritoneum. Osmotic pressure differences at the blood capillary produce a solute free ultrafiltrate via aquaporin 1 that is ~50% of the total filtration. The remainder of the fluid is filtered via inter-endothelial gaps lined with negatively charged glycocalyx, which alters the traditional Starling forces and is easily damaged by inflammation or ischemia. Ultrafiltration failure occurs when intraperitoneal pressure is too high, the inflamed peritoneum dissipates the osmotic agent rapidly because of hyperpermeable angiogenic vessels, or peritoneal scarring lowers the osmotic pressure near the blood supply and there is no force for fluid transport through the scar to the cavity. To remedy problems in net ultrafiltration, lowering the volume lowers the intraperitoneal pressure and often solves the problem of excessive pressure. Preventative measures to decrease inflammation and peritonitis are important for preservation of the barrier. Experimental measures such as peritoneal stem-cell transplants may someday permit reclamation of damaged barrier systems and allow patients to continue the dialytic technique.

The Normal Peritoneal Transport Barrier

Figure 1a illustrates the normal peritoneal barrier, which is made up of blood vessels distributed within a tissue cellular-interstitial matrix and covered

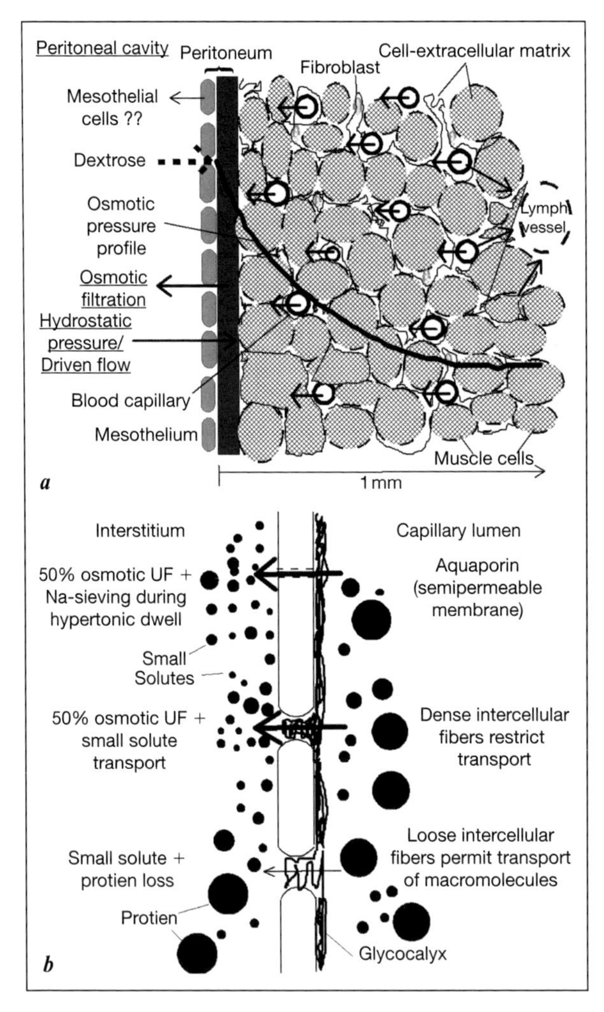

Fig. 1. ***a*** Distributed concept of normal ultrafiltration barrier. Dextrose diffuses from dialysate into tissue and sets up an osmotic pressure profile (thick curved line). Hydrostatic pressure force opposes the osmotic force. ***b*** Pore-matrix concept of endothelial barrier incorporating the luminal glycocalyx.

by a thin layer (<90 μm) of submesothelial connective tissue and a single layer of mesothelial cells, termed the anatomic peritoneum. Hypertonic solutions, used to remove fluid from the body of an anephric patient in peritoneal dialysis, typically contain small solutes such as amino acids or dextrose that diffuse into the tissue adjacent to the peritoneum and establish a decreasing concentration

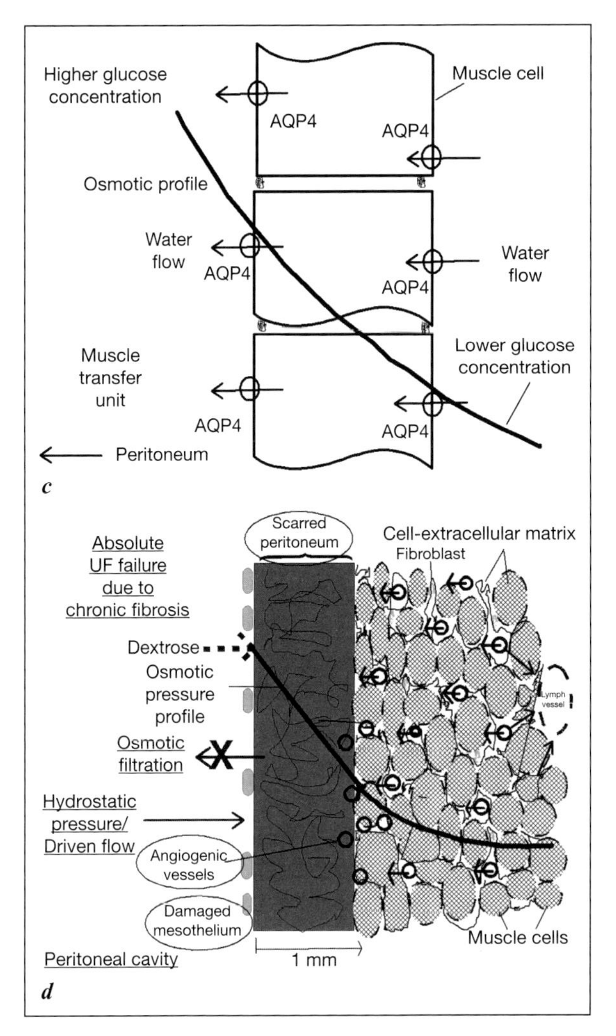

Fig. 1. ***c*** Muscle transfer unit that facilitates transfer of water from the tissue toward the peritoneal cavity. ***d*** Scarred peritoneum that results in marked decrease in osmotic pressure in vicinity of microvessel source of water.

and osmotic pressure profile in the tissue interstitium [1]. Blood vessels are the main source of ultrafiltration, and the water flow from the capillaries to the interstitium will depend on the difference between the capillary luminal pressures and the effective pressure on the interstitial side. The concentration profile results from the diffusion of the small solutes through the tissue interstitium and simultaneous uptake into the blood capillaries. The largest gradients of osmotic pressure will therefore be across blood vessels closest to the peritoneum, while blood vessels that are more than a millimeter away will have almost no role in the process [1].

The three potential barriers to both solute and water flow are the anatomic peritoneum, the cellular-interstitial matrix surrounding the blood vessels, and the capillary endothelium. Recent research has shown that the anatomic peritoneum, although important for the preservation of the overall barrier and for lubrication of the visceral-parietal tissues, has insignificant resistance to both solute diffusion and water flow in peritoneal dialysis [2].

Osmotic Filtration Barrier: Aquaporin and Inter-Endothelial Glycocalyx

Exciting work in the last 15 years has demonstrated that aquaporin 1, the water-only channel in the endothelial membrane (fig. 1b), is responsible for approximately half of the osmotic ultrafiltration from the plasma into the cavity. Definitive experiments in aquaporin I knockout mice showed that 40–50% of the net ultrafiltration due to hypertonic solutions was eliminated by the absence of aquaporin 1 [3, 4].

The other half of ultrafiltration from the blood to the interstitium occurs through interendothelial spaces filled with a luminal glycocalyx that restricts passage of solutes and sets up alternative pressure forces [5] from those calculated with the classic Starling equation: $F = (P_p - P_i) - \sigma(\pi_p - \pi_i)$, where F = sum of convective forces across the endothelium, P_p = capillary hydrostatic pressure, P_i = interstitial hydrostatic pressure, σ = solute reflection coefficient, π_p = plasma oncotic pressure, π_i = interstitial oncotic pressure. The glycocalyx adds an additional layer on the luminal side of the capillary as illustrated in figure 1b and alters the microenvironment near the true, size-selective boundary. Oncotic forces opposing filtration are larger than those that have been estimated from simple differences between the luminal plasma concentration and the concentration in the interstitium [6, 7]. The albumin concentration just below the glycocalyx but above the tight junction is likely much lower than the albumin concentration in the bulk interstitium, due to the inability of albumin to diffuse against the ultrafiltrate flow through the gap in the glycocalyx [8]. In the new form of Starling's equation, the interstitial pressures are replaced by P_g = hydrostatic pressure in the capillary gap, π_g = osmotic pressure in the region just below the glycocalyx: $F = (P_p - P_g) - \sigma(\pi_p - \pi_g)$ [8].

The mechanism is modeled by a pore-matrix theory [9] that incorporates an extracellular coating of anionic polysaccharides on the luminal surface of endothelium and explains many observations in microcirculatory physiology [10]. Perturbations of this layer by electrical stimulation, perfusion with tumor necrosis factor alpha, perfusion with adenosine, perfusion of oxidized low-density lipoproteins, and ischemia reperfusion injury all damage the glycocalix and result in release of larger proteins and albumin from the plasma into the interstitium [11].

Osmotic Filtration Barrier: Integrated Cellular-Interstitial Matrix

The integrated cellular-interstitial matrix makes up the other major part of the barrier. As illustrated in figure 1a, water flows from the lumen of the blood vessel to the tissue interstitium in accordance with the modified Starling law. However, there is no clear mechanism to transport water through the tissue toward the peritoneum. Diffusion of water is too slow to account for the typical flow rate (1 μl/min/cm^2) from mammalian tissue due to a solution with osmotic pressure of ~500 mosm/kg, and calculations have demonstrated that the nearest vessels do not possess the blood flow to support this flux [12]. The answer must lie in the properties of the integrated muscle cell-interstitial matrix.

A large percentage of the cellular structure underneath the is made up of muscle that is a potential network of osmotic transducers. Under normal conditions, these have minimal interstitium, and the cells are linked together by collagen fibers through β1-integrins. Figure 1b illustrates how these muscle cells as well as the blood capillaries are subjected to a decreasing concentration gradient of the osmotic agent. These relatively large cells will therefore be subject to differential forces on opposite sides of each cell. While there is no aquaporin 1 in muscle cells [3], all of these cells are populated by aquaporin 4, which allow cells to regulate their volume when confronted with osmotic stress [13]. The integrated barrier possibly functions as an osmotically powered pump, with water drawn through the side of the cell closest to the peritoneum causing the interior of the cell to increase its osmolality to above the interstitial osmolality at the far side of the cell (fig. 1c). With successive steps through different cells within the tissue, water could move toward the peritoneal cavity. While there is currently no hard data to support this, recent theoretical work by Waniewski et al. [14] has pointed to the fact that the cell-intersitial matrix has to have an ability to present an osmotic barrier.

Net Ultrafiltration: Effect of Intraperitoneal Hydrostatic Pressure

The intraperitoneal pressure may rise to 2–10 mm Hg, depending on the position of the patient and the volume instilled. This pressure exerts force across peritoneal tissue, and in particular, across the abdominal wall. Research has demonstrated that total fluid loss from the peritoneal cavity to the body occurs at 60–90 ml/h in humans [15], with 10–20 ml/h flowing into lymphatics and 50–80 ml/h flowing into the surrounding tissues, in particular, the abdominal wall and the diaphragm, in direct proportion to the intraperitoneal hydrostatic pressure [15]. Therefore, the osmotic flow into the cavity must overcome this 'wrong-way' flow back to the tissue (fig. 1a). The following equation illustrates the principle involved:

$$\textit{Net UF} = \frac{\textit{Volume collected} - \textit{volume instilled}}{\textit{duration of dialysis dwell}} = \textit{Osmotic filtration} - \textit{fluid loss}$$

where osmotic filtration = osmotic pressure-driven fluid flow from the tissue to the cavity, and fluid loss = fluid flow from the cavity to lymphatics and to subperitoneal tissue due to the elevated hydrostatic pressure in the cavity (2–12 mm Hg) relative to the tissue (typically –1 to 0 mm Hg) [15].

Ultrafiltration Failure and Treatment

The peritoneal cavity was not designed by our creator to be a dialysis device. Placing a foreign body (catheter) and instilling dextrose-based solutions 24 hours per day alters the physiology of the peritoneum and its underlying tissue. In some patients this leads to failure to remove sufficient fluid during dialysis. Absolute ultrafiltration (UF) failure is the condition of the patient in which volume collected is less than or equal to the volume instilled over a dwell of 2–4 h (despite using the highest dextrose concentration), and therefore the Net Ultrafiltration will be zero or negative, indicating fluid absorption. Relative UF failure can be defined by less than 400 ml of net ultrafiltration after a 4-hour PET with 4.25% dextrose solution. There are several possible causes for this failure, which will be illustrated by cases.

Fluid Loss Exceeds Osmotic Ultrafiltration due to High Intraperitoneal Pressure

A 47-year-old female with weight of 80 kg and height of 1.37 m, progressed to renal failure and was started on peritoneal dialysis, 2 liters, 4 times per day. From the beginning, she complained of fullness and pressure in her abdomen. The PET at 2 months demonstrated a high rate of dextrose transport and poor ultrafiltration. Her fluid volume was gradually decreased to 1.5

liters, which she could easily tolerate, and net UF rose to >500 ml per day with adequate solute clearance.

Rapid Loss of Small Osmotic Solute from Cavity Leading to Relative UF Failure

After 3 more years and 3 cases of bacterial peritonitis, the patient was unable to ultrafiltrate more than 400 ml, even with rapid cycling overnight. She was placed on one exchange of icodextrin during the long daytime dwell, and she remains on PD but restricts her total daily fluid intake to <1.5 liters.

Inflammatory changes in the peritoneum due to long-term exposure to bio-incompatible solutions or recurrent peritonitis may result in the progressive loss of ultrafiltration. Typical characteristics of this are rapid absorption of glucose from the peritoneal solution, loss of sodium-free water flux during the first hour of a hypertonic dwell, and a markedly diminished loss of net UF over a 4-hour PET test. Angiogenic vessels from inflammation may significantly increase the functional area for solute transfer and therefore increase the overall solute absorption from the tissue interstitium. They typically lack glycocalyx and therefore solute transport in both directions across the endothelium could be quite rapid.

At this time, corrective interventions are unproved, and efforts have focussed on prevention of inflammatory changes. Use of more biocompatible solutions, icodextrin or prophylactic drugs constitute therapy that can be used on these patients to slow the inflammatory change. Decreasing cases of peritonitis is an obvious way to prevent inflammation and scarring of the peritoneum. Detection and reduction of infected catheters and removal of biofilm with specialized solutions would be another therapy.

Loss of Net UF with the Use of a High-Molecular-Weight Starch Solution: Absolute UF Failure

A 59-year-old woman was treated with PD for 3 years. She developed a severe fungal peritonitis that required removal of her catheter. After replacement of the catheter, net UF was negative with dextrose solutions and icodextrin, and she was switched to hemodialysis.

Significant scarring of the peritoneum is often present after 6 or more years of continuous peritoneal dialysis. The peritoneum is typically an acellular, avascular layer of tissue, often as thick as 1 mm. Solute transport rapid across this avascular, acellular layer and uptake into abnormal blood capillaries is rapid.

However, with the loss of the cell interstitial matrix and the increase in the distance of the blood capillaries from the peritoneum, transport of water to the peritoneal cavity will be nearly zero. This is illustrated in figure 1d. Replacing the peritoneum with stem cells early in the course of the inflammatory-fibrotic process may someday be a potential therapy. However, a severely scarred peritoneum may not be susceptible to the binding of mesothelial cells; even if the mesothelial cells bind, they may be lying over a scarred tissue such as in figure 1d.

References

1 Flessner MF, Fenstermacher JD, Dedrick RL, Blasberg RG: A distributed model of peritoneal-plasma transport: tissue concentration gradients. Am J Physiol 1985;248:F425–F435.
2 Flessner MF, Henegar J, Bigler S, Genous L: Is the peritoneum a significant transport barrier in peritoneal dialysis? Perit Dial Int 2003;23:542–549.
3 Ni J, Verbavatz J-M, Rippe A, Boisde I, Moulin P, Rippe B, Verkman AS, Devuyst O: Aquaporin-1 plays and essential role in water permeability and ultrafiltration during peritoneal dialysis. Kidney Int 2006;69:1518–1525.
4 Yang B, Folkesson HG, Yang J, Mattick LR, Ma T, Verkman AS: Reduced osmotic water permeability of the peritoneal barrier in aquaporin-1 knockout mice. Am J Physiol 1999;276:C76–C81.
5 Flessner MF: Endothelial glycocalyx and the peritoneal barrier. Perit Dial Int 2008;28:6–12.
6 Hu X, Adamson RH, Liu B, Curry FE, Weinbaum S: Starling forces that oppose filtration after tissue oncotic pressure is increased. Am J Physiol Heart Circ Physiol 2000;279:H1724–H1736.
7 Adamson RH, Lenz JF, Zhang X, Adamson GN, Weinbaum S, Curry FE: Oncotic pressures opposing filtration across non-fenestrated rat microvessels. J Physiol 2004;557:889–907.
8 Levick JR: Revision of the Starling principle: new views of tissue fluid balance. J Physiol 2004;557:704.
9 Weinbaum S, Zhang X, Han Y, Vink H, Cowin SC: Mechanotransduction and flow across the endothelial glycocalyx. Proc Natl Acad Sci USA 2003;100:7988–7995.
10 Vink H, Duling BR: Identification of distinct luminal domains for macromolecules, erythrocytes, and leukocytes within mammalian capillaries. Circ Res 1996;79:581–589.
11 Nieuwdorp M, Meuwese MC, Vink H, Hoekstra JB, Kastelein JJ, Stroes ES: The endothelial glycocalyx: a potential barrier between health and vascular disease. Curr Opin Lipidol 2005;16:507–511.
12 Flessner MF, Credit K, Li X, Tanksley J: Similitude of transperitoneal permeability in different rodent species. Am J Physiol Renal Physiol 2007;292:F495–F499.
13 Frigeri A, Nicchia GP, Balena R, Nico B, Svelto M: Aquaporins in skeletal muscle: reassessment of the functional role of aquaporin-4. FASEB J 2004;18:905–907.
14 Waniewski J, Dutka V, Stachowska-Pietka J, Cherniha R: Distributed modeling of glucose-induced osmotic flow. Adv Perit Dial 2007;23:2–6.
15 Flessner MF: Net ultrafiltration in peritoneal dialysis: Role of direct fluid absorption into peritoneal tissue. Blood Purif 1992;10:136–147.

Michael F. Flessner, MD, PhD
Division of Nephrology, Department of Medicine, University of Mississippi Medical Center
2500 North State Street
Jackson, MS 39216–4505 (USA)
Tel. +1 601 984 5670, Fax +1 601 984 5765, E-Mail mflessner@umsmed.edu

Ronco C, Crepaldi C, Cruz DN (eds): Peritoneal Dialysis – From Basic Concepts to Clinical Excellence. Contrib Nephrol. Basel, Karger, 2009, vol 163, pp 15–21

Long-Term Changes in Solute and Water Transport

Lily Mushahar[a], *Mark Lambie*[a], *Kay Tan*[a], *Biju John*[a], *Simon. J. Davies*[a,b]

[a]Department of Nephrology, University Hospital of North Staffordshire, and [b]Institute of Science and Technology in Medicine, Keele University, Stoke on Trent, UK

Abstract

The rate of transport of small solutes across the peritoneal membrane is one of the most important measurements in PD patients. Significant between-patient variability is associated with an impact on small solute clearance, ultrafiltration and even survival. Cross-sectional and longitudinal studies show that solute transport generally increases with time on treatment although this is again highly variable between individuals and is likely to represent an increased vascularity of the membrane. Initially, there is a coupled decrease in the ultrafiltration capacity of the membrane that can be explained by the earlier loss of the osmotic gradient leading predominantly to reduced free water transport via aquaporins combined with more fluid reabsorption once the osmotic gradient has dissipated. Subsequently, in some patients a further disproportionate fall in ultrafiltration occurs due to uncoupling of fluid transport from solute transport as a result of a reduction in the osmotic conductance of the membrane. Drivers of this damage appear to be peritonitis, glucose exposure and early loss of residual renal function. Cytokines and growth factors appear to be involved in this process and may prove useful biomarkers of membrane injury in the future.

The rate at which small solutes cross the peritoneal membrane varies considerably between individuals treated with peritoneal dialysis and is as important as body size and residual renal function (RRF) to the clinician when prescribing the most appropriate dialysis regime [1]. This variability affects both the ease with which molecules larger than urea can be cleared (e.g. creatinine) and conversely the rapidity of solute absorption (e.g. glucose), which in turn means that it is inversely coupled to the efficiency of osmotically driven water transport. Patients with high solute transport will have less-efficient overall fluid removal due to a combination of reduced ultrafiltration, especially free water transport via aquaporins in the early part of a glucose dwell combined with more rapid fluid reabsorption in the later part of the dwell once the osmotic gradient has dissipated [2]. This inverse coupling means that any increase in solute transport with time on PD will automatically result in reduced ultrafiltration capacity with osmotic agents. There is, however, evidence that this coupling becomes less apparent with time on therapy due to a disproportionate fall in ultrafiltration [3, 4]. This article will discuss the changes in solute transport that occur with time on treatment, the factors that drive this change and why changes in water transport might become uncoupled from this process. The implications for clinical practice will also be considered.

Solute Transport at the Start of Peritoneal Dialysis

Our understanding of physiology tells us that the variability between patients in solute transport should be a function of the anatomical area of the membrane in contact with PD fluid combined with the density of perfused capillaries within the membrane and their relative blood flow. In reality, there are very few clinical correlates with solute transport rate at the start of PD treatment. On average, men have slightly higher solute transport rates than women, which is in keeping with a larger anatomical area but this accounts for just a very small amount of the variance (1–2%) [1]. Rumpsfeld et al. [5] found an inverse relationship with body mass index on multivariate analysis which is at first sight surprising until we remember that this is inversely related to body surface area. Comorbidity, especially diabetes, has been linked to high transport rates but again this accounts for very little of the variance and generally disappears on multivariate analysis. It is also important to recognise that the timing of the first measurement of membrane function is important as this changes significantly during the first few weeks of treatment. A preliminary unpublished analysis of the GLOBAL fluid study has found a significant centre effect that was explained by the different timing of the first Peritoneal Equilibration Test.

The negative correlation between transport status and plasma albumin at the start of peritoneal dialysis has led to the argument that high transport is a reflection of systemic inflammation. Patients with an IL-6 polymorphism associated with increased systemic and local peritoneal production of IL-6 do have higher solute transport at the start of PD but again this does not account for more than a few percent of the variance [6]. It should also be remembered that the majority of peritoneal protein leak is lost via a different (large) pore system compared to small solutes. The peritoneal membrane can only be considered to be inflamed if there is also a relative increase in large pores; in this context, peritoneal protein loss appears to be a predictor of survival in PD independent of solute transport characteristics, i.e. weakly correlated with inflammation and vascular disease [7, 8].

To summarise, therefore, there are no clear clinical correlates with solute transport at the time of treatment and it should not be considered as a surrogate measure of membrane inflammation until further direct evidence is obtained.

Changes in Solute Transport with Time on Peritoneal Dialysis

The majority of cross-sectional and longitudinal studies, provided they are of significant size and length, have shown that solute transport tends to increase with time on treatment [3, 9, 10]. There is considerable between-patient variation in this and it must be remembered that because changing membrane function is often associated with increased risk for drop-out from therapy that there is an effect of informative censoring when interpreting the data. This effect is illustrated in figure 1 where is can be seen that patients who had prolonged survival on PD tended to start with a lower solute transport and have a more gradual rise with time when compared to patients who had to stop PD earlier due to death or technique failure. As it is unlikely that there is a significant change in the anatomical area of the membrane in contact with dialysate over this number of years it seems most likely that this change in solute transport reflects a pathophysiological change in the membrane itself due to a combination of increased perfusion of existing capillaries or new vessel formation.

Coupling and Uncoupling of Solute and Water Transport with Time on Peritoneal Dialysis

As discussed above, an increase in solute transport will inevitably lead to a reduction in the ultrafiltration capacity of the peritoneal membrane when osmotically active dialysis solutions are being used. This will not be the case for solutions such as icodextrin which are iso-osmotic with plasma, achieving their ultrafiltration through the small pores. There is evidence from both longitudinal and cross-sectional studies that in long-term PD patients, however, that the fall in ultrafiltration capacity of the membrane is greater than can be explained by increases in solute transport alone [3, 10]. This indicates that there is an additional problem occurring leading to a reduction in the osmotic conductance of the membrane. Osmotic conductance is the term used to describe the amount of ultrafiltration achieved for a given osmotic gradient and it will be determined by three membrane properties: membrane area, efficiency of the osmotic agent at the given pore system (reflection coefficient) and the ability

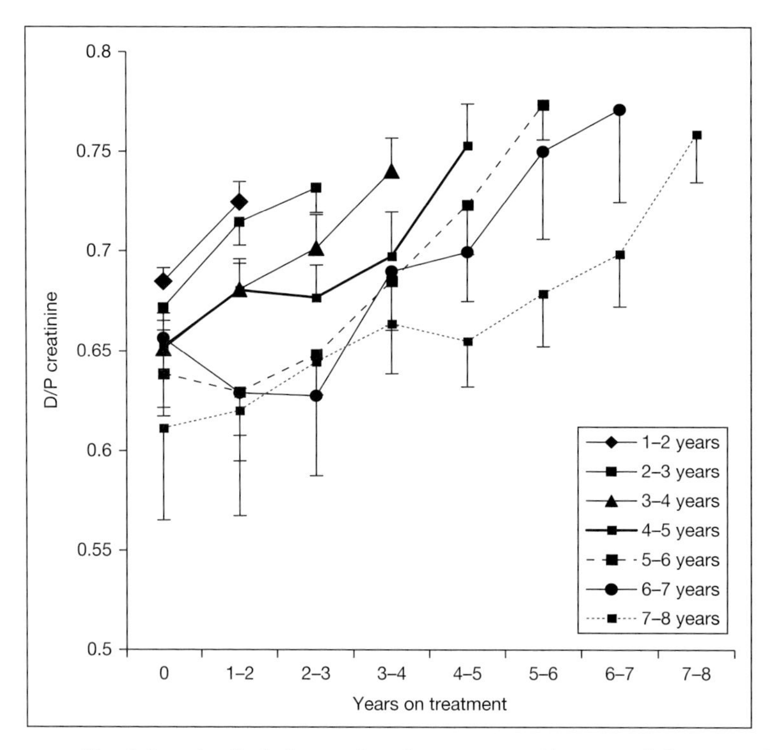

Fig. 1. Longitudinal changes in solute transport (shown as dialysate to plasma creatinine ratio at 4 h taken from the peritoneal equilibration test) according to how many years the patients were on PD. It can be seen that patients spending 1–2 years on PD have a higher transport at the start of treatment that increases more rapidly than patients who stayed on PD for 7–8 years. Data taken from the Stoke PD Study 1990 to 2005. Eventually, all patients develop high transport but there is a drop out effect associated with high transport which leads to informative censoring.

of fluid to pass through the membrane (liquid permeability). By definition, the fall in osmotic conductance of the membrane cannot be explained by a fall in membrane area as increasing solute transport is in itself leading to an increase the effective membrane area. A fall in the overall osmotic conductance could be due to a change in the relative proportion of pore types across which glucose is exerting its osmotic effect – for example, a relative reduction in aquaporins compared to the increase in small pores. Alternatively, increased fibrosis of the interstitium of the peritoneal membrane could be reducing the permeability of water flow whilst having no significant effect on solute transport. The best study measuring ultrafiltration in long term patients with loss of osmotic conductance would suggest that the latter is the most likely explanation [4] and this has been modelled successfully by extending the 3-pore model to include a fibrotic interstitium [11].

What Drives Changes in Membrane Solute Transport?

Early studies of membrane change demonstrated that peritonitis, especially when associated with severe inflammation or recurrent episodes led to increases in solute transport and reduced ultrafiltration capacity [12]. The early use of hypertonic exchanges in the course of time on PD is also associated with more rapid increases in solute transport and loss of ultrafiltration. As would be expected, the earlier use of higher glucose concentrations is required by an earlier loss of RRF and it can be difficult to decide which of the two factors is most important in driving membrane change and damage [3, 13]. Data from the EAPOS study, which was confined to anuric patients treated with automated peritoneal dialysis demonstrated relatively rapid changes in membrane function that could not be attributed to changing RRF [14]. In this study, greater glucose exposure was associated with faster membrane change. The evidence would suggest that all three components, peritonitis, glucose exposure and lack of RRF are potentially synergistic drivers of increasing solute transport.

What Are the Mechanisms of Membrane Change?

Having concluded that there is little evidence to support baseline differences in solute transport as being due to membrane inflammation, it seems very likely that this is important when considering longitudinal change. Potential mediators of this change might include cytokines such as IL-6 which are almost certainly produced locally within the peritoneal cavity. Relatively small studies have supported this showing increased dialysate as opposed to systemic concentrations being associated with solute transport [15, 16]. It is also now well established that epithelial to mesenchymal cell transition of mesothelial cells occurs in the membrane of patients on PD and that this is a likely mechanism, driven by TGF, leading to progressive membrane fibrosis [17]. Whether or not the processes leading to increased solute transport (increased vascularity) and reduction in osmotic conductance (fibrosis) are separate but coincident or represent a single common pathway of injury remain unclear at this point in time.

Conclusions and Clinical Consequences

Increasing solute transport is the first clinical evidence that PD is having an adverse effect on peritoneal membrane function. It usually predates the more serious reduction in ultrafiltration that is due to loss of osmotic conductance. For these reasons, it is good practice to perform relatively simple membrane function tests, e.g. Peritoneal Equilibration Test, on a regular basis, for example 6-monthly. The drivers of membrane damage indicate that where possible the clinician should strive to avoid peritonitis, preserve RRF and avoid excess glucose exposure. The clinico-pathological correlations between membrane function change and morphology are beginning to be understood more and the underlying mechanisms of membrane injury point to several possible interventions. The future evaluation of biomarkers that predate the development of functional and morphological changes will help the clinician pursue the correct strategy in the management of their patients.

References

1 Davies SJ: Mitigating peritoneal membrane characteristics in modern peritoneal dialysis therapy. Kidney Int Suppl 2006;103:S76–S83.
2 Asghar RB, Davies SJ: Pathways of fluid transport and reabsorption across the peritoneal membrane. Kidney Int 2008;73:1048–1053.
3 Davies SJ: Longitudinal relationship between solute transport and ultrafiltration capacity in peritoneal dialysis patients. Kidney Int 2004;66:2437–2445.
4 Parikova A, Smit W, Struijk DG, et al: Analysis of fluid transport pathways and their determinants in peritoneal dialysis patients with ultrafiltration failure. Kidney Int 2006;70:1988–1994.
5 Rumpsfeld M, McDonald SP, Purdie DM, et al: Predictors of baseline peritoneal transport status in Australian and New Zealand peritoneal dialysis patients. Am J Kidney Dis 2004;43:492–501.
6 Gillerot G, Goffin E, Michel C, et al: Genetic and clinical factors influence the baseline permeability of the peritoneal membrane. Kidney Int 2005;67:2477–2487.
7 Heaf JG, Sarac S, Afzal S: A high peritoneal large pore fluid flux causes hypoalbuminaemia and is a risk factor for death in peritoneal dialysis patients. Nephrol Dial Transplant 2005;20:2194–2201.
8 Van Biesen W, Van der Tol A, Veys N, et al: The Personal Dialysis Capacity Test is superior to the peritoneal equilibration test to discriminate inflammation as the cause of fast transport status in peritoneal dialysis patients. Clin J Am Soc Nephrol 2006;1:269–274.
9 Heimburger O, Wang T, Lindholm B: Alterations in water and solute transport with time on peritoneal dialysis. Perit Dial Int 1999;19(suppl 2):S83–S90.
10 Smit W, Parikova A, Struijk DG, et al: The difference in causes of early and late ultrafiltration failure in peritoneal dialysis. Perit Dial Int 2005;25(suppl 3):S41–S45.
11 Rippe B, Venturoli D: Simulations of osmotic ultrafiltration failure in CAPD using a serial three-pore membrane/fiber matrix model. Am J Physiol Renal Physiol 2007;292:F1035–F1043.
12 Davies SJ, Bryan J, Phillips L, et al: Longitudinal changes in peritoneal kinetics: the effects of peritoneal dialysis and peritonitis. Nephrol Dial Transplant 1996;11:498–506.
13 Davies SJ, Phillips L, Naish PF, et al: Peritoneal glucose exposure and changes in membrane solute transport with time on peritoneal dialysis. J Am Soc Nephrol 2001;12:1046–1051.

14 Davies SJ, Brown EA, Frandsen NE, et al: Longitudinal membrane function in functionally anuric patients treated with APD: Data from EAPOS on the effects of glucose and icodextrin prescription. Kidney Int 2005;67:1609–1615.
15 Pecoits-Filho R, Carvalho MJ, Stenvinkel P, et al: Systemic and intraperitoneal interleukin-6 system during the first year of peritoneal dialysis. Perit Dial Int 2006;26:53–63.
16 Rodrigues AS, Martins M, Korevaar JC, et al: Evaluation of peritoneal transport and membrane status in peritoneal dialysis: focus on incident fast transporters. Am J Nephrol 2007;27:84–91.
17 Aroeira LS, Aguilera A, Sanchez-Tomero JA, et al: Epithelial to mesenchymal transition and peritoneal membrane failure in peritoneal dialysis patients: pathologic significance and potential therapeutic interventions. J Am Soc Nephrol 2007;18:2004–2013.

Prof. Simon Davies
Department of Nephrology, University Hospital of North Staffordshire, Royal Infirmary
Princess Road
Stoke on Trent, Staffordshire, ST4 7LN (UK)
Tel. +44 1782 554164, Fax +44 1782 620759, E-Mail simondavies1@compuserve.com

Ronco C, Crepaldi C, Cruz DN (eds): Peritoneal Dialysis – From Basic Concepts to Clinical Excellence. Contrib Nephrol. Basel, Karger, 2009, vol 163, pp 22–26

Fluid Transport with Time on Peritoneal Dialysis: The Contribution of Free Water Transport and Solute Coupled Water Transport

Annemieke M. Coester[a], *Watske Smit*[a,b], *Dirk G. Struijk*[a,b], *Raymond T. Krediet*[a]

[a]Division of Nephrology, Department of Internal Medicine, Academic Medical Center, University of Amsterdam, Amsterdam, and [b]Dianet Foundation Amsterdam, Utrecht, The Netherlands

Abstract

Ultrafiltration in peritoneal dialysis occurs through endothelial water channels (free water transport) and together with solutes across small pores: solute coupled water transport. A review is given of cross-sectional studies and on the results of longitudinal follow-up.

During peritoneal dialysis (PD), the capillary wall and its surrounding tissues serve as a biological membrane for fluid and solute transport and are therefore the major determinants of peritoneal permeability. Also, the intrinsic permeability (the size selectivity of the peritoneal membrane) together with the hydraulic permeability, which is the resistance of the peritoneal membrane to fluid transport, are important.

The peritoneum is a heterogeneous biological membrane consisting of different structures: the mesothelium, the interstitial tissue and the endothelial cells of the microvascular wall. The latter is the most important structure for transport according to the three pore theory [1]. In this theory, it is assumed that small interendothelial pores are especially involved in small solute transport and fluid transport coupled to solutes (SCWT). The number of large pores is so small that their contribution to fluid transport can be ignored, but allow passing of macromolecules. Intraendothelial aquaporins are so small that they only allow fluid: solute-free water transport (FWT). A possible contribution of the peritoneal endothelial glycocalyx is speculative [2, 3].

Together, FWT and SCWT result in total transcapillary ultrafiltration (TCUF). If using glucose as the osmotic agent, determinants of TCUF are the osmotic reflection coefficient (sigma): the resistance of the peritoneal membrane to glucose transport from the peritoneal cavity to the circulation determines the potency of glucose to maintain a crystalloid osmotic gradient. Whereas glucose is so small and can pass the small pores rather easily, sigma is dependent on aquaporin-1 (AQP-1) function. Also the osmotic force for ultrafiltration (osmotic conductance) is important. It is the product of sigma, the hydraulic permeability and the vascular surface area available for transport.

However, the use of the peritoneal dialysis solutions, containing large amounts of glucose and glucose degradation products, induce peritoneal membrane alterations. Especially, alterations of the peritoneal vessels develop: diabetiform neovascularization and vasculopathy [4]. The latter being a risk factor for neovascularisation and fibrosis of the peritoneal membrane.

Resulting from the above, it is quite logical that also alterations in fluid and solute transport occur, eventually resulting in ultrafiltration failure. Mostly cross-sectional studies have been performed in order to study the effects of peritoneal dialysis on fluid transport. The most important finding was a decrease in the contribution of FWT to TCUF in long-term PD patients with ultrafiltration failure [5]. This was due to a decreased reflection coefficient for glucose, although a decreased hydraulic permeability could not be excluded in very long-term patients [6]. At present, there is a paucity in large longitudinal studies on the natural time course of fluid transport and its determinants.

The present review aims to discuss fluid transport with time on PD and especially focuses on the contribution of FWT and SCWT to TCUF. Therefore, we first discuss results of a cross-sectional study designed to evaluate differences in short-term and long-term patients and the presence or absence of long-term PD ultrafiltration failure [7]. Secondly, results of a large longitudinal analysis will be discussed to learn something about the natural time course of fluid transport [8].

Cross-Sectional Study

Free water transport is strongly dependent on the crystalloid osmotic gradient [9] which disappears during the dwell, depending on the magnitude of the effective peritoneal vascular surface area. This can be illustrated by figure 1, left panel. The steepest increase in FWT, and so the fastest transport rates, is present in the first hour of the dwell. After around 120 min, the increase levels

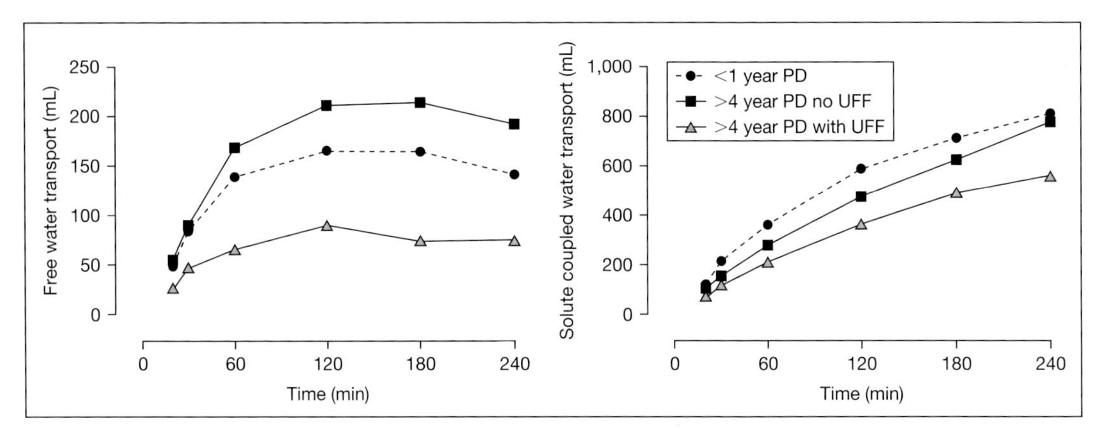

Fig. 1. Left panel: fluid profile of FWT [13] in absolute values during a 4-hour 3.86% dwell of the three different groups. Right panel: fluid profile of SCWT in absolute values during a 4-hour 3.86% dwell of the three different groups.

off, indicating that the osmotic gradient has become smaller. For long-term PD patients this initial increase in FWT is not that steep and already levels off at 60 min of the dwell. The effective vascular surface area is important, as judged from small solute transport, showing fastest values in the long-term group with ultrafiltration failure and slowest values in the long-term group without ultrafiltration failure.

For solute-coupled water transport, the trends during the dwell show a continuous almost linear increase (fig. 1, right panel), indicating that other than osmotic forces must also be involved [6, 9]. The lowest values for the long-term group with ultrafiltration failure can be explained by the fact that in poor ultrafiltration also its components are decreased. The time courses for the short-term group and long-term group without ultrafiltration failure are in line with expectations. The larger effective peritoneal vascular surface area allows more SCWT. This indicates that the osmotic pressure gradient has a limited role in SCWT [9].

The contribution of FWT to TCUF changes in the presence of ultrafiltration failure. This is illustrated in figure 2. The contribution of FWT is obviously less in ultrafiltration failure [5].

Longitudinal Analysis of Fluid Transport: FWT and SCWT

This study [7] was performed in a cohort of 138 incident PD patients without previous renal replacement therapy and with at least a baseline peritoneal

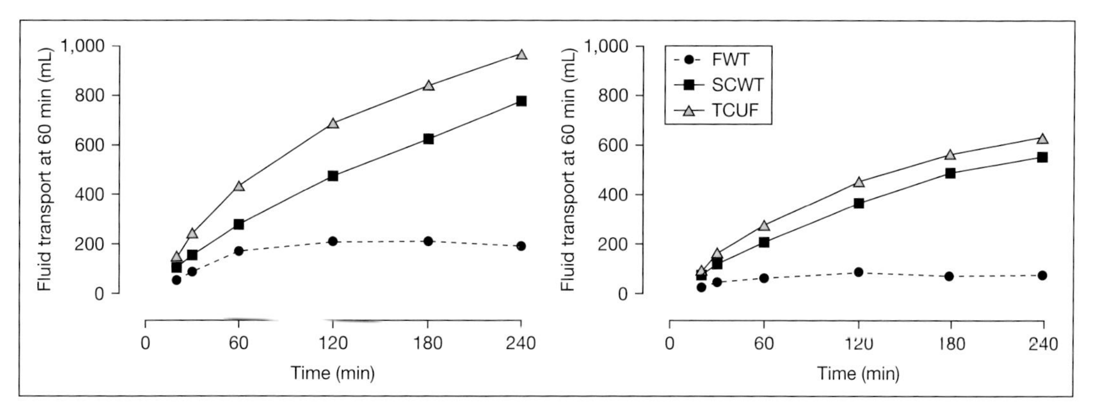

Fig. 2. Left panel: total fluid profiles of TCUF, SCWT and FWT in absolute values during a 4-hour 3.86% dwell of the long-term group without ultrafiltration failure in order to show the relative contribution of FWT and SCWT to TCUF during the whole dwell. Right panel: same, but for the long-term group with ultrafiltration failure.

function test available done during a 4-hour 3.86% glucose dwell. Peritoneal function was monitored in the first year and repeated every year until 5 years of follow-up. The time courses for TCUF, SCWT and FWT (in absolute values) showed a gradual decrease with duration of PD, mainly contributed by the patients who developed ultrafiltration failure. It was remarkable that the contribution of FWT to TCUF increased slightly, especially during the first few years. This can be explained by a slight decrease in the effective peritoneal vascular surface area (reflected by small solute transport) during the same period. This allows a greater amount of glucose in the peritoneal cavity to maintain a crystalloid osmotic gradient. Or it possibly suggests an upregulation of AQP-1 [10]. The contribution of FWT to TCUF continued thereafter, together with an increase in small solute transport, meaning that the contribution of SCWT decreased. Possibly an increase in the resistance of the peritoneal membrane to fluid transport, as judged from a decreasing hydraulic permeability in this study and a hypothetical greater amount of AQP-1s in the increase vascular surface area can be of importance in this finding. AQP-1 function might of course also be impaired [11].

Also, as shown previously [12], an increase in the restriction coefficient for macromolecules can make the assumption of resistance of the peritoneal membrane stronger. The fact that patients were not chosen for absence or presence of ultrafiltration failure, induces the effect of case mix. However, in order to show a natural time course both patient types must be included.

Conclusion

In conclusion, fluid profiles of FWT and SCWT during a dwell can be explained by the current knowledge of the three-pore model. However, as judged from the longitudinal analysis, a thickened interstitium or interstitial fibrosis might also be of importance.

References

1 Rippe B: A three-pore model of peritoneal transport. Perit Dial Int 1993;13(suppl 2):S35–S38.
2 Flessner M: Endothelial glycocalyx and the peritoneal barrier. Perit Dial Int 2008;28:6–12.
3 Rippe B: Does an endothelial surface layer contribute to the size selectivity of the permeable pathways of the three-pore model? Perit Dial Int 2008;28:20–24.
4 Williams JD, Craig KJ, Topley N, et al: Morphologic changes in the peritoneal membrane of patients with renal disease. J Am Soc Nephrol 2002;13:470–479.
5 Smit W, Parikova A, Struijk DG, Krediet RT: The difference in causes of early and late ultrafiltration failure in peritoneal dialysis. Perit Dial Int 2005;25(suppl 3):S41–S45.
6 Parikova A, Smit W, Zweers MM, Struijk DG, Krediet RT: Free water transport, small pore transport and the osmotic pressure gradient. Nephrol Dial Transpl 2008;23:2350–2355.
7 Coester AM, Struijk DG, Smit W, de Waart DR, Krediet RT: The cellular contribution to effluent potassium and its relation to free water transport during peritoneal dialysis. Nephrol Dial Transpl 2007;22:3593–3600.
8 Coester AM, Parikova A, Smit W, Struijk DG, Krediet RT: Longitudinal analysis of fluid transport and its determinants in peritoneal dialysis patients. Perit Dial Int 2009;29(suppl 1):S3–S3.
9 Parikova A, Smit W, Struijk DG, Zweers MM, Krediet RT: The contribution of free water transport and small pore transport to the total fluid removal in peritoneal dialysis. Kidney Int 2005;68:1849–1856.
10 Schoenicke G, Diamant R, Donner A, Roehrborn A, Grabensee B, Plum J: Histochemical distribution and expression of aquaporin 1 in the peritoneum of patients undergoing peritoneal dialysis: relation to peritoneal transport. Am J Kidney Dis 2004;44:146–154.
11 Goffin E, Combet S, Jamar F, Cosyns JP, Devuyst O: Expression of aquaporin-1 in a long-term peritoneal dialysis patient with impaired transcellular water transport. Am J Kidney Dis 1999;33:383–388.
12 Ho-Dac-Pannekeet MM, Koopmans JG, Struijk DG, Krediet RT: Restriction coefficients of low molecular weight solutes and macromolecules during peritoneal dialysis. Adv Perit Dial 1997;13:72–76.
13 La Milia V, Di Filippo S, Crepaldi M, et al: Mini-peritoneal equilibration test: a simple and fast method to assess free water and small solute transport across the peritoneal membrane. Kidney Int 2005;68:840–846.

Annemieke M. Coester, MD
Department of Internal Medicine, Division of Nephrology, Academic Medical Center, University of Amsterdam
Meibergdreef 11
NL–1105 AZ Amsterdam-ZO (The Netherlands)
Tel. +31 205 666 139, E-Mail a.m.coester@amc.uva.nl

Ronco C, Crepaldi C, Cruz DN (eds): Peritoneal Dialysis – From Basic Concepts to Clinical Excellence. Contrib Nephrol. Basel, Karger, 2009, vol 163, pp 27–34

Peritoneal Dialysis: A Biological Membrane with a Nonbiological Fluid

Janusz Witowski[a,b], *Achim Jörres*[a]

[a]Department of Nephrology and Medical Intensive Care, Charité-Universitätsmedizin Berlin, Campus Virchow-Klinikum, Berlin, Germany; [b]Department of Pathophysiology, Poznań University of Medical Sciences, Poznań, Poland

Abstract

During chronic peritoneal dialysis (PD), the peritoneal membrane undergoes structural and functional alterations that, in some patients, may eventually lead to loss of function as the 'dialysis membrane'. These alterations include loss of mesothelial cells, thickening of the submesothelial compact zone, and changes of vascularization. At least some of these effects are thought to be associated with chronic exposure to unphysiological PD fluids. On a cellular level, chronic PD is associated with the phenomenon of epithelial-to-mesenchymal transformation (EMT) as well as premature aging and senescence of mesothelial cells. The present article discusses the potential mechanisms involved in these phenomena and reviews the current information derived from cell culture and animal studies.

Ultrafiltration failure is a serious complication of peritoneal dialysis that may affect even 30% of patients within 4 years on treatment [1–3]. It leads to inadequate fluid removal and extravascular volume expansion and the patients affected may ultimately require transfer to hemodialysis. If not caused by catheter malfunction, poor patient compliance and/or insufficient dialysis regimen, ultrafiltration failure is likely to be caused by peritoneal membrane dysfunction [4]. Dysfunction of the peritoneum as a dialysis organ is thought to be a result of structural changes that develop in the peritoneal membrane with time on PD. They typically include loss of mesothelial cells, thickening of the submesothelial compact zone, and variety of vascular changes [5]. Of those, an increase in vascularization and permeability of the peritoneum is arguably the primary cause of peritoneal membrane dysfunction occurring in up to 75% of patients with ultrafiltration failure [2, 3]. Increased peritoneal vascular area leads to an

Table 1. Effects of GDPs on peritoneal mesothelial cells

Effect	References
Reduced viability	[9, 14–19]
Impaired proliferation/wound healing	[8, 9, 18–22]
Decreased release of IL-6	[9, 14, 19]
Decreased secretion of fibronectin	[14]
Reduced collagen I synthesis	[23]
Increased apoptosis	[15, 22, 24]
Increased generation of reactive oxygen species	[17, 19]
Increased carbonyl stress	[17]
Increased expression of AGE receptors	[25]
Increased VEGF expression and release	[18, 25, 26]
Increased adhesion molecule (VCAM-1) expression	[22]
Increased IL-8 release	[22]
Depletion of cellular glutathione	[20, 27]
Decreased tight junction protein expression	[18]
Mesothelial cell HLA antigen expression	[28]
Induction of epithelial-to- mesenchymal transition	[29]

increase in small solute transport, rapid glucose absorption and dissipation of the osmotic gradient that drives ultrafiltration.

Although it is not clear why the alterations in the peritoneal membrane structure and function develop only in some patients, it has been postulated that chronic exposure to a nonphysiologic dialysate might be a contributing factor. Of PD fluid components, glucose degradation products (GDPs) have been suspected to exert particularly detrimental effects [6]. A number of in vitro and in vivo studies have demonstrated that exposure to GDPs may have adverse consequences to the peritoneal membrane [7]. These effects result partly from direct toxicity of some GDPs. In this respect 3,4-di-deoxyglucosone-3-ene and methylglyoxal appear to be particularly harmful [8, 9]. In addition, GDPs may avidly link with proteins to form advanced glycation end products (AGEs) [10, 11]. With time, AGEs accumulate increasingly in the peritoneum [12] and their presence correlates with a decrease in ultrafiltration [13]. As a result of these observations, the process of PD fluid manufacturing has been modified to compartmentalize dialysate components and minimize GDP formation. New PD solutions contain not only significantly less GDPs, but have also more physiological pH and offer the possibility of using bicarbonate as a buffer.

Unfavorable effects of GDPs have been particularly well documented in mesothelial cells (table 1). These cells were also first to test the potential benefits of new PD solutions.

Table 2. Major clinical trials with new PD solutions

PD solution	Clinical design	Patients	Duration months	Dialysate CA125	Ultra-filtration	Reference
Physioneal (Baxter)	CAPD, prospective, randomized, parallel, 106 patients	106	6	↑	↑	[32]
	APD, prospective, randomized, cross-over, 14 patients,	14	12	↑	↔	[33]
Gambrosol-Trio (Gambro)	CAPD, prospective, randomized, parallel, 80 patients	80	24	↑	↔	[34]
Balance (Fresenius)	CAPD, prospective, randomized, cross-over with parallel arms, 86 patients	86	6	↑	↓	[35]
	CAPD; prospective, randomized, parallel	50	12	↑	↔	[36]
	CAPD, prospective, randomized, parallel	104	12	↑	↑	[37]
Bicavera (Fresenius)	APD, prospective, randomized, cross-over	28	6	↑	↔	[38]

Only studies that investigated CA125 levels and peritoneal ultrafiltration were included. Symbols indicate an increase (↑), a decrease (↓) or no change (↔) in the dialysate concentration of a given mediator compared to conventional PDF.

Indeed, the most consistent effect recorded in clinical trials with new PD fluids was that of increased CA125 concentration in the dialysate (table 2). The phenomenon is intriguing because it was observed in studies of various design, duration, and employing solutions with different buffers [7, 30]. Intraperitoneal levels of CA125 are thought to reflect mesothelial cell mass [31], thus increased CA125 concentrations are usually interpreted as reflecting better preservation of the peritoneal mesothelium. In contrast, the impact of new solutions on ultrafiltration lacked the consistency, which seemed to indicate that mesothelial cells did not have a major role in the maintenance or loss of ultrafiltration.

However, there appear to be serious limitations in the value of dialysate CA125 as a clinical marker of mesothelial cell or membrane integrity [39]. It appears that the magnitude of CA125 release by mesothelial cells may depend not only on their number but also functional properties modulated by age and exposure to high glucose [40]. Moreover, it is not clear how dialysate CA125 levels can be affected by epithelial-to-mesenchymal transition (EMT) of mesothelial cells, a process strongly implicated in peritoneal membrane dysfunction

[41]. It has been observed that with time on PD mesothelial cells isolated from dialysate effluent increasingly show the fibroblast-like phenotype rather than the epithelial cobblestone morphology [42]. The transformation is accompanied by a loss of epithelial cell markers E-cadherin and cytokeratins, and an acquisition of a myofibroblast marker α-smooth muscle actin. Such cells lose polarity, disassemble tight junctions, display great motility and invade the underlying stroma. In this respect, fibroblast-like cells were detected in the submesothelial interstitium of PD patients [43].

Transforming growth factor-β (TGF-β) and a transcription factor snail have been identified as important mediators of EMT [44]. It has been demonstrated that TGF-β can induce EMT both in mesothelial cells in vitro [45] and in the peritoneal mesothelium in vivo [46]. Importantly, it has recently been demonstrated that mesothelial cells undergoing EMT show increased expression of genes associated with fibrosis and angiogenesis, including vascular endothelial growth factor (VEGF) [47]. In this respect, mesothelial cells with fibroblast-like appearance isolated from dialysate effluent were found to express more VEGF [48] and cyclo-oxygenase-2 (COX-2), a key enzyme involved in inflammation-associated prostaglandin production [49]. Importantly, the donors of such cells tended to have increased peritoneal solute transport and decreased ultrafiltration. More recently, it has been reported that the type of PD solution may impact on the phenotype of mesothelial cells found in the effluent [49, 50]. Mesothelial cells with fibroblast-like morphology and high expression of snail were isolated more frequently from patients dialyzed with conventional solutions than from those receiving new PD fluids low in GDPs. In keeping with this observation, methylglyoxal was found to induce snail and TGF-β in rat mesothelial cells in vitro and in vivo [29]. The presence of these cells in the peritoneum was associated with its fibrotic thickening and increased expression of VEGF. These results suggest that mesothelial cells undergoing EMT as a result of exposure to GDP-containing solutions may contribute to high peritoneal transport rates by producing more vasoactive mediators, such as VEGF. Although patients receiving fluids with either low or high GDPs were found to have similar concentrations of VEGF in the dialysate [35, 38, 50], they might have differed in a more biologically important local presence of VEGF within the peritoneal membrane. Interestingly, new data indicate that several options recommended to improve declining ultrafiltration may be associated with amelioration of EMT in mesothelial cells. They include the use of icodextrin [41] and peritoneal rest [51].

Besides the induction of EMT, peritoneal dialysis may also be associated with premature ageing and accelerated cellular senescence of mesothelial cells. The first report of this phenomenon is from Gotloib et al. [52, 53] who detected senescent mesothelial cells in rodents treated with PD fluids. We have recently found that human peritoneal mesothelial cells treated with a PD solution exhibit increased expression of senescence-associated β-galactosidase (SA-β-Gal) [54]. Accelerated mesothelial senescence may be of potential relevance in the context of peritoneal dialysis as it is associated with decreased cell proliferation [55] and increased production of TGF-β, VEGF and fibronectin [56, 57], all factors implicated in the pathogenesis of chronic peritoneal membrane alterations.

The cellular mechanisms of mesothelial senescence are not completely understood. It appears that HPMC undergo senescence in culture predominantly as a result of nontelomeric DNA damage [58], a process which is associated with mitochondrial dysfunction and increased production of reactive oxygen species (ROS) [55, 59] and which can be exacerbated by exposure to high glucose [59, 60]. Moreover, TGF-β is not only induced by high glucose but conversely is also capable of inducing the senescence response in mesothelial cells [56]. Human peritoneal mesothelial cells treated with TGF-β show increased activity of SA-β-Gal and increased secretion of VEGF [54].

Taken together it is tempting to speculate that accumulation of senescent mesothelial cells in PD patients may result in increased peritoneal generation of ROS, TGF-β and VEGF, all factors which may contribute to peritoneal neoangiogenesis and fibrosis and, ultimately, peritoneal membrane failure.

References

1 Slingeneyer A, Canaud B, Mion C: Permanent loss of ultrafiltration capacity of the peritoneum in long-term peritoneal dialysis: an epidemiological study. Nephron 1983;33:133–138.
2 Ho-Dac-Pannekeet MM, Atasever B, Struijk DG, Krediet RT: Analysis of ultrafiltration failure in peritoneal dialysis patients by means of standard peritoneal permeability analysis. Perit Dial Int 1997;17:144–150.
3 Heimburger O, Waniewski J, Werynski A, et al: Peritoneal transport in CAPD patients with permanent loss of ultrafiltration capacity. Kidney Int 1990;38:495–506.
4 Margetts PJ, Churchill DN: Acquired ultrafiltration dysfunction in peritoneal dialysis patients. J Am Soc Nephrol 2002;13:2787–2794.
5 Williams JD, Craig KJ, Topley N, et al: Morphologic changes in the peritoneal membrane of patients with renal disease. J Am Soc Nephrol 2002;13:470–479.
6 Wieslander AP, Kjellstrand PT, Rippe B: Heat sterilization of glucose-containing fluids for peritoneal dialysis: biological consequences of chemical alterations. Perit Dial Int 1995;15:S52–S59.
7 Witowski J, Jörres A: Effects of peritoneal dialysis solutions on the peritoneal membrane: clinical consequences. Perit Dial Int 2005;25(suppl 3):S31–S34.
8 Morgan LW, Wieslander A, Davies M, et al: Glucose degradation products (GDP) retard remesothelialization independently of D-glucose concentration. Kidney Int 2003;64:1854–1866.
9 Witowski J, Korybalska K, Wisniewska J, et al: Effect of glucose degradation products on human peritoneal mesothelial cell function. J Am Soc Nephrol 2000;11:729–739.
10 Lamb EJ, Cattell WR, Dawnay AB: In vitro formation of advanced glycation end products in peritoneal dialysis fluid. Kidney Int 1995;47:1768–1774.
11 Tauer A, Knerr T, Niwa T, et al: In vitro formation of N(epsilon)-(carboxymethyl)lysine and imidazolones under conditions similar to continuous ambulatory peritoneal dialysis. Biochem Biophys Res Commun 2001;280:1408–1414.

12 Nakayama M, Kawaguchi Y, Yamada K, et al: Immunohistochemical detection of advanced glycosylation end-products in the peritoneum and its possible pathophysiological role in CAPD. Kidney Int 1997;51:182–186.
13 Honda K, Nitta K, Horita S, et al: Accumulation of advanced glycation end products in the peritoneal vasculature of continuous ambulatory peritoneal dialysis patients with low ultra-filtration. Nephrol Dial Transplant 1999;14:1541–1549.
14 Witowski J, Wisniewska J, Korybalska K, et al: Prolonged exposure to glucose degradation products impairs viability and function of human peritoneal mesothelial cells. J Am Soc Nephrol 2001;12:2434–2441.
15 Santamaria B, Ucero AC, Reyero A, et al: 3,4-Dideoxyglucosone-3-ene as a mediator of peritoneal demesothelization. Nephrol Dial Transplant 2008;23:3307–3315.
16 Tomo T, Okabe E, Yamamoto T, et al: Synergistic cytotoxicity of acidity and 3,4-dideoxyglucosone-3-ene under the existence of lactate in peritoneal dialysis fluid. Ther Apher Dial 2005;9:182–187.
17 Alhamdani MS, Al Kassir AH, Abbas FK, et al: Antiglycation and antioxidant effect of carnosine against glucose degradation products in peritoneal mesothelial cells. Nephron Clin Pract 2007;107:c26–c34.
18 Leung JC, Chan LY, Li FF, et al: Glucose degradation products downregulate ZO-1 expression in human peritoneal mesothelial cells: the role of VEGF. Nephrol Dial Transplant 2005;20:1336–1349.
19 Breborowicz A, Witowski J, Polubinska A, et al: L-2-Oxothiazolidine-4-carboxylic acid reduces in vitro cytotoxicity of glucose degradation products. Nephrol Dial Transplant 2004;19:3005–3011.
20 Yamamoto T, Tomo T, Okabe E, et al: Glutathione depletion as a mechanism of 3,4-dideoxyglucosone-3-ene-induced cytotoxicity in human peritoneal mesothelial cells: role in biocompatibility of peritoneal dialysis fluids. Nephrol Dial Transplant 2009;4:1436–1442.
21 Boulanger E, Wautier MP, Gane P, et al: The triggering of human peritoneal mesothelial cell apoptosis and oncosis by glucose and glycoxydation products. Nephrol Dial Transplant 2004;19:2208–2216.
22 Welten AG, Schalkwijk CG, ter Wee PM, et al: Single exposure of mesothelial cells to glucose degradation products (GDPs) yields early advanced glycation end-products (AGEs) and a proinflammatory response. Perit Dial Int 2003;23:213–221.
23 Witowski J, Jörres A, Korybalska K, et al: Glucose degradation products in peritoneal dialysis fluids: do they harm? Kidney Int 2003;(suppl):S148–S151.
24 Lee DH, Choi SY, Ryu HM, et al: 3,4-Dideoxyglucosone-3-ene induces apoptosis in human peritoneal mesothelial cells. Perit Dial Int 2009;29:44–51.
25 Lai KN, Leung JC, Chan LY, et al: Differential expression of receptors for advanced glycation end-products in peritoneal mesothelial cells exposed to glucose degradation products. Clin Exp Immunol 2004;138:466–475.
26 Inagi R, Miyata T, Yamamoto T, et al: Glucose degradation product methylglyoxal enhances the production of vascular endothelial growth factor in peritoneal cells: role in the functional and morphological alterations of peritoneal membranes in peritoneal dialysis. FEBS Lett 1999;463:260–264.
27 Korybalska K, Wisniewska-Elnur J, Trominska J, et al: The role of the glyoxalase pathway in reducing mesothelial toxicity of glucose degradation products. Perit Dial Int 2006;26:259–265.
28 Pajek J, Kveder R, Bren A, et al: Short-term effects of a new bicarbonate/lactate-buffered and conventional peritoneal dialysis fluid on peritoneal and systemic inflammation in CAPD patients: a randomized controlled study. Perit Dial Int 2008;28:44–52.
29 Hirahara I, Ishibashi Y, Kaname S, et al: Methylglyoxal induces peritoneal thickening by mesenchymal-like mesothelial cells in rats. Nephrol Dial Transplant 2009;24:437–447.
30 Locatelli F, La MV: Preservation of residual renal function in peritoneal dialysis patients: still a dream? Kidney Int 2008;73:143–145.
31 Krediet RT: Dialysate cancer antigen 125 concentration as marker of peritoneal membrane status in patients treated with chronic peritoneal dialysis. Perit Dial Int 2001;21:560–567.
32 Jones S, Holmes CJ, Krediet RT, et al: Bicarbonate/lactate-based peritoneal dialysis solution increases cancer antigen 125 and decreases hyaluronic acid levels. Kidney Int 2001;59:1529–1538.

33 Fusshoeller A, Plail M, Grabensee B, Plum J: Biocompatibility pattern of a bicarbonate/lactate-buffered peritoneal dialysis fluid in APD: a prospective, randomized study. Nephrol Dial Transplant 2004;19:2101–2106.
34 Rippe B, Simonsen O, Heimburger O, et al: Long-term clinical effects of a peritoneal dialysis fluid with less glucose degradation products. Kidney Int 2001;59:348–357.
35 Williams JD, Topley N, Craig KJ, et al: The Euro-Balance Trial: the effect of a new biocompatible peritoneal dialysis fluid (balance) on the peritoneal membrane. Kidney Int 2004;66:408–418.
36 Szeto CC, Chow KM, Lam CW, et al: Clinical biocompatibility of a neutral peritoneal dialysis solution with minimal glucose-degradation products–a 1-year randomized control trial. Nephrol Dial Transplant 2007;22:552–559.
37 Choi HY, Kim DK, Lee TH, et al: The clinical usefulness of peritoneal dialysis fluids with neutral pH and low glucose degradation product concentration: an open randomized prospective trial. Perit Dial Int 2008;28:174–182.
38 Haas S, Schmitt CP, Arbeiter K, et al: Improved acidosis correction and recovery of mesothelial cell mass with neutral-pH bicarbonate dialysis solution among children undergoing automated peritoneal dialysis. J Am Soc Nephrol 2003;14:2632–2638.
39 Topley N, Michael D, Bowen T: CA125: holy grail or a poisoned chalice. Nephron Clin Pract 2005;100:c52–c54.
40 Breborowicz A, Breborowicz M, Pyda M, et al: Limitations of CA125 as an index of peritoneal mesothelial cell mass. Nephron Clin Pract 2005;100:c46–c51.
41 Aroeira LS, Aguilera A, Sanchez-Tomero JA, et al: Epithelial to mesenchymal transition and peritoneal membrane failure in peritoneal dialysis patients: pathologic significance and potential therapeutic interventions. J Am Soc Nephrol 2007;18:2004–2013.
42 Yanez-Mo M, Lara-Pezzi E, Selgas R, et al: Peritoneal dialysis and epithelial-to-mesenchymal transition of mesothelial cells. N Engl J Med 2003;348:403–413.
43 Jimenez-Heffernan JA, Aguilera A, Aroeira LS, et al: Immunohistochemical characterization of fibroblast subpopulations in normal peritoneal tissue and in peritoneal dialysis-induced fibrosis. Virchows Arch 2004;444:247–256.
44 Zavadil J, Bottinger EP: TGF-beta and epithelial-to-mesenchymal transitions. Oncogene 2005;24:5764–5774.
45 Yang AH, Chen JY, Lin JK: Myofibroblastic conversion of mesothelial cells. Kidney Int 2003;63:1530–1539.
46 Margetts PJ, Bonniaud P, Liu L, et al: Transient overexpression of TGF-{beta}1 induces epithelial mesenchymal transition in the rodent peritoneum. J Am Soc Nephrol 2005;16:425–436.
47 Zhang J, Oh KH, Xu H, Margetts PJ: Vascular endothelial growth factor expression in peritoneal mesothelial cells undergoing transdifferentiation. Perit Dial Int 2008;28:497–504.
48 Aroeira LS, Aguilera A, Selgas R, et al: Mesenchymal conversion of mesothelial cells as a mechanism responsible for high solute transport rate in peritoneal dialysis: role of vascular endothelial growth factor. Am J Kidney Dis 2005;46:938–948.
49 Aroeira LS, Lara-Pezzi E, Loureiro J, et al: Cyclooxygenase-2 mediates dialysate-induced alterations of the peritoneal membrane. J Am Soc Nephrol 2009;20:582–592.
50 Do JY, Kim YL, Park JW, et al: The association between the vascular endothelial growth factor-to-cancer antigen 125 ratio in peritoneal dialysis effluent and the epithelial-to-mesenchymal transition in continuous ambulatory peritoneal dialysis. Perit Dial Int 2008;28(suppl 3):S101–S106.
51 Yu MA, Shin KS, Kim JH, et al: HGF and BMP-7 ameliorate high glucose-induced epithelial-to-mesenchymal transition of peritoneal mesothelium. J Am Soc Nephrol 2009;20:567–581.
52 Gotloib L, Shostak A, Wajsbrot V, Kushnier R: High glucose induces a hypertrophic, senescent mesothelial cell phenotype after long in vivo exposure. Nephron 1999;82:164–173.
53 Gotloib L, Wajsbrot V, Shostak A: Icodextrin-induced lipid peroxidation disrupts the mesothelial cell cycle engine. Free Radic Biol Med 2003;34:419–428.
54 Witowski J, Ksiazek K, Jörres A: Glucose-induced mesothelial cell senescence and peritoneal neoangiogenesis and fibrosis. Perit Dial Int 2008;28(suppl 5):S34–S37.
55 Ksiazek K, Piwocka K, Brzezinska A, et al: Early loss of proliferative potential of human peritoneal mesothelial cells in culture: the role of p16INK4a-mediated premature senescence. J Appl Physiol 2006;100:988–995.

56 Ksiazek K, Korybalska K, Jörres A, Witowski J: Accelerated senescence of human peritoneal mesothelial cells exposed to high glucose: the role of TGF-beta1. Lab Invest 2007;87:345–356.
57 Ksiazek K, Jörres A, Witowski J: Senescence induces a proangiogenic switch in human peritoneal mesothelial cells. Rejuvenation Res 2008;11:681–683.
58 Ksiazek K, Passos JF, Olijslagers S, et al: Premature senescence of mesothelial cells is associated with non-telomeric DNA damage. Biochem Biophys Res Commun 2007;362:707–711.
59 Ksiazek K, Passos JF, Olijslagers S, von Zglinicki T: Mitochondrial dysfunction is a possible cause of accelerated senescence of mesothelial cells exposed to high glucose. Biochem Biophys Res Commun 2008;366:793–799.
60 Ksiazek K, Breborowicz A, Jörres A, Witowski J: Oxidative stress contributes to accelerated development of the senescent phenotype in human peritoneal mesothelial cells exposed to high glucose. Free Radic Biol Med 2007;42:636–641.

Prof. Achim Jörres, MD
Department of Nephrology and Medical Intensive Care
Charité-Universitätsmedizin Berlin, Campus Virchow-Klinikum
DE–13353 Berlin (Germany)
Tel. +49 30 450 553 423, Fax +49 30 450 553 916, E-Mail achim.joerres@charite.de

Ronco C, Crepaldi C, Cruz DN (eds): Peritoneal Dialysis – From Basic Concepts to Clinical Excellence. Contrib Nephrol. Basel, Karger, 2009, vol 163, pp 35–44

Mechanisms of Cell Death during Peritoneal Dialysis

A Role for Osmotic and Oxidative Stress

Lazaro Gotloib

Laboratory of Experimental Nephrology, Ha Emek Medical Center, Afula, Israel

Abstract

Aims: To offer a condensed description of the modes of cell death and the involved mechanisms behind them as detected in the different layers of the peritoneal tissue during experimental and clinical, long-term peritoneal dialysis (PD). **Main Remarks:** Several types of cell death have been observed in the mesothelial monolayer: apoptosis, anoikis, secondary necrosis, pure necrosis and mitotic catastrophe. Death of mesothelial cells exposed to glucose-enriched solutions derives mainly from a degree of oxidative insult leading to DNA damage, provoked by glucose itself and/or its degradation products. Use of icodextrin is associated with a higher degree of oxidative injury that also leads to genomic damage and consequently to cell death. Peritoneal leukocytes exposed to glucose-enriched PD solutions share the fate of mesothelial cells. Endothelial cells treated in vitro with high glucose concentrations have higher rates of apoptosis induced by a degree of oxidative stress. Endothelial apoptosis plays an important role in remodeling the vascular network, since this development has been observed at the beginning of neo-angiogenesis and at the branching and regression of microvessels of neoformation. Acute osmotic stress results in increased proportions of mesothelial cells dying in apoptosis, anoikis, secondary necrosis as well as in pure necrosis. After long-term exposure, cells apparently adapt to the new hypertonic environment even though the eventual presence of functional changes cannot be ruled out. **Conclusions:** All cells lining the peritoneal cavity or living near its immediate environment, exposed most of the time to nonphysiological fluids, undergo changes that lead to their death. This problem is behind the poor regenerative capabilities showed by the mesothelial monolayer, the microvascular changes that lead to neoangiogenesis, and, probably, to a defective response to peritoneal infection. Therapeutic modulation of apoptotic cell death could inhibit its progress. Since liberation of oxygen radicals is thereby causally involved in apoptosis, reduced levels of GDPs and/or use of antioxidants appear to be indicated in order to prevent mesothelial, endothelial cells and leukocyte death.

The scientist, by the very nature of his commitment, creates more and more questions, never fewer.

G. W. Allport. In: Becoming

At a certain point of their life, cells reach a situation of terminal replicative senescence and are destined to die in apoptosis without further division. This situation of terminal growth arrest is characterized by the differential expression of cell-cycle-regulated genes, and a failure to complete the mitogen-stimulated cascade of signaling events that lead to DNA synthesis. This development appears to derive from critical shortening of telomeres that occurs due to incomplete DNA replication of the chromosome ends. This mechanism is fueled and mediated by the p53 tumor suppressor protein and its downstream effector, $p21^{Cip1}$ [1], which shows cells the way to die through the apoptotic pathway. At any given time, the mesothelium of intact unexposed animals shows about 2% of the cell population to be undergoing the apoptosis.

Pathways to Cellular Death

Different types of cell death are often defined by morphological criteria without a clear reference to precise biochemical mechanisms. An assortment of modes of mesothelial cell death during peritoneal dialysis has been identified: necrotic, apoptotic, a merge of apoptosis and necrosis, and death associated with mitosis. This phenomenon occurs with and without the involvement of nucleases or of distinct classes of proteases, such as caspases, calpains, cathepsins and transglutaminases. Functionally, these mechanisms can be classified as programmed or accidental, physiological or pathological [2].

During necrotic death (fig. 1a), cells swell, cytosolic and nuclear structures show structural abnormalities, but the arrangement of chromatin is preserved. The cytoplasmic membrane breaks down and becomes permeable to vital dyes, but on the other hand allows the cytoplasmic content reach the extracellular space. Then, cytosolic enzymes, notably those of mitochondrial origin, create a situation of local inflammation. This is a relevant difference between necrosis and apoptosis.

The course of noninflammatory apoptotic death, a central mechanism of which is activation of proteases of the caspase (cysteine aspartases) family [3] in most cases, goes through three phases of structural changes [4]. During the first step, cells keep their normal morphology even though at times they appear disengaged from surrounding cells, losing specialized membrane structures like microvilli. This first period lasts from 2 to 44 h. During the next step that covers from 10 min to about 40 h, electron microscopy can detect subtle changes in the

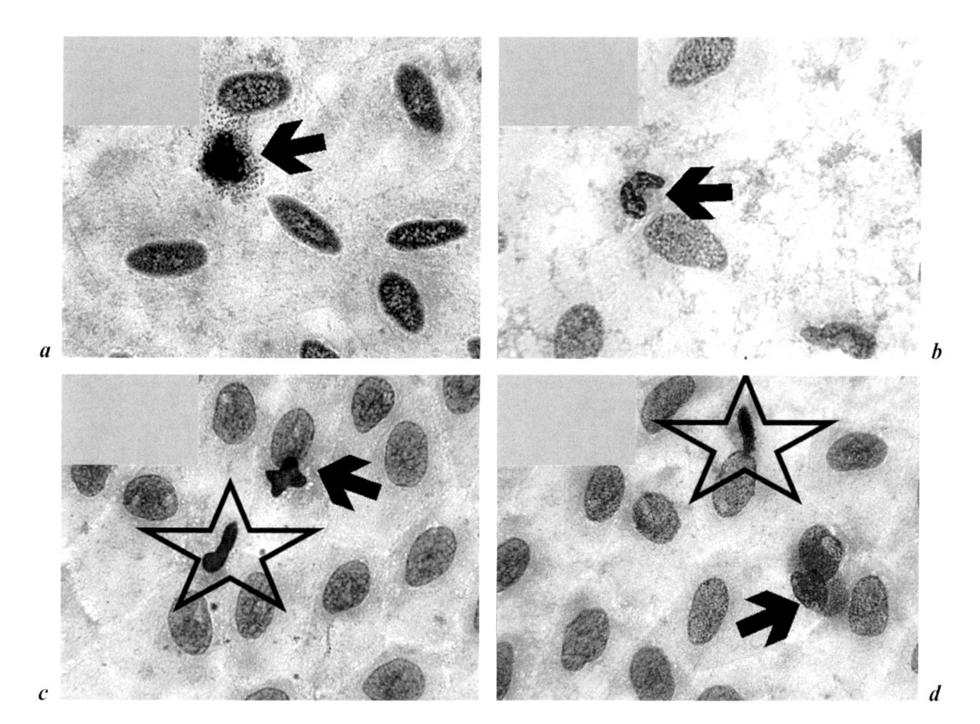

Fig. 1. ***a*** Mesothelial cell undergoing necrosis (arrow). The sample was exfoliated from a mouse 24 h after intraperitoneal infusion of 7.5% icodextrin PD solution. HE. ×1,000. ***b*** Typical aspect of a mesothelial cell undergoing apoptosis showing a shrunken cytoplasm and a kidney-shaped nucleus (arrow). The rat was treated with 4.25% glucose, healt-sterilized, single-bag PD fluid during 30 days. HE. ×1,000. ***c*** Phagocytosis of one apoptotic body by a neighboring mesothelial cell (arrow). Circle surrounds one apoptotic body. HE. ×1,000. ***d*** Apoptotic cell being extruded from the peritoneal surface (arrow). This is the usual appearance of anoikis. The mesothelial imprint was recovered from a mouse exposed during for 30 days to one daily intraperitoneal injection of 4.25%, heat-sterilized, single-bag dialysis fluid. Star encloses one mesothelial cell in the course of apoptosis. HE. ×1,000.

cell membrane, namely kind off blebbing. Then cells enter into the execution phase the duration of which is around 90 h, showing besides the blebbing of the plasma membrane, fast shrinkage of the cytoplasm that becomes denser, as well as condensation of the nuclear chromatin. Typically, nuclei appear as compact, kidney-shaped structures (fig. 1b). Cells undergoing apoptosis can also be identified by means of immunohistochemistry (TUNEL or annexin staining).

Finally, the common pathway of most apoptotic cells is phagocytosis by healthy, younger neighbors or by phagocytes, without any inflammatory reaction (fig. 1c).

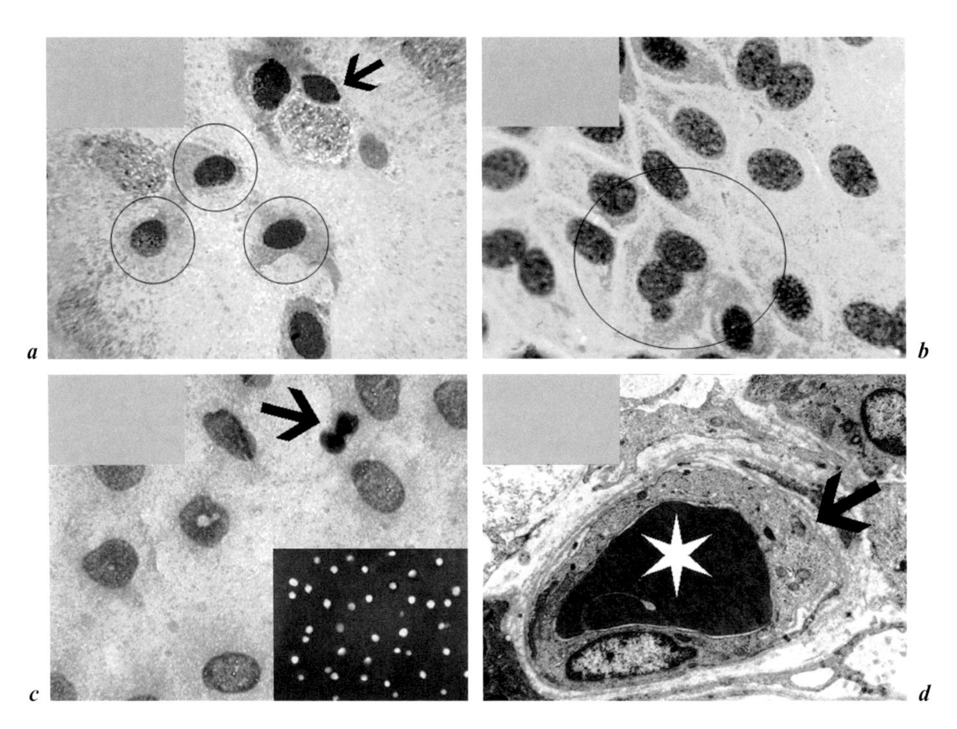

Fig. 2. ***a*** Arrow points at one binucleated mesothelial cell traversing the path of secondary necrosis. Notice the presence of apoptotic bodies (circles). This fragment of the monolayer was taken from a mouse after being exposed to 4.25% glucose PD fluid during a period of 3 h. HE. ×1,000. ***b*** Trinucleated cell displaying one micronucleus (circle). This image was photographed from an imprint of the monolayer recovered from a mouse after 30 days' exposure to a 4.25% glucose, heat-sterilized, single-bag PD solution. HE. ×1,000. ***c*** Apoptotic death of one mesothelial cell at the time of mitosis (arrow) representing the mode of dead defined as mitotic catastrophe. The sample was obtained from a rat exposed during 24 h to a 1.5% glucose, heat-sterilized, single-bag solution for peritoneal dialysis. HE. ×1,000. ***b*** Small inset: Mesothelial cells observed in one imprint obtained from a rat after 30 days of one daily intraperitoneal injection of 7.5% icodextrin PD solution. Numerous cells show positive immunostaining for 8-hydroxy-2'-deoxyguanosine. 8-Hydroxy-2'-deoxyguanosine staining. Sigma, Israel, Holon. ×400. ***d*** Section of parietal peritoneum recovered from a patient after 17 months of maintenance peritoneal dialysis, performed with 1.5 and 4.25% glucose, heat-sterilized, single bag PD solutions. This electron micrograph shows a substantially thickened endothelial cell of a submesothelial capillary (arrow). White star = Erythrocyte. ×12,600.

An unknown proportion of apoptotic mesothelial cells, in addition to leaving behind the normally observed links with surrounding cells, lose their adhesion to the extracellular matrix and are rejected from the tissue surface. This phenomenon, a variation of the dismissal of apoptotic cells in epithelia, is called anoikis [5] and has been detected in the peritoneal mesothelium of animals exposed to peritoneal dialysis (PD) fluids (fig. 1d). Besides its specific form of induction, the molecular mechanisms of anoikis-associated cell death match those activated during classical apoptosis.

Occasionally, cells going through the apoptotic pathway, an energy requiring process, are faced with severe adenosine triphosphate (ATP) depletion, and are forced to die in necrosis. This situation, defined as secondary necrosis, has also been detected in mesothelium of rodents exposed to icodextrin-based PD fluids (fig. 2a).

Mitotic catastrophe is a cell death mode occurring either during or shortly after a dysregulated failed mitosis. It can be accompanied by morphological alterations including micronucleation and multinucleation deriving from a deficient separation during cytokinesis, and can be effected through necrosis or apoptosis [2] (fig. 2b).

Autophagic cell death is morphologically defined as a type of cell death that occurs in the absence of chromatin condensation, but is accompanied by massive autophagic vacuolization of the cytoplasm effecting digestion of the cellular content by their own lysosome machinery. In contrast to apoptotic cells, those dying because of autophagy are generally not taken away by phagocytes. This process entails the formation of membrane-bound autophagosomes that collect cytoplasmic materials making them available to lysosomes, and then digested by acid hydrolases completing the protein degradation pathway. Several studies have offered evidence supporting the concept that ROS (reactive oxygen species) are endowed with the capability of activating the mechanism of cell death through autophagy [6]. This mode of death has been recently identified by electron microscopy in human mesothelium recovered from peritoneal biopsies of patients on long-term peritoneal dialysis. Besides, its presence in human vascular endothelial cells has been well documented [7].

Additionally, apoptosis is a key element in detecting deviations of the normal cell cycle derived from DNA damage or growth dysregulation that could become precursors of malignant clones. So, apoptosis becomes a critical regulatory component developed by nature to keep within the homeostatic range age, number as well as quality of cells from any given tissue.

Causes of Mesothelial Cell Death during Long-Term Peritoneal Dialysis

PD generates a situation in which cells lining the peritoneal cavity are exposed during a substantial part of their life span to a hyperosmotic, glucose-enriched unfriendly environment. A 3 h in vivo exposure to high glucose concentration induces marked shrinkage of mesothelial cells, noticeable widening of intercellular spaces, as well as a substantially increased proportion of cells displaying the morphological aspect of apoptosis, whereas in samples obtained from animals treated with icodextrin images of cells dying either by apoptosis or necrosis are detected [Gotloib, unpubl. obs.]. These observations coincide with an investigation performed in peripheral blood monocytes incubated with high glucose or icodextrin. Cells shrunk immediately after 1 min incubation, whereas the increase in apoptosis was evident after a 120-min interval [8]. The fact that these acute alterations were also observed in experiments done by using a radical scavenger like mannitol in equimolar concentration suggests that oxidative stress, resulting from the hypertonic insult was not behind the observed increased rate of cell death, even though it has been shown that osmotic stress can generate an oxidative additional injury [9]. The question of whether this apoptotic effect is related or not to caspase-3 activity is still debated.

Sustained osmotic contraction of the cytoplasma triggers changes of gene expression, that result in an increased synthesis and accumulation of organic osmolytes like sorbitol, myo-inositol, betaine, which drive water back into the dehydrated cell, setting in motion the mechanism of regulatory volume increase that in order to survive, cells launch after shrinkage [10]. This path is followed by cells to maintain the steadiness of their volume, but requires spending large amounts of energy that, leading to ATP depletion, may well derive in early apoptosis or even in secondary necrosis (fig. 2a).

This compensatory mechanism is evident 24 h after the initial hyperosmolar insult reaching a peak after 48 h as shown in rat mesothelial cells in culture [11], and explains the observed adaptation of the mesothelium, both, in vitro as well as in vivo, to a new environment after long-term exposure to hyperosmolar, glucose-enriched PD fluids [12, 13]. Even though the exposed cells became larger after 30 days' contact with the experimental solution, their regenerative capabilities remained unaffected. It may be assumed that microvascular endothelial cells share the changes displayed by the mesothelial monolayer when facing a hypertonic environment.

In vitro and in vivo studies provided evidence showing that exposure of the mesothelium to glucose induced a marked and dose-dependent acceleration of the cell cycle that exhausted after four rounds of replication, results in a monolayer populated by large, hypertrophic, multinucleated, prematurely senescent cells [14] (fig. 2c), the final outcome of which is death in apoptosis. The speeding up of the cell cycle results from a degree of oxidative stress induced by glucose itself, glucose auto-oxidation as well as by glucose degradation products present in different concentration in commercially available dialysis solutions [14]. More to the point, in vitro and in vivo studies have demonstrated that the presence of glucose in PD fluids is associated to DNA damage of the exposed mesothelium [15], adding one more reason to put in motion p53, the sentinel of the G1/S checkpoint, in order to launch the apoptotic pathway to prevent further progression of these cells along the cell cycle.

Acute exposure of intact rats to the 7.5% icodextrin solution resulted in a linear increase in the concentration of thiobarbituric-reactive substances (basically MDA), during a follow-up period of up to 4 h [16]. These findings brought to light the fact that the osmotic agent and/or large and medium-sized polymers derived from its intra-abdominal degradation-induced lipid peroxidation at the level of the mesothelial cell membrane, resulting in DNA damage, as shown by the presence of mesothelial cells having rod-like and semilunar nuclei, micronuclei (fig. 2b), increased numbers of heterogeneous nucleoli [16], and, not less relevant, positive staining for 8-hydroxy-2-deoxyguanosine (fig. 2c, small inset), a marker of cellular oxidative DNA damage, leading the exposed monolayer to commit protective cellular suicide that reaches a peak after 30 days of exposure. Besides, images of secondary necrosis have also been detected. This fact in addition to the absence of early acceleration of the cell cycle, as observed with glucose, suggests that the oxidative insult induced by icodextrin appears to be more marked than that derived from the former osmotic agent. This notion is supported by published evidence indicating that increasing levels of oxidative stress can stimulate cell proliferation, block cell growth at the G1-S checkpoint, bring about genotoxicity, apoptosis, or even cause necrosis, all these reactions being the result of a dose-related effect [17].

It should be noticed that studies performed in humans also put in evidence the presence of carbonyl stress during the long dwell commonly used in clinical grounds [18].

The increased proportion of mesothelial cells undergoing apoptosis in the experimental set up has also been detected in effluent of human patients on long-term PD showing a large proportion of mesothelial cells expressing p53 [Gotloib, unpubl. obs.].

Apoptosis of Endothelial Cells Exposed to High Glucose Concentrations

Experiments performed in vitro have demonstrated that endothelial cells treated with high glucose concentrations have higher rates of apoptosis induced by a degree of oxidative stress [19]. Studies done in bovine endothelial cells have shown that advanced glycation end products (AGEs) trigger the oxidative reaction that accelerates endothelial cell apoptosis, this reaction being a critical event in the process of microvascular complications [20]. The effect of this oxidative injury upon the peritoneal microvasculature has not yet been fully investigated. In vitro studies have demonstrated that when human umbilical vein endothelial cells (HUVEC) are incubated with deoxyglucosone, this dicarbonyl compound is internalized by cells in a time-dependent manner, leading to an increase in ROS levels in the cells that subsequently undergo apoptosis [21]. It should be noticed, however, that methylglyoxal, in combination with high glucose in concentrations comparable to those seen in the blood circulation of diabetic patients, has been found to induce apoptosis or necrosis of HUVEC in vitro as a result of oxidative injury [22].

It is known from investigations performed in other microvascular beds that endothelial apoptosis plays an important role in remodeling the vascular network, since this development has been observed at the beginning of neoangiogenesis and at the branching and regression of microvessels of neoformation. The fact that high-dose glucose induces apoptosis in human retinal capillary endothelial cells, an outcome that at least appears to contribute to the development of diabetic retinopathy, suggests that the same mechanism may be involved in the development of the neoangiogenesis observed in the peritoneal microcirculation during long-term PD. Besides, peritoneal biopsies taken from patients on maintenance PD revealed a marked hypertrophy of microvascular endothelial cells that at times occupied a substantial part of the capillary lumen (fig. 2d), suggesting that they may well have been in a situation of terminal replicative senescence.

Apoptosis of Leukocytes Exposed to High Glucose Concentration

In vitro experiments have shown that neutrophils and peripheral blood mononuclear cells exposed to PD heat-sterilized, single-chambered, 4.25% glucose solutions containing high concentrations of GDPs undergo a substantially increased rate of caspase-dependent apoptosis that appeared related to the presence of a high concentration of 3,4-di-deoxyglucosone [23]. This substance appears to be the most active of the α-oxoaldehydes that, together with glyoxal and methylglyoxal, derived from the nonenzymatic degradation of glucose induce, as stated before, oxidative injury.

It may well be assumed that the increased rate of leukocyte death induced by glucose-enriched PD solutions negatively affects the peritoneal defense mechanisms facing infection.

Relevance of Homeostatic Death in Long-Term Peritoneal Dialysis

It becomes evident that all cells lining the peritoneal cavity or living near its immediate environment, exposed most of the time to nonphysiological fluids, undergo changes that lead to apoptotic death or worse, to necrosis. This problem is behind the poor regenerative capabilities showed by the mesothelial monolayer, the microvascular changes that lead to neoangiogenesis, as well as to a defective response to peritoneal infection.

It has been proposed that therapeutic modulation of apoptotic cell death targeted on a single cell population by modifying the activity of elements building the apoptotic pathway could inhibit its progress [24]. On the other hand, experimental work has shown that the use of more biocompatible 3-bag PD fluids containing considerably lower concentrations of GDP than those present in one-bag, heat-sterilized, commonly used solutions, effectively limited the occurrence of apoptosis [25]. Since liberation of oxygen radicals is thereby causally involved in apoptosis, reduced levels of GDPs and/or use of antioxidants appear to be indicated in order to prevent mesothelial, endothelial and leukocyte cell death.

References

1 Ben Porath I, Weinberg RA: The signals and pathways activating cellular senescence. Int J Biochem Cell Biol 2005;37:961–976.

2 Kroemer G, Galluzzi L, Vandenabeele PJ, Abrams J, Alnemri ES, Baehrecke EH, Blagosklonny MV, El-Deiry WS, Golstein P, Green DR, Hengartner M, Knight RA, Kumar S, Lipton SA, Malorni W, Nun G, Peter ME, Tschopp J, Yuan J, Piacentini M, Zhivotovsky B, Melino G: Classification of cell death: recommendations of the Nomenclature Committee on Cell Death 2009. Cell Death Diff 2009;16:3–11.

3 Patel T, Gores GJ, Kaufmann SH. The role of proteases during apoptosis. FASEB J 1996;10:587–559.

4 Messam CA and Pittman RN: Asynchrony and commitment to die during apoptosis. Exp Cell Res. 1998;238:389–398.

5 Frisch SM, Francis H: Disruption of epithelial cell-matrix interactions induces apoptosis. J Cell Biol 1994;124:619–626.

6 Vellai T: Autophagy genes and ageing. Cell Death Diff 2009;16:94–102.

7 Patschan S, Goligorsky MS: Autophagy: the missing link between non-enzymatically glycated proteins inducing apoptosis and premature senescence of endothelial cells? Autophagy 2008;4:521–523.

8 Gastaldello K, Husson C, Dondeigne JP, Vangerweghem JL, Thielemans C: Cytotoxicity of mononuclear cells as induced by peritoneal dialysis fluids: insight into mechanisms that regulate osmotic stress-related apoptosis. Perit Dial Int 2008;28:655–666.

9 Obrosova IG, Fathallah L, Lang HJ: Interaction between osmotic and oxidative stress in diabetic pre-cataractous lens: studies with a sorbitol dehydrogenase inhibitor. Biochem Pharmacol 1999;58:1945–1954.

10 Alexander RT, Grinstein S: Activation of kinases upon volume changes: role in cellular homeostasis. Contrib Nephrol. Basel, Karger, 2006, vol 152, pp 105–124.

11 Matsuoka Y, Yamaguchi A, Nakanishi T, Sugiura T, Kitamura H, Horio M, Takimitsu Y, Ando A, Imai E, Hori M: Response to hypertonicity in mesothelial cells: role of Na / myo-inositol transporter. Nephrol Dial Tansplant 1999;14:1217–1223.

12 Breborowicz A, Polubinska A, Oreopoulos DG. Changes in volume of peritoneal cells exposed to osmotic stress. Perit Dial Int 1999;19:119–123.

13 Gotloib L, Wajsbrot V, Shostak A, Kushnier R: Effect of hyperosmolarity upon the mesothelial monolayer exposed in-vivo and in-situ to a mannitol-enriched dialysis solution. Nephron 1999;81:301–309.
14 Gotloib L, Shostak A, Wajsbrot V, Kushnier R: High glucose induces an hypertrophic senescent mesothelial cell phenotype after long, in-vivo exposure. Nephron 1999;82:164–173.
15 Ishibashi Y, Sugimoto T, Ichikawa Y, Akatsuka A, Miyata T, Nangaku M, Tagawa H, Kurokawa K: Glucose dialysate induces mitochondrial DNA damage in peritoneal mesothelial cells. Perit Dial Int 2002;22:11–21.
16 Gotloib L, Wajsbrot V, Shostak A: Mesothelial dysplastic changes and lipid peroxidation induced by 7.5% icodextrin. Nephron 2002;92:142–155.
17 Ziegler- Skylakakis K, Andrae U: Mutagenicity of hydrogen peroxide in V79 Chinese hamster cells. Mutat Res 1987;192:65–67.
18 Ueda Y, Miyata T, Goffin E, Yoshino A, Inagi R, Ishibashi Y, Izuhara Y, Salito A, Kurokawa K, Van Ypersele De Strihou C: Effect of dwell time on carbonyl stress using icodextrin and amino acid peritoneal dialysis fluids. Kidney Int 2000;58:2518–2524.
19 Du X, Stocklauser-Farber K, Rosen P: Generation of reactive oxygen intermediates, activation of NF-kappaB, and induction of apoptosis in human endothelial cells by glucose: role of nitric oxide synthase? Free Radic Biol Med 1999;27:752–763.
20 Xiang M, Yang M, Zhou M, Li W, Qian Z: Crocetin prevents AGEs-induced vascular endothelial cell apoptosis. Pharmacol Res 2006;54:268–274.
21 Harukiko S, Motoko T, Toshihiro Y, Tadashi T, Ho LS, Yasuhide M, Misonou Y, Naoyuki T: The internalization and metabolism of 3-deoxyglucosone in human umbilical vein endothelial cells. J Biochem 2006;2:245–253.
22 Chan WH, Wu HJ: Methylglyoxal and high glucose co-treatment induces apoptosis or necrosis in human umbilical vein endothelial cells. J Cell Biochem 2008;103:1144–1157.
23 Catalan MP, Santamaria B, Reyero A, Ortiz A, Ejido J, Ortiz A: 3,4-Di-deoxyglucosone-3-ene promotes leukocyte apoptosis. Kidney Int 2005;68:1303–1311.
24 Ortiz O, Catalan MP: Will modulation of cell death increase technique survival? Perit Dial Int 2004;24:105–114.
25 Catalan MP, Revero A, Egido J, Ortiz A: Acceleration of neutrophil apoptosis by glucose-containing peritoneal dialysis solutions: role of caspases. J Am Soc Nephrol 2001;12:2442–2449.

Lazaro Gotloib, MD
Laboratory for Experimental Nephrology, Ha'Emek Medical Center
POB 2886
Menahem Ussishkin 74
Afula 18248 (Israel)
Tel./Fax +972 4 6591537, E-Mail gotloib@012.net.il

Ronco C, Crepaldi C, Cruz DN (eds): Peritoneal Dialysis – From Basic Concepts to Clinical Excellence. Contrib Nephrol. Basel, Karger, 2009, vol 163, pp 45–53

Different Aspects of Peritoneal Damage: Fibrosis and Sclerosis

Guido Garosi

UOC Nefrologia, Dialisi e Trapianto, Azienda Ospedaliera Universitaria Senese, Siena, Italy

Abstract

Peritoneal sclerosis is very common in peritoneal dialysis patients. It can vary from the mild, clinically silent simple sclerosis always present after years of peritoneal dialysis, to rare but dramatic and often fatal cases of encapsulating peritoneal sclerosis. Their frequency, pathology, animal models, pathogenesis, clinical manifestations, therapy, and prevention are reviewed. Peritoneal fibrosis and sclerosis is a complex phenomenon. At one end of the spectrum we have simple sclerosis: the mild, clinically silent sclerosis always present after years of peritoneal dialysis. At the other end, we have encapsulating peritoneal sclerosis, which is rare but dramatic and fatal.

Simple Sclerosis

Frequency

SS can be demonstrated in all peritoneal dialysis (PD) patients [1–3]; it is often present even during chronic kidney failure before dialysis [2, 3].

Pathology

Table 1 shows our data on the histologic characteristics of 180 patients with simple sclerosis (SS) and 44 cases of encapsulating peritoneal sclerosis (EPS). SS seems to be just one of the morphological alterations always associated with PD; it is mainly a parietal alteration [4]. There is very scarce evidence of inflammation and calcification, whereas a slight vasculopathy is present. The histological picture is quite monotonous.

Table 1. Pathology of SS and EPS (median and range; number of cases)

	SS (n = 180)	EPS (n = 44)	p
Thickness of sclerosis, μm	45 (10–70)	750 (250–4,000)	<0.01
Inflammation	5/180	44/44	<0.01
Parvicellular infiltration	5/180	40/44	<0.01
Mild	5/180	0/44	
Severe	0/180	40/44	
Microabscesses	0/180	17/44	<0.05
Giant cells	0/180	39/44	<0.01
Granulation tissue	0/180	39/44	<0.01
Vascular alterations	19/180	44/44	<0.01
Arterial thickening	19/180	44/44	<0.01
Mild	19/180	0/44	
Severe	0/180	44/44	
Arterial occlusion	0/180	41/44	<0.01
Arterial calcification	0/180	26/44	<0.01
Arterial ossification	0/180	9/44	
Tissue calcification	1/180	13/44	<0.01
Tissue ossification	0/180	4/44	
Presence of bone marrow	0/180	2/44	

Statistical analysis: Mann-Whitney test (thickness of sclerosis), χ^2 test (other variables).

Animal Models

Rats and rabbits treated with PD show all the typical mesothelial and sub-mesothelial morphological modifications induced by PD in humans, including SS [5, 6], with no need for the action of any other agent on the peritoneum. These phenomena are constant and reproducible, making these models suitable for comparing the biocompatibility of different PD solutions. No case of SS has ever been described in animals in the absence of PD.

Pathogenesis

There is substantial agreement that the poor biocompatibility of PD (related to glucose, hyperosmolarity, low pH and lactate buffer) is the main reason why SS is constant in PD: many studies demonstrated the consequences of PD on mesothelial cells, macrophages, neutrophils, lymphocytes, and fibroblasts. This research provides biochemical documentation of functional changes in cells exposed to PD solutions [7, 8].

Clinical Manifestations

SS has no clinical manifestation [1–4]: this is not surprising in view of the minor anatomical alterations. It is generally agreed that SS is associated with a loss of peritoneal dialytic capacity [9].

Encapsulating Peritoneal Sclerosis

Frequency

EPS is rare, with a low prevalence (0.5–3.5%) and an incidence of less than 5 cases per thousand patient years; its frequency increases with time of PD and membrane permeability [10]. Recently, an increased incidence of EPS has been reported after kidney transplantation [11].

Pathology

Our data (table 1) show striking pathologic differences between SS and EPS. In EPS sclerosis involves the whole thickness of the peritoneum; its thickness shows a bimodal distribution, with a huge difference between SS and EPS, and a clear gap between the two. In EPS, sclerosis is more evident in the visceral peritoneum (visceral peritoneum median = 1,200 μm, range 600–4,000 μm; parietal peritoneum median = 450 μm, range = 250–2,000 μm; Wilcoxon test $p < 0.05$), at variance with SS (visceral peritoneum median = 25 μm, range 10–40 μm; parietal peritoneum median = 50 μm, range 30–70 μm; Wilcoxon test $p < 0.05$).

Inflammation is present in all our patients with EPS, as opposed to SS, with aspects related to both acute and chronic inflammation.

Milky spots (submesothelial clusters of lymphoid tissue considered the site of origin of peritoneal resident white cells) are not influenced by SS [12], while EPS is associated with a heavy decrease in milky spots: of 44 cases, we observed just 3 milky spots in 2 patients. In EPS, milky spots are not near the mesothelium, but deep under the sclerosis, maybe with different functional characteristics.

In our data [13], vasculopathy and calcifications in EPS are of another order of magnitude with respect to SS. In EPS, vessel occlusion is the hallmark, often coupled with vessel calcification and even ossification. At high magnification real osteoclasts can be observed. In 2 cases we noticed even bone marrow inside islands of ossification. The calcifications are often mixed with inflammatory infiltrate and giant cells, and can therefore be considered dystrophic.

However, the descriptions of EPS pathology by Japanese authors [10, 14, 15] usually do not mention significant inflammation, calcification and vasculopathy, or state that these changes can be found only in the advanced stage of clinical EPS. In these descriptions, there is no clear difference between SS and EPS. Therefore, according to this point of view, EPS pathology could be considered as exaggerated SS alterations.

Recently, Nakayama's group [16] compared the peritoneal pathology in 12 EPS patients Vs 23 non-EPS PD cases. The EPS group showed a significantly increased thickness of degenerated submesothelial layer and fibrin deposition, whereas no differences were found in the incidence of mesothelial detachment, new membrane formation, compact zone thickness, and vascular alterations. The conclusion suggests that angiogenesis, vasculopathy, new membrane formation, fibrosis, and degenerative changes of the compact zone are common findings in both EPS and SS, and are not unique characteristics for EPS. In this study the groups were comparable on age, gender and PD duration. Nevertheless, none of the 12 EPS patients withdrew PD because of intestinal problems (ultrafiltration failure in 8 cases, increased salt intake in 2, patient preference in 2) and just 5 of them were treated (with prednisone).

Animal Models

A spontaneous pathologic process similar to EPS with vomiting, abdominal pain, ascites and palpable intestinal masses has been described [17] in dogs and cats, unrelated to PD. EPS has never been reproduced in animals by means of PD, whereas it has been induced without PD [18] by introducing various substances into the abdomen: chlorhexidine, household bleach, silica, talc, crocidolite, glass fibers, polypropylene, kevlar, asbestos.

Pathogenesis

Some authors consider SS and EPS as two stages of the same disorder [15]. In this view, the time of PD and the degree of PD bioincompatibility are crucial in determining the onset of progressive sclerosis. So, EPS is considered as the final evolution of SS, which will be reached sooner or later by any PD patient.

In our opinion, this is not a correct view: we believe that SS and EPS are two different nosological entities [1, 4]. Unlike for SS, no single causal factor has been identified which is alone sufficient to generate EPS. Only a series of risk factors have been identified, which may or may not be related to PD. A link between duration of dialysis and incidence of EPS [15] is well established, but it is not mandatory: EPS may arise very early in PD [1, 10], and often develops only after kidney transplantation [11]. The poor biocompatibility of PD is an obvious risk factor for EPS [1, 15], but the better biocompatibility of recent PD solutions does not seem to be associated with a reduction in the frequency of EPS. Even peritonitis, the most commonly invoked [15] pathogenetic factor for EPS, is far from mandatory. Overall, PD is a serious risk factor for EPS, but not an etiological factor.

In human beings there is no evidence of SS unrelated to kidney failure. On the contrary, EPS in nonrenal patients is a well known nosological entity, first described by Owtschinnikow [19] in 1907. The number of cases of spontaneous EPS greatly exceeds that of the PD-related form [1, 4]. These non-dialytic forms may be associated with the use of ß-blockers (by decrease in ultrafiltration and inhibition of surfactant release), the presence of tumors (as a paramalignant phenomenon), or may be idiopathic. In these idiopathic cases two considerations are important. The first is their association with a general connective tissue impairment, particularly of the serous membranes [20], suggesting an immune pathogenesis. The second is genetic predisposition, suggested by the high frequency in women from subtropical areas [21], and familial forms such as familial multifocal fibrosclerosis [22].

Finally, a two-hit hypothesis has been proposed [23]: SS predisposes the patient to a second hit (peritonitis, discontinuations of PD, genetic predisposition, etc.) that triggers the process towards EPS.

Only prospective studies in a significant number of PD patients can establish without any doubt if SS and EPS are different disorders or different stages of the same disorder. These studies should include clinical characteristics, genetic profiles and close monitoring of the patients. A very interesting proposal for an international EPS registry and DNA bank was made in 2006 [24], and endorsed by the ISPD International Studies Committee [25].

Clinical Manifestations

The symptoms associated with EPS [1, 3, 4, 10, 15, 23] include anorexia, nausea, vomiting, diarrhea, constipation, abdominal distension, fever, weight loss, abdominal pain, palpable abdominal mass, partial or complete bowel obstruction, hemorrhagic effluent, and ascites. Onset is often insidious, with vague bowel symptoms. In this phase, the concomitant finding of loss of dialytic efficiency, weight loss and hemorrhagic effluent is suggestive of the diagnosis. In other cases onset may be acute, manifesting directly as bowel obstruction. Ascites is an important diagnostic finding in patients developing EPS after suspension of PD. The mortality of EPS is high (26–93%).

Diagnosis

Plain X-ray and contrast studies such as barium meal and small bowel follow-through examination [1, 3, 4, 10, 15, 23] may be completely negative. Signs suggesting EPS include dilated small bowel loops with gas-fluid levels, calcifications in the peritoneum and intestinal wall and, sometimes, bowel wall thickening. In contrast studies, signs suggesting EPS are motility disturbances with delayed transit, varying degrees of obstruction accompanied by hypermobility of some loops, and separation of small bowel loops which appear fixed and rigid.

Ultrasound studies [1, 3, 4, 10, 15, 23, 26] are more specific: besides the usual signs of obstruction and loculated ascites, the typical thickening of the intestinal wall can be observed. This can show a sandwich appearance, with two outer echogenic layers and one central echopoor layer.

Computed tomography [1, 3, 4, 10, 15, 23] is more accurate, providing more detail of obstruction, loculated ascites, calcifications and thickening of the peritoneum. Nevertheless, it has been proven valid and reliable in the diagnosis of EPS, rather than as a screening tool [27].

Recently, positron emission tomography has been suggested as a useful adjunct in the diagnosis of EPS [28].

Although histological examination is an invasive method, it is often the only way to obtain a definite diagnosis, because many bowel diseases may mimic EPS in clinical manifestations, sonography and radiology. This is also the only reliable method to check the degree of inflammation associated with EPS [1, 3, 4, 10, 15, 23].

Therapy

Surgery is usually reserved for intestinal obstruction [1, 3, 4, 10, 15, 23]. The elective method is membrane resection; alternatively enterolysis with partial excision of the membrane or intestinal resection can be performed. The high mortality is mainly due to post-operative complications, typically the opening of intestinal anastomoses. Surgery has recently been proposed also for prevention of recurrence [29].

Interruption of PD and catheter removal [1, 3, 4, 10, 15, 23] often result in considerable improvement in symptoms and some regression of the anatomical lesions. In other cases, however, these measures seem to precipitate a worsening of the situation. PD removes fibrin from the peritoneum and keeps the bowel loops apart so they cannot adhere. When there is a high rate of fibrin production, suspension of PD may be counterproductive.

Many cases of EPS have been treated with steroids and immunosuppressants [1, 3, 4, 10, 15, 23]. Steroids have been used on their own or with cyclophosphamide, azathioprine or colchicine. Most authors report success with good long-term prognosis or at least longer survival and well-being; a minority of cases fails to improve with the treatment. Recently, rapamicin [30], everolimus [31] and mycophenolate mofetil [32] have been successfully used.

Tamoxifen is widely used with good results in EPS patients on the basis of its antagonism towards TGFβ-1 [33].

Other therapies described include total parenteral nutrition [34], progesterone [35], antibiotics [1], and intraperitoneal administration of phosphatidylcholine [36].

Prevention

The poor prognosis of EPS implies that prevention should be very important. Unfortunately, our lack of knowledge about etiology impairs this task. No real prevention of EPS exists at the moment [1, 3, 4, 10, 15, 23]. Nevertheless it could be useful to reduce as much as possible the bioincompatibility of PD by using solutions with bicarbonate buffer and osmotic agents different from glucose (amino acids, glucose polymers). Prevention and treatment of peritonitis should be an absolute priority. Patients should be periodically evaluated [9] for peritoneal transport and ultrafiltration, with a thorough sonographical, radiological and histological study of any dubious case.

References

1 Garosi G, Di Paolo N, Sacchi G, Gaggiotti E: Sclerosing peritonitis: a nosological entity. Perit Dial Int 2005;25(S3):S110–S112.

2 Honda K, Hamada C, Nakayama M, Miyazaki M, Sherif AM, Harada T, Hirano H: Impact of uremia, diabetes, and peritoneal dialysis itself on the pathogenesis of peritoneal sclerosis: a quantitative study of peritoneal membrane morphology. Clin J Am Soc Nephrol 2008;3:720–728.

3 Williams JD, Craig KJ, Topley N, Von Ruhland C, Fallon M, Newman GR, MacKenzie RK, Williams GT: Morphologic changes in the peritoneal membrane of patients with renal disease. J Am Soc Nephrol 2002;13:470–479.

4 Garosi G, Di Paolo N: Peritoneal sclerosis: one or two nosological entities? Semin Dial 2000; 13:297–308.

5 Gotloib L, Wajsbrot V, Shostak A: A short review of experimental peritoneal sclerosis: from mice to men. Int J Artif Organs 2005;28:97–104.
6 Garosi G, Gaggiotti E, Monaci G, Brardi S, Di Paolo N: Biocompatibility of a peritoneal dialysis solution with amino acids: histological study in the rabbit. Perit Dial Int 1998;18:610–619.
7 De Vriese AS, Tilton RG, Mortier S, Lameire NH: Myofibroblast transdifferentiation of mesothelial cells is mediated by RAGE and contributes to peritoneal fibrosis in uremia. Nephrol Dial Transplant 2006;21:2549–2555.
8 Kolesnyk I, Dekker FW, Noordzij M, le Cessie S, Struijk DG, Krediet RT: Impact of ACE inhibitors and AII receptor blockers on peritoneal membrane transport characteristics in long-term peritoneal dialysis patients. Perit Dial Int 2007;27:446–453.
9 Sampimon DE, Coester AM, Struijk DG, Krediet RT: Time course of peritoneal transport parameters in peritoneal dialysis patients who develop peritoneal sclerosis. Adv Perit Dial 2007;23:107–111.
10 Kawaguchi Y, Kawanishi H, Mujais S, Topley N, Oreopoulos DG: Encapsulating peritoneal sclerosis: definition, etiology, diagnosis, and treatment. International Society for Peritoneal Dialysis ad Hoc Committee on Ultrafiltration Management in Peritoneal Dialysis. Perit Dial Int 2000;20(S4):S43–S55.
11 Fieren MWJA, Betjes MGH, Korte MR, Boer WH: Posttransplant encapsulating peritoneal sclerosis: a worrying new trend? Perit Dial Int 2007;27:619–624.
12 Di Paolo N, Sacchi G, Garosi G, Sansoni E, Bargagli L, Ponzo P, Tanganelli P, Gaggiotti E: Omental milky spots and peritoneal dialysis: review and personal experience. Perit Dial Int 2005;25:48–57.
13 Garosi G, Di Paolo N: Inflammation and gross vascular alterations are characteristic histological features of sclerosing peritonitis. Perit Dial Int 2001;21:417–418.
14 Honda K, Nitta K, Horita S, Tsukada M, Itabashi M, Nihei H, Takashi A, Oda H: Histologic criteria for diagnosing encapsulating peritoneal sclerosis in continuous ambulatory peritoneal dialysis patients. Adv Perit Dial 2003;19:169–175.
15 Nakayama M, Maruyama Y, Numata M: Encapsulating peritoneal sclerosis is a separate entity: con. Perit Dial Int 2005;25(S3):S107–S109.
16 Sherif AM, Yoshida H, Maruyama Y, Yamamoto H, Yokoyama K, Hosoya T, Kawakami M, Nakayama M: Comparison between the pathology of encapsulating sclerosis and simple sclerosis of the peritoneal membrane in chronic peritoneal dialysis. Ther Apher Dial 2008;12:33–41.
17 Hardie EM, Rottman JB, Levy JK: Sclerosing encapsulating peritonitis in four dogs and a cat. Vet Surg 1994;23:107–114.
18 Komatsu H, Uchiyama K, Tsuchida M, Isoyama N, Matsumura M, Hara T, Fukuda M, Kanaoka Y, Naito K: Development of a peritoneal sclerosis rat model using a continuous-infusion pump. Perit Dial Int 2008;28:641–647.
19 Owtschinnikow PJ: Peritonitis chronica fibrosa incapsulata. Arch Klin Chir 1907;83:623–634.
20 Narayanan R, Bhargava BN, Kabra SG, Sangal BC: Idiopathic sclerosing encapsulating peritonitis. Lancet 1989;ii:127–129.
21 Dehn TC, Lucas MG, Wood RF: Idiopathic sclerosing peritonitis. Postgrad Med J 1985;61:841–842.
22 Comings DE, Skubi KD, Van Eyes J, Motulsky AG: Familial multifocal fibrosclerosis. Ann Intern Med 1967;66:884–892.
23 Augustine T, Brown PW, Davies SD, Summers AM, Wilkie ME: Encapsulating peritoneal sclerosis: clinical significance and implications. Nephron Clin Pract 2009;111:c149–c154.
24 Summers AM, Brenchley PEC: An international encapsulating peritoneal sclerosis registry and DNA bank: why we need one now. Perit Dial Int 2006;26:559–563.
25 Topley N: Encapsulating peritoneal sclerosis: time to act. Perit Dial Int 2006;26:564–565.
26 Duman S, Ozbek SS, Gunay ES, Bozkurt D, Asci G, Sipahi H, Kircelli F, Ertilav M, Ozkahya M, Ok E: What does peritoneal thickness in peritoneal dialysis patients tell us? Adv Perit Dial 2007;23:28–33.
27 Tarzi RM, Lim A, Moser S, Ahmad S, George A, Balasubramaniam G, Clutterbuck EJ, Gedroyc W, Brown EA: Assessing the validity of an abdominal CT scoring system in the diagnosis of encapsulating peritoneal sclerosis. Clin J Am Soc Nephrol 2008;3:1702–1710.

28 Tarzi RM, Frank JW, Ahmad S, Levy JB, Brown EA: Fluorodeoxyglucose positron emission tomography detects the inflammatory phase of sclerosing peritonitis. Perit Dial Int 2006;26:224–230.
29 Kawanishi H, Ide K, Yamashita M, Shimomura M, Moriishi M, Tsuchiya S, Dohi K: Surgical techniques for prevention of recurrence after total enterolysis in encapsulating peritoneal sclerosis. Adv Perit Dial 2008;24:51–55.
30 Rajani R, Smyth J, Abbs I, Goldsmith JA: Differential effect of sirolimus vs prednisolone in the treatment of sclerosing encapsulating peritonitis. Nephrol Dial Transplant 2002;17:2278–2280.
31 Ricciatti AM, Goteri G, D'Arezzo M, Sagripanti S, Bibiano L, Petroselli F, Fabris G, Frascà GM: Peritonite sclerosante trattata con successo con associazione di steroidi, everolimus e tamoxifene (abstract). Atti XIV Convegno Gruppo di Studio di Dialisi Peritoneale, Modena, Italy, 7–9 febbraio 2008, p 20.
32 Lafrance JP, Létourneau I, Ouimet D, Bonnardeaux A, Leblanc M, Mathieu N, Michette V: Successful treatment of encapsulating peritoneal sclerosis with immunosoppressive therapy. Am J Kidney Dis 2008;51:e7–e10.
33 Wong CF: Clinical experience with tamoxifen in encapsulating peritoneal sclerosis. Perit Dial Int 2006;26:183–184.
34 Pusateri R, Ross R, Marshall R, Meredith JH, Hamilton RW: Sclerosing encapsulating peritonitis: report of a case with small bowel obstruction managed by long-term home parenteral hyperalimentation, and a review of the literature. Am J Kidney Dis 1986;8:56–60.
35 Mazure R, Marty FP, Niveloni S, Pedreira S, Vazquez H, Smecuol E, Kogan Z, Boerr L, Maurino E, Bai JC: Successful treatment of retractile mesenteritis with oral progesterone. Gastroenterology 1998;114:1313–1317.
36 Struijk DG, van der Reijden HJ, Krediet RT, Koomen GC, Arisz L: Effect of phosphatidylcholine on peritoneal transport and lymphatic absorption in a CAPD patient with sclerosing peritonitis. Nephron 1989;51:577–578.

Guido Garosi, MD
UOC Nefrologia, Dialisi e Trapianto, Azienda Ospedaliera Universitaria Senese
Viale Bracci, 16
IT–53100 Siena (Italy)
Tel. +39 0577 586307, Fax +39 0577 586149, E-Mail g.garosi@ao-siena.toscana.it

Ronco C, Crepaldi C, Cruz DN (eds): Peritoneal Dialysis – From Basic Concepts to Clinical Excellence. Contrib Nephrol. Basel, Karger, 2009, vol 163, pp 54–59

Biological Markers in the Peritoneal Dialysate Effluent: Are They Useful

Raymond T. Krediet[a], *Denise E. Sampimon*[a], *Anniek Vlijm*[a], *Annemieke M. Coester*[a], *Dirk G. Struijk*[a], *Watske Smit*[a,b]

[a]Division of Nephrology, Department of Medicine, Academic Medical Centre University of Amsterdam, and [b]Dianet Foundation Utrecht, Amsterdam, The Netherlands

Abstract

A review is given on biomarkers in peritoneal effluent. It comprises methods to distinguish between diffusion and local production. This is followed by examples of various biomarkers. Their potential use is discussed in 4 situations: inherent fast transporters, longitudinal follow-up of patients, biocompatibility testing of new dialysis solutions, and their potential use in the detection of patients who are likely to develop encapsulating peritoneal sclerosis.

Peritoneal effluent of peritoneal dialysis (PD) patients contains many substances. The origin of these can either be diffusion from the circulation, local peritoneal production or both. In the present review, a survey will be given on methods to distinguish between local production and diffusion, on examples of the various biomarkers, and on their usefulness in clinical practice.

Local Production or Diffusion

Diffusion of a substance can be neglected when the effluent concentration of a high-molecular-weight substance approaches or exceeds the serum concentration, for instance cancer antigen-125 (CA125) or interleukin-6 (IL-6). For other solutes, attempts should be made to distinguish between the two. The peritoneal diffusion rate of a substance is dependent on its diffusion in a watery

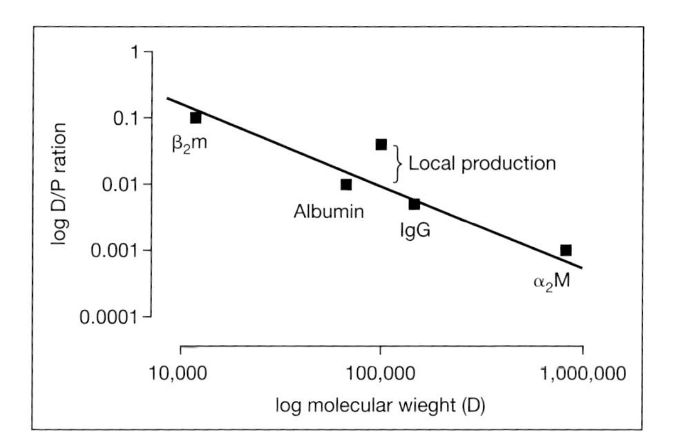

Fig. 1. The principle of assessment of local production of a substance. A transport line is constructed by plotting the D/P ratio of 4 proteins against their molecular weight on a double logarithmic scale. Interpolation of the substance of interest allows the calculation of the amount locally produced. β_2m = β_2-Microglobulin, IgG = immunoglobulin G; α_2M = α_2-macroglobulin.

solution (D_W), which is determined by molecular weight (MW) and shape. Power relationships, that means linear when plotted on a double logarithmic scale, have been shown for correlations between peritoneal clearances or mass transfer area coefficients (MTAC) of various solutes and their MW [1, 2] or D_W [3]. When peritoneal transport rates of various solutes known to be transported only are plotted against their MW or D_W on a double logarithmic scale, a linear transport line can be made in every patient. Interpolation of the MW or D_W of the solute under investigation on the transport line, predicts the expected effluent concentration by diffusion only. The difference with the actually measured concentration is the concentration due to local production or release. This is shown graphically in figure 1.

In case the MW of a solute is very similar to that of creatinine, local production can also be analyzed by dividing the dialysate/plasma (D/P) ratio of that solute by D/P creatinine. These 3 approaches discussed above are only valid for solutes that are not bound to proteins. For instance, transforming growth factor-β (TGF-β) in the circulation is bound to α_2-macroglobulin, which makes a precise analysis of local production impossible.

Examples of Various Biomarkers

Cross-sectional studies have shown that effluent concentrations of the following biomarkers are explained by diffusion only: tumor necrosis factor-α in stable patients [2], nitrate [4], secretory phospholipase A_2 [4], hydroxyproline [5] and various complement factors [6]. Results for complement 3d are equivocal [6, 7].

Local production or release in the peritoneal cavity has been shown for various cytokines and prostanoids [8], i.e. growth factors like vascular endothelial growth factor (VEGF) [9] which is a marker for neoangiogenesis. Also, connective tissue growth factor [unpubl.], coagulation and fibrinolytic factors [10], and various tissue markers [11] are released locally. These include CA125 which is a marker of mesothelial cell mass and/or turnover [12], and various glycosaminoglycans like hyaluronan [11]. Also, matrix metalloproteinase 2 and its tissue inhibitor-1 are produced locally [unpubl.]. In addition to all the above-mentioned factors, K^+ is released from cells during a hypertonic exchange [13]. The amount of K^+ efflux was related to free water transport. This means that an increase in free water transport caused by the hypertonicity of the dialysis fluid was accompanied by an increase in K^+ efflux from cells. Therefore, the phenomenon is probably caused by hypertonic cell shrinkage. This leads to efflux of potassium through K^+ channels to prevent too high intracellular concentrations of potassium. The relationship of K^+ efflux with free water transport was absent in long-term patients with ultrafiltration failure. Apoptosis is accompanied by an increased efflux of K^+ from cells. Therefore, these findings suggest a reduced apoptosis and thereby a tendency for cell proliferation.

How Useful Are Biomarkers?

Most biomarkers have been analyzed using cross-sectional observations. Some have also been studied during peritonitis. These results are of interest, but are not required for the diagnosis or treatment. Taking all the data together, CA125, IL-6 and VEGF in effluent seem the most useful ones at the moment. Their application will be discussed in 4 situations: patients with an inherent fast transport status, during longitudinal follow-up, as marker of biocompatibility of dialysis solutions, and as diagnostic markers for the development of encapsulating peritoneal sclerosis (EPS).

Inherent Fast Transport Status

Two types of this condition can be distinguished. One is associated with comorbidity. These patients have high effluent concentrations of IL-6 and VEGF [14]. The other type is associated with high concentrations of effluent CA125 and VEGF [15, 16]. It is likely that a large mesothelial cell mass leads to a high production rate of VEGF and thereby causes peritoneal vasodilation. Measurement of IL-6 and CA125 might therefore be useful to distinguish between the two types.

Longitudinal Follow-Up

Effluent CA125 decreases during follow-up [17] while VEGF has a tendency to increase, although the interindividual variation is large [18]. Longitudinal follow-up of incident PD patients showed that the time course of D/P creatinine (which reflects the vascular peritoneal surface area) was both associated with effluent VEGF and IL-6, but the effect of IL-6 was somewhat bigger than that of VEGF [19].

Biocompatibility of Dialysis Solutions

The application of all new dialysis solutions with a reduced content of glucose degradation products and a higher pH than in conventional solutions have been associated with increased concentrations of effluent CA125. This is the case for Gambrosol trio® [20], Physioneal® [21], Balance® [22] and Bicaveta® [23]. It suggests an increase in mesothelial cell mass in prevalent patients who had been treated with conventional solutions before. These data make it likely that effluent CA125 is a good marker for biocompatibility of dialysis solutions.

Effluent Markers in the Development of Encapsulating Peritoneal Sclerosis (EPS)

The diagnosis of EPS is made when clinical symptoms of bowel obstruction are present, confirmed by CT scanning [24] or at laparotomy. Once the diagnosis has been made, most patients have a poor prognosis. Therefore, it would be useful to have tests available that can predict its development. Almost all EPS patients have ultrafiltration failure, but fortunately EPS occurs only in a minority of these patients. Recently, we found trends in effluent IL-6 and CA125 2–3 years preceding the clinical diagnosis of EPS [unpubl.]. Effluent CA125 decreased, while IL-6 increased as shown in figure 2. In a case-control analysis, a low CA125 combined with a high IL-6 had a sensitivity of 70% and a specificity of 89%. It may be that changes in peritoneal fluid and solute transport, in combination with time trends in biomarkers, are useful to predict the development of EPS.

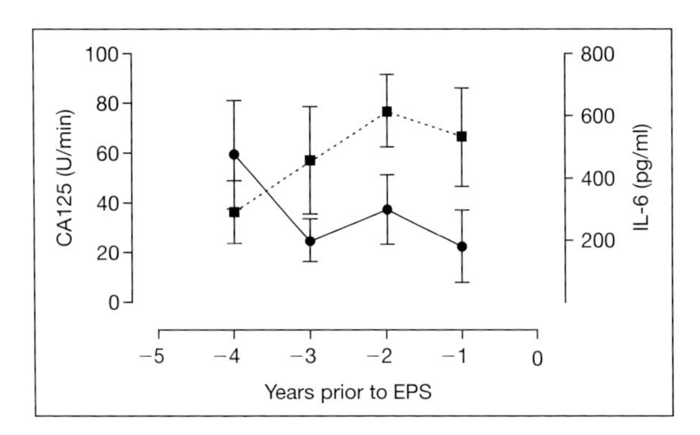

Fig. 2. The time-course of the effluent appearance rates of CA125 and IL-6 during the years preceding the clinical diagnosis of encapsulating peritoneal sclerosis (time point zero). Drawn line = CA125; dotted line = IL-6.

Conclusion

Biomarkers in effluent that are produced or released locally are the most interesting ones. Especially CA125, IL-6 and to a lesser degree VEGF are useful for study purposes and in clinical practice.

References

1 Lasrich M, Maher JM, Hirszel P, Maher JF: Correlation of peritoneal transport rates with molecular weight: a method for predicting clearances. ASAIO J 1079;2:107–113.

2 Zemel D, Imholz ALT, de Waart DR, Dinkla C, Struijk DG, Krediet RT: Appearance of tumor necrosis factor α and soluble TNF-receptors I and II in peritoneal effluent of CAPD. Kidney Int 1994;46:1422–1430.

3 Imholz ALT, Koomen GCM, Struijk DG, Arisz L, Krediet RT: Effect of dialysate osmolarity on the transport of low-molecular weight solutes and proteins during CAPD. Kidney Int 1993;43:1339–1346.

4 Douma CE, de Waart DR, Struijk DG, Krediet RT: Are phospholipase A_2 and nitric oxide involved in the alterations in peritoneal transport during CAPD peritonitis? J Lab Clin Med 1998;132:329–340.

5 Vlijm A, de Waart DR, Zweers MM, Krediet RT: Effluent hydroxyproline in experimental peritoneal dialysis. Perit Dial Int 2007;27:210–213.

6 Reddingius RE, Schröder CH, Daha MR, Willems JL, Koster AM, Monnens LAH: Complement in serum and dialysate in children on continuous ambulatory peritoneal dialysis. Perit Dial Int 1995;15:49–53.

7 Krediet RT, Asghar SS, Koomen GCM, Venneker GT, Struijk DG, Arisz L: Effects of renal failure on complement C3d levels. Nephron 1991;59:41–45.

8 Zemel D, Koomen GCM, Hart AAM, ten Berge RJM, Struijk DG, Krediet RT: Relationship of TNF-α, interleukin-6 and protaglandins to peritoneal permeability for macromolecules during

longitudinal follow-up of peritonitis in continuous ambulatory peritoneal dialysis. J Lab Clin Med 1993;122:686–696.
9 Zweers MM, de Waart DR, Smit W, Struijk DG, Krediet RT: The growth factors VEGF and TGF-beta in peritoneal dialysis. J Lab Clin Med 1999;134:124–132.
10 De Boer AW, Levi M, Reddingius RE, Willems HL, van den Bosch S, Schröder CH, Monnens LAH: Intraperitoneal hypercoagulation and hypofibrinolysis is present in childhood peritonitis. Pediatr Nephrol 1999;13:281–287.
11 Pannekeet MM, Zemel D, Koomen GCM, Struijk DG, Krediet RT: Dialysate markers of peritoneal tissue during peritonitis and in stable CAPD. Perit Dial Int 1995;15:217–225.
12 Krediet RT: Dialysate cancer antigen 125 concentration as marker of peritoneal membrane status in patients treated with chronic peritoneal dialysis. Perit Dial Int 2001;21:560–567.
13 Coester AM, Struijk DG, Smit W, de Waart DR, Krediet RT: The cellular contribution to effluent potassium and its relation with free water transport during peritoneal dialysis. Nephrol Dial Transplant 2007;22:3593–3600.
14 Pecoits-Filho R, Araujo MRT, Lindholm B, et al: Plasma and dialysate IL-6 and VEGF concentrations are associated with high peritoneal solute transport rate. Nephrol Dial Transplant 2002;17:1480–1486.
15 Rodrigues A, Martins M, Santos MJ, et al: Evaluation of effluent markers cancer antigen 125, vascular endothelial growth factor and interleukin-6. Relationship with peritoneal transport. Adv Perit Dial 2004;20:8–12.
16 Van Esch S, Zweers MM, Jansen MAM, de Waart DR, van Manen JG, Krediet RT: Determinants of peritoneal solute transport rates in newly started nondiabetic peritoneal dialysis patients. Perit Dial Int 2004;24:554–561.
17 Ho-dac-Pannekeet MM, Hiralall JK, Struijk DG, Krediet RT: Longitudinal follow-up of CA-125 in peritoneal effluent. Kidney Int 1997;51:888–893.
18 Zweers MM, Struijk DG, Krediet RT: Vascular endothelial growth factor in peritoneal dialysis. J Lab Clin Med 2001;137:125–132.
19 Rodrigues AS, Martins M, Korevaar JC, et al: Evaluation of peritoneal transport and membrane status in peritoneal dialysis: focus on incident fast transporters. Am J Nephrol 2007;27:84–91.
20 Rippe B, Simonsen O, Heimburger O, et al: Long-term clinical effects of a peritoneal dialysis fluid with less glucose degradation products. Kidney Int 2001;59:348–357.
21 Jones S, Holmes CJ, Krediet RT, et al: Bicarabonate/lactate-based peritoneal dialysis solution increases cancer antigen 125 and decreases hyaluronic acid levels. Kidney Int 2001;59:1329–1338.
22 Williams JD, Topley N, Craig KJ, et al: The Euro-Balance trial: the effect of a new biocompatible peritoneal dialysis fluid (balance) on the peritoneal membrane. Kidney Int 2004;66:408–418.
23 Haas S, Schmitt P, Arbeites K, et al: Improved acidosis correction and recovery of mesothelial cell mass with neutral pH bicarabonate dialysis solution among children undergoing automated peritoneal dialysis. J Am Soc Nephrol 2003;14:2632–2638.
24 Vlijm A, Stoker J, Bipat S, et al: CT findings characteristics for encapsulating peritoneal sclerosis: a case-control study. Perit Dial Int 2009;in press.

R.T. Krediet, MD, PhD
Room F4–215, Academic Medical Center, Division of Nephrology, Department of Medicine
PO Box 22700
NL–1100 DE Amsterdam (The Netherlands)
Tel. +31 20 566 5990, Fax +31 20 691 4904, E-Mail C.N.deboer@amc.uva.nl

Ronco C, Crepaldi C, Cruz DN (eds): Peritoneal Dialysis – From Basic Concepts to Clinical Excellence. Contrib Nephrol. Basel, Karger, 2009, vol 163, pp 60–66

Peritoneal Angiogenesis in Response to Dialysis Fluid

Bengt Rippe

Department of Nephrology, University Hospital of Lund, Lund, Sweden

Abstract

In patients on peritoneal dialysis (PD) ultrafiltration failure (UFF) often develops after 3 or 4 years of treatment. Increased angiogenesis, leading to an increase in small solute transport, is a key feature of UFF. Among the pathophysiological mechanisms responsible for an increased angiogenesis, peritonitis, or just 'low-grade inflammation', such as in the uremic condition per se seem to be of importance. However, also the interaction of the peritoneal membrane with peritoneal dialysis fluids (PDF), especially those containing glucose and glucose degradation products (GDP) may be crucial in triggering an increased angiogenesis. This brief review summarizes some recent experimental evidence that PDF low in glucose and GDP, or with an alternative buffer, pyruvate, may partly prevent angiogenesis in long-term PD. The fact that rat models of PD, in which catheters are used for instillation of the PDF, usually show an exaggerated neoangiogenesis, compared to rat models in which the PDF was administered by daily intraperitoneal injections, is also commented upon. To prevent angiogenesis in PD all precautions should be taken to provide a peritonitis-free PD. Although not directly proven, the use of low-GDP solutions should be preferred to GDP-containing solutions. Furthermore, ACE inhibitors have recently been shown to be of value in preventing increases in small solute transport in long-term PD. These findings await confirmation in randomized controlled trials.

More than 30% of patients treated with long-term peritoneal dialysis (PD) eventually develop peritoneal membrane failure (PMF), resulting in a reduced ability of removing excess fluid from the body, so-called ultrafiltration failure (UFF) [1]. Morphologically, PMF is usually manifested by a thickening of the submesothelial (compact) zone, that becomes fibrotic, and an increased angiogenesis, resulting in an increased number of peritoneal microvessels [2, 3]. Furthermore, there is usually partial denudation of the mesothelial cell layer. The increased angiogenesis is usually associated with diabetiform reduplications of capillary basement membranes and with subendothelial hyalinosis, especially in the venules [2]. Theoretically, an increased number of vessels would lead to an increased effective peritoneal vascular surface area, which in turn would result in a more rapid dissipation of the glucose osmotic gradient during each dwell, and therefore, in UFF. It should, however, be noted that for development of severe UFF, both increases in effective vascular surface area and peritoneal development of fibrosis have to be present [4]. The morphological and functional changes leading to UFF develop rather slowly in PD patients and usually are not seen until after 3 or 4 years of treatment. However, in animal models, especially in catheter-bearing rats, marked changes, including an increased angiogenesis and submesothelial fibrosis, usually develop within 12–16 weeks [2, 5].

Pathophysiological Aspects

Peritonitis, or just 'low-grade inflammation', such as in the uremic condition per se, but, particularly in combination with the exposure of the peritoneal membrane to high concentrations of glucose and GDP in conventional PDF seem to trigger PMF. These factors can lead to an increased release of vascular endothelial growth factor (VEGF), basic fibroblast growth factor (bFGF), both pro-angiogenic, and the production of reactive carbonyl compounds (RCOs) and advanced glycation end products (AGEs). These mediators, in turn, may contribute to the increased angiogenesis and fibrogenesis of PMF. Not the least the presence of GDP in conventional PDF will expose the peritoneum to RCOs resulting from conventional heat sterilization of glucose. However, RCOs also enter the peritoneum from the uremic circulation. The RCOs promote AGE-formation from proteins, which can initiate a range of cellular responses, including stimulation of monocytes, secretion of inflammatory cytokines, proliferation of vascular smooth muscle, stimulation of growth factors and secretion of matrix proteins [6]. RCOs may also increase VEGF expression directly in endothelial and mesothelial cells [7]. Interestingly, both nitric oxide (NO) and endothelial NO synthase (eNOS) are required for VEGF-driven angiogenesis. While VEGF is able to upregulate eNOS and NO-production, NO also can modulate, and even suppress, the hypoxic induction of VEGF creating a negative feedback between NO and VEGF induction. The intricate interrelations between the induction of pro-angiogenic factors following exposure to PDF have been discussed at some length earlier [6], and some of the known endogenous pro- and anti-angiogenic factors modulating endothelial cell growth are shown in table 1 [8].

Another interesting hypothesis behind the increased angiogenesis in PMN is the pseudohypoxia hypothesis, much discussed to explain the increased

Table 1. Endogenous factors that modulate endothelial cell growth

Proangiogenic factors	Antiangiogenic factors
Vascular endothelial growth factor	Angiostatin
Basic fibroblast growth factor (bFGF, FGF2)	Endostatin
Transforming growth factor-β	Thrombospondin-1
Transforming growth factor-α	SPARC (secreted protein, acidic and rich in cysteine)
Hypoxia inducible factor-1β	Vascular endothelial growth inhibitor
Hepatocyte growth factor	
Prostaglandin E_2	
Platelet-derived endothelial cell growth factor	
Angiogenin	
Interleukin-8	
Angiopoietin	

angiogenesis seen in patients with diabetes mellitus. During the cellular degradation of glucose, NADH is formed from NAD^+, both via glycolysis and the polyol pathway. This results in an increase in the $NADH/NAD^+$ ratio, which is reversed by the conversion of pyruvate into lactate by lactate dehydrogenase. NAD^+ is also regenerated in the citric acid cycle. The presence of very high lactate concentrations in conventional PD solutions leads to an increase in intracellular lactate, which tends to inhibit the NAD^+ regeneration, thereby leading to an increased intracellular $NADH/NAD^+$ ratio. This mimics intracellular hypoxia, the most important stimulus for the release of VEGF and HIF-1α, the major drivers of angiogenesis. This would speak in favor for the use of pyruvate instead of lactate, or at least for partial replacement of lactate with pyruvate, as buffer in the PDF.

Role of Glucose and GDP in the Angiogenesis of Peritoneal Membrane Failure

Indications of detrimental effects of glucose on the peritoneal membrane mainly come from clinical studies. In a cohort study on 303 patients at one single dialysis center, Davies et al. [9] reported that of 22 patients who were treated continuously for five years (or more) in CAPD, 13 had stable transport ($D/P_{creatinine}$ = 0.67, stable for 5 years), whereas 9 had a sustained increase in small solute transport ($D/P_{creatinine}$ = 0.56 at start, increasing to 0.77 after 5 years). Compared with the stable patients, those with increasing transport had an earlier loss in residual renal function and were exposed to significantly more hypertonic glucose solutions for the first 2 years of treatment. Further increases in glucose exposure were observed due to reductions in UF, as solute transport continued to rise. Although being an observational study, which could not prove cause and effect, these data support the contention that hypertonic glucose (and high concentrations of GDPs) may be detrimental to the peritoneal membrane and cause PMF in the long term.

Along with the hypothesis that GDP may be of key importance for the increased angiogenesis in long-term PD, Mortier et al. [5] exposed catheter-bearing rats to conventional lactate-buffered 3.86% PDF at pH 5, in comparison with PDF containing low levels of GDP, being bicarbonate/lactate-buffered at pH 7.4 (Physioneal®), or a lactate-buffered amino acid PDF at pH 6.7 (Nutrineal®), all infused twice daily (2 × 10 ml) for 12 weeks. With standard PDF UF was reduced after 12 weeks, but not with low-GDP (bicarbonate/lactate buffered) or amino acid PDF. Furthermore, rats exposed to standard PDF showed an increased VEGF expression, more microvascular proliferation and submesothelial fibrosis, not observed in the other groups. Also, there was an increased accumulation of AGEs and upregulation of the receptor for AGE. In another attempt to reduce not only glucose (and GDP) exposure, but also, along the pseudohypoxia hypothesis, to reduce the lactate concentration, Krediet and co-workers in a similar long-term peritoneal rat exposure model have reported some very promising results of the combination of glycerol 1.4%, amino acids 0.5%, and dextrose 1% in a lactate/bicarbonate or pyruvate buffered PDF, the so-called GLAD-study [10]. Replacing lactate with pyruvate as buffer was associated with a reduction in the number of vessels after 20 weeks of PDF exposure. Furthermore, also reducing the GDP exposure, there was a further reduction in angiogenesis, and, in particular, in submesothelial fibrosis. It was concluded that GDP might be even more relevant for the development of fibrotic alterations than for affecting the number of vessels in this chronic rat exposure model.

The mentioned experimental studies performed in catheter-bearing rats lend strong support for the role of GDP (and/or glucose) in producing PMF. There are, however, reasons to believe that the presence of an indwelling catheter (a Rat-o-Port system) may have acted as a catalyst in producing the marked angiogenesis and fibrosis noted in these animal models, since in rats devoid of an implanted catheter, changes have been found to be much less pronounced [11, 12]. Thus, in a study from our own group chronic peritoneal exposure of rats to lactate-buffered GDP-containing or GDP-free solutions was studied using daily i.p. injections (10 + 10 ml) of the PDF for 12 weeks. In all animals exposed to lactate-buffered PDF, irrespective of GDP content, there were increases in submesothelial tissue thickness and in vascular density, paralleling increases in the expression of mRNA for VEGF and TGF-β. However, the VEGF and TGF-β-1 expression at 12 weeks actually tended to be higher in animals exposed to GDP-free solutions than in those treated with standard solution. Only the presence of mesothelial cell damage could be connected to the presence of GDP. All the other changes observed were similar for all fluids. Furthermore, in a clinical study of a commercial PDF having low GDP (Gambrosol-trio®), compared to standard PDF (Gambrosol®), we were not able to show any differences in transport parameters or UF after two years, although markers of mesothelial integrity (CA125) were considerably higher in the low-GDP PDF [13]. Furthermore, in the Euro Balance Study, testing a lactate-buffered, low-GDP PDF with neutral pH, Balance®, patients on the low-GDP solution actually demonstrated a reduced UF capacity [14]. Thus, in clinical studies, but also in animal studies in which indwelling catheters have been avoided, there may be much less of peritoneal changes due to GDP. Could animal models with chronically implanted catheters produce an accelerated angiogenesis and fibrosis, not observed in human PD or in animals exposed to PDF by daily needle injections? Still, the data from the previously mentioned clinical cohort PD study provides evidence that exposure to high levels of glucose (and GDP) may be detrimental for the peritoneal membrane in the long term [9].

How Can Peritoneal Angiogenesis Be Avoided?

First, training should aim at teaching the patient completely aseptic techniques in order to make the treatment 'peritonitis-free'. Second, overhydration should be avoided, since overhydration implies the use of solutions containing more glucose (and GDP). Dietary salt restriction and/or the use of icodextrin (once daily) to manage volume overload should be considered [15]. Although definite evidence is still lacking from randomized controlled trials, low-GDP PDF should be preferred to conventional solutions. It should thus be noted that not a single case of sclerosing encapsulating peritonitis has yet been reported in patients on low-GDP solutions. Clinical trials using anti-angiogenic therapy are lacking. However, in animal studies there is evidence that ACE inhibitors, and also octreotide, an anti-VEGF inhibitor, could markedly reduce fibrogenesis and angiogenesis [16, 17]. Furthermore, ACE inhibitors have, in a recently published (observational) clinical study, been shown to be of value in preventing increases in small solute transport ($D/P_{creatinine}$) in long-term PD [18]. Concerning more specific anti-VEGF therapy, caution should be taken not to provoke undesirable side effects. This is because VEGF, bFGF and TGF-β are all necessary growth factors involved in proper wound healing and tissue repair. Antagonizing VEGF activity may be detrimental, not the least for the residual renal function, since reduced, not increased, VEGF expression in podocytes and tubular cells has been reported to be an important feature in many progressive kidney diseases [8].

References

1 Heimbürger O, Waniewski J, Werynski A, Tranaeus A, Lindholm B: Peritoneal transport in CAPD patients with permanent loss of ultrafiltration capacity. Kidney Int 1990;38:495–506.

2 Williams JD, Craig KJ, Topley N, Von Ruhland C, Fallon M, Newman GR, Mackenzie RK, Williams GT: Morphologic changes in the peritoneal membrane of patients with renal disease. J Am Soc Nephrol 2002;13:470–479.

3 Mateijsen MA, van der Wal AC, Hendriks PM, Zweers MM, Mulder J, Struijk DG, Krediet RT: Vascular and interstitial changes in the peritoneum of CAPD patients with peritoneal sclerosis. Perit Dial Int 1999;19:517–525.

4 Rippe B, Venturoli D: Simulations of osmotic ultrafiltration failure in CAPD using a serial three-pore membrane/fiber matrix model. Am J Physiol Renal Physiol 2007;292:F1035–1043.

5 Mortier S, Faict D, Schalkwijk CG, Lameire NH, De Vriese AS: Long-term exposure to new peritoneal dialysis solutions: effects on the peritoneal membrane. Kidney Int 2004;66:1257–1265.

6 Devuyst O: New insights in the molecular mechanisms regulating peritoneal permeability. Nephrol Dial Transplant 2002;17:548–551.

7 Inagi R, Miyata T, Yamamoto T, Susuki D, Urakami K, Saito A: Glucose degradation product methylglyoxal enhances the production of vascular endothelial growth factor in peritoneal cells: role in the functional and morphological alterations of peritoneal membranes in peritoneal dialysis. FEBS Lett 1999;463:260–264.

8 Kang DH, Kanellis J, Hugo C, Truong L, Anderson S, Kerjaschki D, Schreiner GF, Johnson RJ: Role of the microvascular endothelium in progressive renal disease. J Am Soc Nephrol 2002;13:806 816.

9 Davies SJ, Phillips L, Naish PF, Russell GI: Peritoneal glucose exposure and changes in membrane solute transport with time on peritoneal dialysis. J Am Soc Nephrol 2001;12:1046–1051.

10 Krediet RT, Zweers MM, van Westrhenen R, Zegwaard A, Struijk DG: Effects of reducing the lactate and glucose content of PD solutions on the peritoneum: is the future GLAD? Nephrol Dial Transplant Plus 1 2008;(suppl 4):iv56–iv62.

11 Musi B, Braide M, Carlsson O, Wieslander A, Albrektsson A, Ketteler M, Westenfeld R, Floege J, Rippe B: Biocompatibility of peritoneal dialysis fluids: long-term exposure of nonuremic rats. Perit Dial Int 2004;24:37–47.

12 Flessner MF, Credit K, Henderson K, Vanpelt HM, Potter R, He Z, Henegar J, Robert B: Peritoneal changes after exposure to sterile solutions by catheter. J Am Soc Nephrol 2007;18:2294–2302.

13 Rippe B, Simonsen O, Heimbürger O, Christensson A, Haraldsson B, Stelin G, Weiss L, Nielsen FD, Bro S, Friedberg M, Wieslander A: Long-term clinical effects of a peritoneal dialysis fluid with less glucose degradation products. Kidney Int 2001;59:348–357.

14 Williams JD, Topley N, Craig KJ, Mackenzie RK, Pischetsrieder M, Lage C, Passlick-Deetjen J: The Euro-Balance Trial: the effect of a new biocompatible peritoneal dialysis fluid (balance) on the peritoneal membrane. Kidney Int 2004,66:408–418.

15 Davies SJ, Woodrow G, Donovan K, Plum J, Williams P, Johansson AC, Bosselmann HP, Heimburger O, Simonsen O, Davenport A, Tranaeus A, Divino Filho JC: Icodextrin improves the fluid status of peritoneal dialysis patients: results of a double-blind randomized controlled trial. J Am Soc Nephrol 2003;14:2338–2344.

16 Duman S, Gunal AI, Sen S, Asci G, Ozkahya M, Terzioglu E, Akcicek F, Atabay G: Does enalapril prevent peritoneal fibrosis induced by hypertonic (3.86%) peritoneal dialysis solution? Perit Dial Int 2001;21:219–224.

17 Günal AI, Celiker H, Akpolat N, Ustundag B, Duman S, Akcicek F: By reducing production of vascular endothelial growth factor octreotide improves the peritoneal vascular alterations induced by hypertonic peritoneal dialysis solution. Perit Dial Int 2002;22:301–306.

18 Kolesnyk I, Noordzij M, Dekker FW, Boeschoten EW, Krediet RT: A positive effect of AII inhibitors on peritoneal membrane function in long-term PD patients. Nephrol Dial Transplant 2009;24:272–277.

Prof. Bengt Rippe
Department of Nephrology, University Hospital of Lund, Lund University
SE–211 85 Lund (Sweden)
Tel. +46 171247, Fax +46 46 2114 356, E-Mail Bengt.Rippe@med.lu.se

Ronco C, Crepaldi C, Cruz DN (eds): Peritoneal Dialysis – From Basic Concepts to Clinical Excellence. Contrib Nephrol. Basel, Karger, 2009, vol 163, pp 67–73

Tailoring Peritoneal Dialysis Fluid for Optimal Acid-Base Targets

Mariano Feriani

Department of Nephrology and Dialysis, Dell'Angelo Hospital, Mestre-Venezia, Italy

Abstract

Mild derangements of acid-base status are common features in peritoneal dialysis patients, metabolic acidosis being the most frequent alteration. One of the main tasks of dialysis is to correct these derangements and the target is the normalization of the acid-base parameters since they affect several organs and functions. Since factors affecting acid-base homeostasis are intrinsic characteristics of the individual patient (metabolic acid production, distribution space for bicarbonate, dialytic prescription, etc.), it is not surprising that only relatively few patients achieve the normal range. Only a certain modulation of buffer infusion by using different buffer concentrations in the dialysis fluid may ensure a good correction in a large percentage of patients.

Clinical Consequences of Metabolic Acidosis

The majority of patients treated with continuous ambulatory peritoneal dialysis (CAPD) show a mild-to-moderate metabolic acidosis without acidemia [1]. Although Uribarri et al. [2] demonstrated that patients in CAPD were in day-to-day acid-base balance, this type of metabolic acidosis probably results in several clinical consequences.

The relationship between metabolic acidosis, parathyroid hormone (PTH) dysfunction, and metabolic bone disease is complex and somewhat controversial. In experimental animals, acute induction of metabolic acidosis stimulates PTH secretion [3] but whether secretion of this hormone remains increased with sustained acidosis is unclear. The data in support of such an effect comes from patients receiving chronic hemodialysis therapy. In one controlled prospective study, patients with serum [HCO_3^-] restored to normal by adding additional HCO_3^- to the bath solution had less of an increase in PTH levels over an 18-month period of observation when compared to a group of patients with no correction of acidosis [4]. Strikingly, correction of acidosis not only decreased bone turnover in high-turnover bone disease (i.e. hyperparathyroidism) osteodystrophy (documented by bone biopsies and osteocalcin measurements), but also improved bone turnover in low-turnover bone disease (unrelated to hyperparathyroidism). In a second study, correction of acidosis in hemodialysis patients improved the sensitivity of PTH to calcitriol [5].

In patients receiving peritoneal dialysis, the use of a low calcium (1.25 mmol/l) dialysate with a higher lactate concentration (40 mmol/l) is associated with a fall in plasma PTH levels. It was postulated that this beneficial outcome was a consequence of the better control of serum phosphate due to the increased supplementation of calcium carbonate [6]. It is interesting to note that serum [HCO_3^-] was higher in the low calcium bath group in this study. It is conceivable that serum PTH decreased because of the improvement in acid-base status rather than from any effect on serum phosphate.

Numerous studies in humans and in experimental animals have shown that metabolic acidosis promotes protein catabolism [7–16]. This catabolic process appears to be dependent on stimulation of glucocorticoid secretion and is directly attributable to acidosis as opposed to other effects of chronic renal insufficiency [7, 8]. Acidosis-induced protein degradation is associated with increased rates of branched-chain amino acid oxidation [9]. In humans with end-stage renal disease, net uptake by muscle of branched-chain amino acids is directly correlated with steady-state serum [HCO_3^-] [15]. In normal human subjects given ammonium chloride to induce metabolic acidosis, albumin synthesis is inhibited and negative nitrogen balance develops within 7 days [17]. This effect is associated with a suppression of insulin-like growth factor, free thyroxine and tri-iodothyronine, and it is possible that these changes are contributing factors to the catabolic state.

In both hemodialysis and peritoneal dialysis patients there is evidence that catabolic effects of metabolic acidosis on proteins and amino acids can only be reversed by a full correction of this condition by increasing serum [HCO_3^-] from 17–18 to 25–26 mEq/l) [4–7, 18, 19]. In a randomized study on 200 CAPD patients [8], normal venous bicarbonate (27.2 mmol/l) was associated with nutritional benefits such as increase in body weight, midarm circumference and decreased morbidity as compared to mildly low venous bicarbonate (23.0 mmol/l).

General Principles of Acid Balance and Acid-Base Balance Homeostasis in End-Stage Renal Disease

Maintenance of a normal serum [HCO_3^-] and pH requires day-to-day replenishment of the alkali consumed in neutralizing the acids produced by endogenous metabolic processes. In patients with functioning kidneys, alkali stores are replenished by renal acid excretion, a process that generates new HCO_3^- in the body. In patients without functioning kidneys, alkali replenishment is accomplished by the addition of either HCO_3^- itself or a metabolic precursor of this anion, such as lactate or acetate. Regardless of the type of renal replacement therapy, however, a new equilibrium almost certainly develops once the amount of dialysis and the alkali concentration of the dialysis bath are fixed, in which steady-state serum [HCO_3^-] is determined primarily by endogenous acid production [20, 21].

During renal replacement therapy, the rate of net alkali addition during treatment is dependent on the transmembrane concentration of HCO_3^-. Thus, when extracellular [HCO_3^-] is lower, net alkali addition will be greater during any given treatment, and when extracellular [HCO_3^-] is higher, less alkali will be added. Using a HCO_3^--containing bath, the transmembrane concentration gradient regulates the amount of HCO_3^- added, while with peritoneal dialysis using a lactate-containing bath, or with hemodialysis using an acetate-containing bath, the transmembrane gradient regulates the amount of HCO_3^- lost into the bath, and therefore net alkali addition. In all patients receiving renal replacement therapy, the prevailing pH and [HCO_3^-] are determined by the characteristics of the dialysis treatment and by endogenous acid production. Because the bath alkali concentration and the dialyzability of HCO_3^- are both fixed once the dialysis prescription is set, the only variable component of these determinants is endogenous acid production. Given these fixed and unvarying alkali replacement conditions, it is not surprising that steady-state serum [HCO_3^-] varies as a function of endogenous acid production much more than in individuals with functioning kidneys, who can vary acid excretion in response to variations in acid production.

The implication of this analysis is that the acids produced by body metabolism do not continually accumulate in patients with end-stage renal disease receiving dialysis treatment. If true, then continued consumption of bone buffers by retained H^+ should not be occurring. The deleterious effects of chronic metabolic acidosis, described earlier, are still likely to be evident, however, because they appear to be related more to the prevailing serum [HCO_3^-] and pH than to the state of acid balance.

Normal acid-base status parameters are reported in table 1 [22].

Table 1. Acid-base normal values [from ref. 22]

	pH	pCO_2	Total CO_2	HCO_3
Arterial blood	7.40	41	26.0	24.8
Capillary blood	7. 42	39	25.1	24.0
Venous blood	7.37	49	28.5	27.0

Lactate-Based Peritoneal Dialysis Fluid

In clinical practice normal values of serum bicarbonate are achieved only in about 25% of CAPD patients by using a solution containing 35 mmol/l lactate [23]. Following the general concept of the alkali transfer through a dialytic membrane, when a higher buffer content in the dialysis fluid is applied a more positive body buffer balance is achieved. This is confirmed in the largest published study on acid-base status in CAPD patients [19]. Two hundred new CAPD patients were randomly allocated to receive either a 40-mmol/l lactate-buffered solution or a 35-mmol/l lactate-buffered CAPD solution. The target venous bicarbonate in the first group was 30 mmol/l and if this was not achieved an oral supplementation with calcium carbonate and sodium bicarbonate was added. Although the target was not reached in all patients, nevertheless a normal acid-base status (venous serum bicarbonate 27.2 ± 0.3 mmol/l) was achieved in the majority of patients. The correction of metabolic acidosis led to greater increases in body weight and midarm circumference and decreased morbidity in terms of number of admissions and days in the hospital in the first year of CAPD. These data support the view that the target for the acid-base correction in dialysis patients should be the normal value for the healthy population.

In a previous study, however, the use of a solution containing 40 mmol/l lactate increased the percentage of patients with a normal acid base to about 40% as compared to the 35-mmol/l solution, but also increased the number of patients with serum bicarbonate concentration in the range of alkalosis [24].

The large interpatient variability is due to a number of individual factors influencing the base requirements in different patients [20]. Among them protein intake and catabolism, distribution space for bicarbonate, dialytic fluid removal and peritoneal membrane permeability are probably the most important.

Even though a daily supplement of oral alkalinizing salts could increase serum bicarbonate in dialyzed patients [19], the intestinal absorption of calcium-containing phosphate binders (calcium carbonate and acetate) is variable and the alkalinizing effect moderate [25], and sodium bicarbonate exposes patients further to the risk of sodium and fluid retention. Moreover, patient compliance is often not optimal.

Bicarbonate-Based Peritoneal Dialysis Fluid

From the acid-base point of view, a peritoneal dialysis fluid based on bicarbonate should be ideal since bicarbonate is the true buffer of the body, while lactate is only an intermediate of the glucose metabolism. In addition, the bicarbonate absorption from solution is dependent on the individual serum bicarbonate concentration so that a feedback between blood and dialysis fluid bicarbonate concentration occurs, thus preventing severe acidosis and alkalosis [26].

Different bicarbonate-based solutions for CAPD have recently been introduced [27, 28]. We reported the results of a multicenter randomized study in which a conventional solution containing 35 mmol/l lactate was compared with a new solution containing 34 mmol/l bicarbonate [29]. Besides the high tolerability reflected by reduction of infusion pain and discomfort, an unexpected result was a small, but significant, increase in serum bicarbonate. This finding was also recorded by another study in which a differently formulated bicarbonate fluid (bicarbonate/lactate) was used and compared with the conventional lactate-based solution [30].

Despite this improvement, a substantial number of patients still had a serum bicarbonate concentration lower than normal.

In a subsequent pilot study [31], we tested a solution with a higher bicarbonate concentration (39 mmol/l) in a few patients. As expected, serum bicarbonate rose in all patients but in those patients with normal acid-base status at baseline, a mild metabolic alkalosis developed.

In order to optimize the patient's acid-base status, a study was designed to assess the possibility of normalizing the serum bicarbonate level in CAPD patients by individualized application of CAPD solutions with two different bicarbonate concentrations [32].

At baseline, 61 patients were stratified for acid-base parameters into normal or alkalotic (31%) and acidotic (69%) groups. The first group was treated with a bicarbonate solution containing 34 mmol/l and the other with a 39 mmol/l bicarbonate solution.

Body weight, body surface area, blood urea nitrogen, serum creatinine, $K_{pr}t/V$, and PNA were significantly higher in the acidotic group.

During the 9-month study, if a patient in the low-bicarbonate group became acidotic he/she was switched to the high-bicarbonate group, whereas, in case of metabolic alkalosis in the high-bicarbonate group, the patient was switched to the low-bicarbonate group.

At the end of the study, 64% of patients had normal acid-base values, 12.7% were alkalotic and 23.4% were still acidotic.

Our study demonstrated that, for achieving a normal acid-base status in a large percentage of CAPD patients, buffer infusion with dialysis fluid should be adjusted to the individual patient characteristics. A fluid containing the high-bicarbonate concentration should be employed in patients with larger body size and relatively high protein intake, while a solution containing the low-bicarbonate concentration is useful in patients with smaller body size and low protein intake thus avoiding alkalosis.

References

1 Feriani M: Adequacy of acid base correction in Continuous Ambulatory Peritoneal Dialysis. Perit Dial Int 1984;14(suppl 3):S133–S138.
2 Uribarri J, Buquing J, Oh MS: Acid-base balance in peritoneal dialysis patients. Kidney Int 1995;47:269–273.
3 Bichara M, Mercier O, Borensztein P, Paillard M: Acute metabolic acidosis enhances circulating parathyroid hormone, which contributes to renal response against acidosis in the rat. J Clin Invest 1990;86:430–443.
4 Lefebvre A, de Verneoul MC, Gueris J, Goldfarb B, Graulet AM, Morieux C: Optimal correction of acidosis changes progression of dialysis osteodystrophy. Kidney Int 1989;36:1112–1118.
5 Graham KA, Hoenich NA, Tarbit M, Ward MK, Goodship THJ: Correction of acidosis in hemodialysis patients increases the sensitivity of the parathyroid glands to calcium. J Am Soc Nephrol 1997;8:127–631.
6 Hutchison AJ, Freemont AJ, Boulton HF, Gokal R: Low calcium dialysis fluid and oral calcium carbonate in CAPD: a method of controlling hyperphosphatemia whilst minimizing aluminum exposure and hypercalcaemia. Nephrol Dial Transplant 1992;7:1219–1225.
7 May RC, Kelly RA, Mitch WE: Mechanisms for defects in muscle protein metabolism in rats with chronic uremia: the influence of metabolic acidosis. J Clin Invest 1987;79:1099–1103.
8 May RC, Kelly RA, Mitch WE: Metabolic acidosis stimulates protein degradation in rat muscle by a glucocorticoid-dependent mechanism. J Clin Invest 1986;77:614–621.
9 Hara Y, May RC, Kelly RA, Mitch WE: Acidosis, not azotemia, stimulates branched-chain amino acid catabolism in uremic rats. Kidney Int 1987;32:808–814.
10 Mitch WE, Clark AS: Specificity of the effect of leucine and its metabolites on protein degradation in skeletal muscle. Biochem J 1984;222:579–586.
11 Papadoyannakis NJ, Stefanides CJ, Mc Geown M: The effect of the correction of metabolic acidosis on nitrogen and protein balance of patients with chronic renal failure. Am J Clin Nutr 1984;40:623–627.
12 Reaich D, Channon SM, Scrimgeour CM, Daley SE, Wilkinson R, Goodship THJ: Correction of acidosis in humans with chronic renal failure decreases protein degradation and amino acid oxidation. Am J Physiol 1993;265:E230–E235.
13 Jenkins D, Burton PR, Bennet SE, Baker F, Walls J: The metabolic consequences of the correction of acidosis in uraemia. Nephrol Dial Transpl 1989;4:92–95.
14 Williams B, Hattersley J, Layward E, Walls J: Metabolic acidosis and skeletal muscle adaptation to low protein diets in chronic uremia. Kidney Int 1991;40:779–786.
15 Bergström J, Alvestrand A, Furst P: Plasma and muscle free amino acids in maintenance hemodialysis patients without protein malnutrition. Kidney Int 1990;38:108–114.
16 Garibotto G, Russo R, Sofia A, Sala MR Robaudo C, Moscatelli P, Deferrari G, Tizianello A: Skeletal muscle protein synthesis and degradation in patients with chronic renal failure. Kidney Int 1994;45:1432–1439.

17 Ballmer PE, Mc Nurlan MA, Hulter HN, Anderson SE, Gatlick PJ, Krapf R: Chronic metabolic acidosis decreases albumin synthesis and induces negative nitrogen balance in humans. J Clin Invest 1995;95:39–451994.
18 Graham KA, Reaich D, Channon SM, Downie S, Goodship THJ: Correction of acidosis in hemodialysis decreases whole-body protein degradation. J Am Soc Nephrol 1997;8:632–637.
19 Graham KA, Reaich D, Channon SM, Downie S, Gilmour E, Passlick-Deetjen J, Goodship THJ: Correction of acidosis in CAPD decreases whole-body protein degradation. Kidney Int 1996;49:1396–1400.
20 Gennari FJ: Acid-base homeostasis in end-stage renal disease. Semin Dialy 1996;9:404–411.
21 Gennari FJ: Acid-base balance in dialysis patients. Kidney Int 1985;28:678–688.
22 Gennari FJ: Normal acid base values; in Cohen JJ, Kassirer JP (eds): Acid Base. Boston, Little, Brown, 1982, pp 107–110.
23 Feriani M: Buffers: bicarbonate, lactate, pyruvate. Kidney Int 1996;50(suppl 56):S75–S80.
24 Nolph KD, Prowant B, Serkes KD, et al: Multicenter evaluation of a new peritoneal dialysis solution with a high lactate and low magnesium concentration. Perit Dial Bull 1983;3:63–65.
25 Stein A, Moorhouse J, Iles-Smith H, et al: Role of an improvement in acid-base status and nutrition in CAPD patients. Kidney Int 1997;52:1089–1095.
26 Feriani M, Passlick-Deetjen J, La Greca G: Factors affecting bicarbonate transfer with bicarbonate-containing CAPD solution. Perit Dial Int 1995;15:336–341.
27 Feriani M, Biasioli S, Borin D, et al: Bicarbonate buffer for CAPD solution. Trans Am Soc Artif Intern Organs 1985;31:668–672.
28 Traneus A, for the Bicarbonate/Lactate Study Group: A long-term study of a bicarbonate/lactate-based peritoneal dialysis solution: clinical benefits. Perit Dial Int 2000;20:516–523.
29 Feriani M, Kirchgessner J, La Greca G, Passlick-Deetjen J, Bicarbonate CAPD Cooperative Group: A randomized multicenter long-term clinical study comparing a bicarbonate buffered CAPD solution with the standard lactate buffered CAPD solution. Kidney Int 1998;54:1731–1735.
30 Otte K, Gonzalez MT, Bajo MA, et al: Clinical experience with a new bicarbonate (25 mmol/L)/ lactate (10 mmol/L) peritoneal dialysis solution. Perit Dial Int 2003;23:138–145.
31 Feriani M, Carobi C, La Greca G, Buoncristiani U, Passlick-Deetjen J: Clinical experiences with a bicarbonate buffered (39 mmol/l) peritoneal dialysis solution. Perit Dial Int 1997;17:17–21.
32 Feriani M, Passlick-Deetjen J, Jaeckle-Meyer I, La Greca G: Individualize bicarbonate concentration in the peritoneal dialysis fluid to optimize acid-base status in CAPD patients. Nephrol Dial Transplant 2004;19:195–202.

Mariano Feriani
Department of Nephrology and Dialysis, Dell'Angelo Hospital
IT–30174 Mestre-Venezia (Italy)
Tel. +39 041 9657367, Fax +39 041 9657382, E-Mail mferiani@goldnet.it

Ronco C, Crepaldi C, Cruz DN (eds): Peritoneal Dialysis – From Basic Concepts to Clinical Excellence. Contrib Nephrol. Basel, Karger, 2009, vol 163, pp 74–81

Preservation and Modulation of Peritoneal Function

A Potential Role of Pyruvate and Betaine

Lazaro Gotloib

Laboratory for Experimental Nephrology, Ha'Emek Medical Center, Afula, Israel

Abstract

Aims: To review interventions aimed at protecting the peritoneal membrane from the detrimental effects of dialysis solutions, and to analyze proposed pharmacological interventions aiming to modulate peritoneal permeability. **Main Remarks:** Sustained oxidative stress appears as the most relevant factor bringing about substantial damage to all the components of the peritoneal membrane. In vivo and in vitro studies suggest that pyruvate, a natural radical scavenger, may well neutralize or substantially reduce the oxidative insult. Additional research suggest that trimetazidine and the glutathione precursor L-2-oxothiazolidine-4-carboxylic (OTZ), both showing antioxidant capabilities, are also promising agents that deserve to be further explored. Acute osmotic stress provokes different modes of cell death. In the long-term mesothelial cells adapt to the hypertonic environment, at least from a morphological point of view, by accumulating compatible osmolytes like betaine, launching the mechanism of regulatory volume increase. Addition of betaine to PD formulation is proposed in order to attenuate cells shrinkage and facilitate adjustment to a new, non physiological environment. Modulation of peritoneal function as a reusable dialyzing, biological membrane has been investigated using a variety of drugs in different experimental set ups: heparin, sulodexide, condroitin sulphate, phosphatidylcholine, indomethacin, and vasoactive drugs like verapamil, captopril, nicardipine, diltiazem, enalapril, valsartan and lisinopril, offering at times conflicting results. **Conclusions:** Efforts focused in improving the dialyzing capabilities of the peritoneal membrane by means of pharmacological manipulation, have not yet offered clinically relevant innovations. Conversely, interventions designed to preserve the peritoneum as a reusable dialyzing living membrane are more promising, especially those aimed at neutralizing oxidative and osmotic injury.

Imagination is more important than knowledge
Albert Einstein. In: Science and Religion.

A large body of literature supports the concept that commercially available PD solutions have detrimental effects upon peritoneal structure and, consequently, affect its performance as a reusable dialysis membrane. These changes basically result from the almost continuous exposure of the peritoneal tissue to the commonly used osmotic agents that induce, per se or through substances derived from their degradation, substantial injury to the mesothelial monolayer, the subjacent connective tissue as well as the submesothelial microvessels. Premature senescence of the monolayer and increased prevalence of different modes of mesothelial cell death are the leading mechanisms involved behind these unwanted effects that result, in the case of regularly used osmotic agents, glucose and icodextrin, in different degrees of oxidative insult [1]. Besides, in vivo acute exposure of the membrane to both osmotic agents results in substantial morphological alterations like shrinkage of the mesothelial cells and, consequently, widening of the intercellular spaces typically detected at the time of acute osmotic stress, as well as increased proportions of cells undergoing apoptosis (fig. 1), [Gotloib, unpubl. obs]. Experimental studies have shown evidence indicating that after long-term exposure to hyperosmolar fluids, mesothelial cells have the capability of adaptation to the new and unusual environment [2, 3], putting in motion the mechanisms of cell volume regulation, the more relevant one is accumulation of nonperturbing osmolytes like betaine. This reaction results in a more than 50% increase in cell volume [2].

On the other hand, the peritoneum being a biological membrane, research focused on the eventual relevance of pharmacological modulation of its dialyzing capabilities appears justified.

Ways Aimed at Reducing Oxidative Injury

Use of one-bag glucose PDF results in the development of a dose-related senescent mesothelial cell phenotype, not linked to the acid pH or the presence of lactate buffer. This complication derives mostly from oxidative stress induced by both, glucose itself and glucose degradation products (GDPs). New glucose formulations are offered in dual or three-chambered bags, keeping glucose at a low pH. This results in a reduced presence but not complete elimination of GDPs. These formulations show a potential to slow down the oxidative injury of the peritoneal membrane resulting from the presence of products derived from the nonenzymatic degradation of glucose [4].

Another approach aimed at reducing the oxidative insult, still in the experimental stages, is based on the inclusion of pyruvate to the dialysis formulation. This agent, part of the alpha-ketoacids-pyruvate system, is a natural H_2O_2 (hydrogen peroxide) scavenger. It has been shown that oxygen radicals

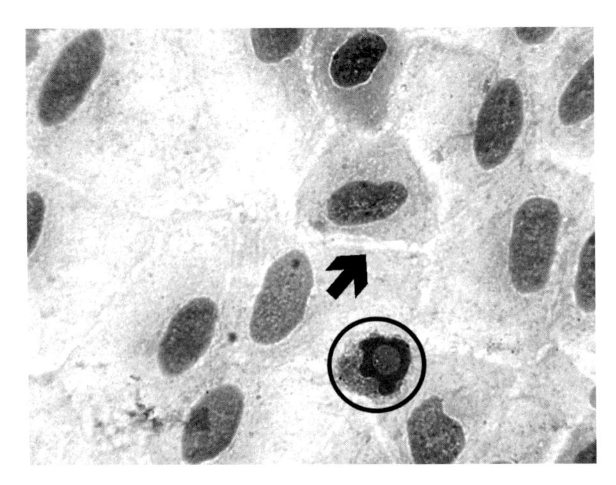

Fig. 1. Sample of mesothelium obtained from a mouse 3 h after one intraperitoneal injection of 7.5% icodextrin dialysis solution. Circle surrounds one mesothelial cell undergoing apoptosis. Notice widening of the intercellular spaces. HE. ×1,000.

at a concentration of 1 mM provoke necrotic death of mammalian cells in culture. On the other hand, in vitro experiments revealed that viability of mesothelial cells incubated in medium containing 2 mM of H_2O_2, a substantial vital challenge, was preserved by adding 2 mM of pyruvate [5]. These observations as well as other in vitro studies stimulated additional experiments testing the benefit to be drawn from the inclusion of pyruvate as a natural scavenger in the formulation of dialysis solutions [6]. A thoroughly designed recently published study delivered evidence indicating that rats exposed during 20 weeks to a 35 mM pyruvate-buffered, 4.25% glucose dialysis fluid showed less angiogenesis and submesothelial fibrosis than animals exposed to a high-glucose, lactate-buffered solution [7]. The influence on the mesothelium of incorporating pyruvate to the dialysis fluid, not explored in this study, has been evaluated in rats exposed during a period of 30 consecutive days to a 4.25% glucose-enriched solution buffered with pyruvate at concentrations of 35 mM [8]. Observation of mesothelial cell imprints sequentially recovered during and at the end of the follow-up period revealed the absence of the early acceleration of the mesothelial cell's life cycle that regularly results from the oxidant effect, observed in animals treated with high-glucose PD solutions. After 30 days' exposure, the cell population showed normal density, normal mitotic activity, normal prevalence of senescent cells and of nonviable cells, as well as prevalence of apoptosis not significantly different from those seen in normal unexposed controls [8] (fig. 2). All this evidence suggests that

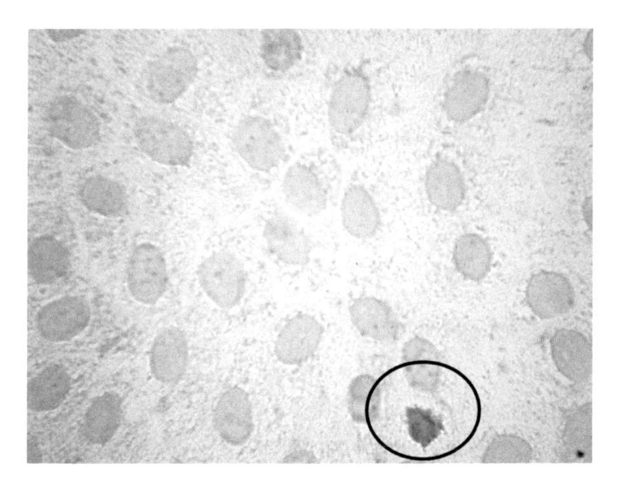

Fig. 2. Imprint recovered from a rat after one intraperitoneal injection of a pyruvate-buffered, 4.25% glucose solution. Only one cell appears stained by the vital dye, indicating a situation of nonviability. Trypan blue. ×1,000.

the addition of pyruvate to PD formulations appears a promising area that deserves to be further explored.

Other agents showing antioxidant capabilities have been recently proposed. Oral administration of trimetazidine to rats has been shown to inhibit the oxidative stress induced by high-glucose PD fluids [9].

An interesting approach to prevent oxidative injury to the peritoneum has been developed by Breborowicz et al. [10]. In vitro experiments using the glutathione precursor L-2-oxothiazolidine-4-carboxylic (OTZ) acid increased the cytoplasmic concentration of glutathione in spite of being incubated with increasing concentrations of glucose. In addition, use of OTZ substantially neutralized liberation of free radicals and reversed inhibition of mesothelial cell growth resulting from exposure to GDPs. In vivo experiments performed on rats showed some attenuation of the morphological changes induced by standard PD solutions [11].

Could Addition of Betaine Attenuate the Acute Osmotic Stress?

Cells challenged by a hyperosmotic environment respond launching compensatory molecular adaptations that allow to reinstate the homeostatic steady state of osmotically altered aspects of cell structure and function. Failure of these mechanisms leads to substantial reduction of growth and increased rate of apoptotic death. Cells are endowed with a host of automatic mechanisms that work against influences tending toward disruption of the homeostatic pathway. During osmotic stress cells react accumulating organic compatible osmolytes that will eventually neutralize the osmotic gap, recovering the conditions of the physiological steady state. Betaine is one of the more relevant compatible osmolytes synthesized by cells in situations of hyperosmotic insult. It has been shown that addition of this trimethyl derivative, in concentrations ranging between 10 and 25 mM to cultured S-3T3 cells incubated in a hyperosmotic medium (0.5 osm), produced dramatic and consistent effects. Basically, prevention of the 90% inhibition of cell proliferation that occurred in its absence [12], and safeguarding endothelial cells from dying in apoptosis, is expected to occur as a consequence of the environmental hypertonicity [13]. It should be noticed that the osmolarity used in this experiment is not far from that of a 4.25% glucose dialysis solution (0.483 osm). This information suggests that given the magnitude of the detected changes derived from the acute exposure of the peritoneum to the osmotic agents, inclusion of betaine to dialysis formulations could help to protect cells from the hyperosmotic insult. This topic is now being investigated.

Heparin

It has been proposed that heparin, particularly in the presence of inflammation, prevents peritoneal fibrosis, whereas when used in humans it can increase ultrafiltration [14]. However, a carefully designed study clearly demonstrates that heparin or low-molecular-weight heparin administration fail to prevent the peritoneal membrane damage induced by exposure to a high-glucose bioincompatible PD solution [15]. Additionally, in vitro experiments offered evidence indicating that heparin has inhibitory effects upon mesothelial cell growth and protein synthesis [16]. Hence, the relevance of using heparin in PD is still unclear.

Hyaluronan has a relevant role maintaining the negative electric mesothelial and microvascular charges that regulate the permselectivity of both endothelia facing negatively charged proteins. Additionally, evidence obtained from studies performed on cultured fibroblasts indicate that hyaluronan has a quite effective antioxidant capability [17].

Breborowicz and coworkers [reviewed in 8] showed that the addition of hyaluronan to PDF results in increased net UF, increased creatinine clearance and lower permeability for protein. Conversely, Moberly's group [reviewed in 8] reported no significant differences in solute clearances and net UF in a small series of PD patients treated with intraperitoneal hyaluronan. Observations reported by Fracasso and coworkers [reviewed in 8] who studied the effect of increasing doses of the glycosaminoglycan sulodexide on peritoneal transport in 6 CAPD patients over a 6-month period, add more uncertainty upon the effectiveness of this compounds. They observed a significant increase in D/P of urea and D/P of creatinine, a reduction in D/P of albumin and no statistically significant effect on UF.

Phosphatidylcholine

The peritoneal effluent contains surface active material that acts as a lubricant. The composition of this material is mostly phospholipids and more specifically phosphatidylcholine. Evidence obtained from clinical and experimental studies suggest that the role of this compound upon UF is controversial. However, its main task as a surfactant lining promotes very high permeability to lipid-soluble drugs and a barrier to water-soluble solutes. Krack's group [reviewed in 8] observed that intraperitoneal phosphatidylcholine resulted in a significant enhancement of UF but no changes in peritoneal permeability as assessed by the PET. Conversely, other investigators have failed to show a significant effect of this agent on UF [De Vecchi et al., reviewed in 8].

Other Possible Agents that Require Further Investigation

Table 1 illustrates several substances which have been proposed to improve peritoneal performance.

Intraperitoneal injection of indomethacin to rabbits has been shown to inhibit the cavitary production of prostanoids associated with a reduced peritoneal permeability to proteins. This outcome, however, was not detected in patients on CAPD.

Use of vasoactive drugs has been explored in different experimental setups, at times offering conflicting results (table 1).

Final Considerations

From this review, it may be concluded that to date efforts focused on improving the dialyzing capabilities of the peritoneal membrane by means of pharmacological manipulation have not yet offered clinically relevant innovations. Conversely, and on the basis of experimental work, it appears that interventions designed to preserve the peritoneum as a reusable dialyzing liv-

Table 1. Additional pharmacological interventions proposed to modulate peritoneal permeability

Agent	Effect	Investigator
Indomethacin, rabbits	Inhibition of prostanoids production and lower peritoneal permeability to protein	Peng et al: Kidney Int 2001;59:44–51
Patients	No effect	Douma et al: Nephrol Dial Transplant 2001;16:803–808
Sulodexide patients	increased D/P urea and D/P creatinine and reduction in D/P albumin	Fracasso et al: Perit Dial Int 2003;23:595–599
L-2-Oxothiazolidine-4-carboxylic acid (OTZ) In vitro	increases the intracellular concentration of glutathione in mesothelial cells	Breborowicz et al : Int J Artif Organs 1996;19:268–275
Oral verapamil, patients	significantly increased CrCl, Kt/V urea, and 24-hour drained dialysate volume	Rojas-Campos et al: Perit Dial Int 2005;25:576–582
Captopril, nicardipine, diltiazem, and verapamil, rats	decreased UF	Kumano et al: Adv Perit Dial 1996;12:27–32
Enalapril, rats	preservation of UF	Duman et al: Perit Dial Int 2001;21:219–224
Oral enalapril, patients	no significant change in small solutes clearances and UF	Ripley et al: Perit Dial Int 1994;14:378–383
Valsartan-lisinopril, rats	attenuation of glucose induced UF reduction	Duman et al : Int J Artif Org 2005;28:156–163
Oral tranexamic acid, patients	increased UF	Kuriyama et al : Perit Dial Int 1999;19:38–44

ing membrane are promising, especially those aimed at neutralizing oxidative and osmotic injury derived from the use of currently prescribed dialysis solutions.

References

1 Gotloib L, Wajsbrot V, Shostak A: Ecology of the peritoneum: a substantial role for the osmotic agents resulting ion apoptosis of mesothelial cells. Contrib Nephrol. Basel, Karger, 2003, vol 140, pp 10–17.

2 Gotloib L, Wajsbrot V, Shostak A, Kushnier R: Effect of hyperosmolality upon the mesothelial monolayer exposed in vivo and in situ to a mannitol-enriched dialysis solution. Nephron 1999; 81:301–309.

3 Breborowicz A, Polubinska A, Oreopoulos DG: Changes in volume of peritoneal mesothelial cells exposed to osmotic stress. Perit Dial Int 1999;19:119–123.

4 Zareie M, Keuning ED, ter Wee PM, Schalkwijk CG, Beelen RH, van den Born J: Improved biocompatibility of bicarbonate-lactate-buffered PDF is not related to pH. Nephrol Dial Transplant 2006;21:208–216.
5 Shostak A, Gotloib L, Kushnier R, Wajsbrot V: Protective effect of pyruvate upon cultured mesothelial cells exposed to 2 mM hydrogen peroxide. Nephron 2000;84:362–366.
6 Brunkhorst R, Mahiout A: Pyruvate neutralizes peritoneal dialysate cytotoxicity: maintained integrity and proliferation of cultured human mesothelial cells. Kidney Int 1995;48:177–181.
7 van Westrhenen R, Zweers MM, Kunne C, de Waart DR, van der Wal AC, Krediet RT: A pyruvate-buffered dialysis fluid induces less peritoneal angiogenesis and fibrosis than a conventional solution. Perit Dial Int 2008;28:487–496.
8 Diaz Buxo J, Gotloib L: Agents that modulate peritoneal membrane structure and function. Perit Dial Int 2007;27:16–30.
9 Gunal AL, Celiker H, Ustundag B, Akpolat N, Dogukan A, Akcicek F: The effect of oxidative stress inhibition with trimetazidine on the peritoneal alterations induced by hypertonic peritoneal dialysis solution. J Nephrol 2003;16:225–230.
10 Breborowicz A, Witowski J, Polubinska A, Pyda M, Oreopoulos DG: L-2-oxothiazolidine-4-carboxylic acid reduces in vitro cytotoxicity of glucose degradation products. Nephrol Dial Transplant. 2004;19:3005–3011.
11 Styszinki A, Wieczorowska-Tobis K, Podkowka R, Breborowicz A, Oreopoulos DG: Effects of glutathione supplementation during peritoneal dialysis. Adv Perit Dial 2006;22:88–93.
12 Petronini PG, De Angelis EM, Borghetti P, Borghetti AF, Wheeler KP: Modulation by betaine of cellular responses to osmotic stress. Biochem J 1992;282:69–73.
13 Alfieri R, Cavazzoni A, Petronini PG, Bonelli M, Caccamo AE: Compatible osmolytes modulate the response of porcine endothelial cells to hypertonicity and protect them from apoptosis. J Physiol 2002;540:499–508.
14 Sjoland JA, Smith PR, Jespersen J, Gram J: Intraperitoneal heparin reduces peritoneal permeability and increases ultrafiltration in peritoneal dialysis patients. Nephrol Dial Transplant 2004;19:1264–1268.
15 Schilte MN, Loureiro J, Keuning FD, ter Wee PM, Celie JWAM, Beelen RHJ, van der Born J: Long- term intervention with heparins in a rat model of peritoneal dialysis. Perit Dial Int 2009;29:26–35.
16 Manalaysay MT, Kumano K, Hyodo T, Sakai T: Inhibition of mesothelial cell growth and protein synthesis by heparin. Adv Perit Dial 1995;11:239–242.
17 Campo GM, Avenoso A, Campo S, D'Ascola A, Ferlazzo AM, Calatroni A: Reduction of DNA fragmentation and hydroxyl radical production by hyaluronic acid and chondroitin-sulphate in iron plus ascorbate-induced oxidative stress in fibroblast cultures. Free Radic Res 2004;38:601–611.

Lazaro Gotloib, MD
Laboratory for Experimental Nephrology, Ha'Emek Medical Center
POB 2886
Menahem Ussishkin 74
Afula 18284 (Israel)
Tel./Fax +972 4 659 1537, E-Mail gotloib@012.net.il

Ronco C, Crepaldi C, Cruz DN (eds): Peritoneal Dialysis – From Basic Concepts to Clinical Excellence. Contrib Nephrol. Basel, Karger, 2009, vol 163, pp 82–89

How to Assess Peritoneal Transport: Which Test Should We Use?

Olof Heimbürger

Divisions of Renal Medicine, Department of Clinical Science, Intervention and Technology, Karolinska Institutet, Karolinska University Hospital, Stockholm, Sweden

Abstract

The transport characteristics of the peritoneal membrane will differ between patients, and will change over time in individual patients, which in turn will affect the fluid and solute transport kinetics of peritoneal dialysis (PD) as well as the dialysis efficiency. There are several different simplified tests available for the evaluation of peritoneal transport. The most widely used are the peritoneal equilibration test (PET) and the personal dialysis capacity test (PDC), which may give relatively detailed information about the peritoneal transport characteristics, and computer software is available for the calculation of transport coefficients as well as for simulations of therapy changes. However, for some purposes, in particular the evaluation of patients with poor ultrafiltration, other tests such as the double mini-PET will give additional information. A rough estimate of the diffusive transport can also be made from a 24-hour dialysate collection in patients on continuous ambulatory dialysis, which also of course gives the possibility to calculate different adequacy parameters. In summary, which test to use is dependent on the question and the clinical situation.

Peritoneal dialysis (PD) utilizes dialysis fluid infused into the peritoneal cavity to remove 'uremic toxins' (including water) from the patient through the peritoneal membrane. Whereas the artificial hemodialysis membranes have well-characterized, reproducible, solute and fluid transport characteristics, the peritoneum is a complex structure of living tissues with different transport characteristics. The transport characteristics of the peritoneal membrane will differ between patients, and will change over time in individual patients, which in turn will affect the fluid and solute transport kinetics of PD as well as the dialysis efficiency. It has also been demonstrated that the different transport characteristics of the membrane are related to the patients clinical outcome [1].

The Peritoneal Transport Process

The capillary wall is considered to be the main transport barrier for diffusion and convection through the peritoneal barrier (though it is likely that the interstitium may be a significant transport barrier in pathological conditions with thickening and fibrosis of the peritoneal membrane). The peritoneal capillaries behave functionally as having a heteroporous structure, with a large number of 'ultra-small' water pores (radius 4–6 Å), a large number of 'small pores' (radius 40–65 Å), and a small number of large pores (radius 200–400 Å) through which macromolecules are filtered due to convective flow [1–3]. The anatomical correlates of the water channels are aquaporin-1, and of the small pores the interendothelial clefts, but the anatomical correlate of the large pores is not established. Whereas only water may pass through the aquaporins, the small pores do not restrict the passage of small solutes but are impermeable for macromolecules larger than albumin [1–3]. In addition to the transcapillary exchange between plasma and dialysate, there is a peritoneal absorptive flow of fluid and solutes, comprising of two different pathways [1, 3]: direct lymphatic absorption, and fluid absorption into interstitial tissues.

Ultrafiltration in peritoneal dialysis is achieved by the application of high concentration of an osmotic agent (usually glucose) in the dialysate, resulting in a high osmotic pressure gradient across the peritoneal barrier [1, 2]. When glucose is used as osmotic agent, it will have a high osmotic pressure over the water pores (as they are impermeable for glucose) and about half of the water flow from the capillaries will pass through these pores. In contrast, glucose can easily diffuse through the small pores and the osmotic force over the small pores will be relatively low, and only about half of the ultrafiltration will occur over the small pores in spite of the large total area [1, 2]. However, the osmotic pressure gradient decreases rapidly due to the absorption of the osmotic agent, when small solutes like glucose are used as osmotic agents.

Small solutes are transported through the peritoneum mainly by diffusion [1, 2]. The diffusive solute transfer rate is proportional to the concentration gradient between dialysate and plasma, and the solutes diffusive mass transport coefficient (K_{BD}), also denoted permeability surface area product (PS), mass transfer coefficient (MTC) or mass transfer area coefficient (MTAC) [1, 2]. Thus, the speed of diffusion for a solute will be dependent on the concentration gradient and the K_{BD}/PS/MTC for the particular solute [1, 2].

The convective transport of small solutes is dependent on the ultrafiltration through the small pores, whereas the ultrafiltration through the aquaporins will not result in convective transport as all solutes will be sieved at the aquaporins. Therefore, the sieving of small solutes will be dependent of the fraction of ultrafiltration that passes through small pores in relation to the total ultrafiltration flow. (The large pores are so rare that the flow through this pathway may be neglected except for the transport of large macromolecules from blood to dialysate.)

Different Peritoneal Transport Tests

There are several tests available for the assessment of peritoneal transport characteristics. A detailed evaluation can be done using a single dwell study with a macromolecular tracer and frequent dialysate and plasma sampling making it possible to assess the intraperitoneal volume-over-time curve, the convective fluid absorption rate as well as the changes in dialysate concentration of different solutes over time [1]. From these data, different transport coefficients can be calculated in detail. However, the single dwell study is difficult to apply in the routine clinical setting and several simplified tests have been developed such as the peritoneal equilibration test (PET), the personal dialysis capacity test (PDC), and the double mini-PET. Based on these tests, there are commercial computer programs available to assess basic peritoneal transport parameters and to predict effects of various treatment schedules on peritoneal small solute clearances and ultrafiltration [4–7]. In general, the results will be closely dependent on the quality of data used for calculations or put into the computer. The lab methods are also important for the results, and, in particular, creatinine levels in dialysate measured with the Jaffé method must be corrected for the interference with high concentrations of glucose in dialysate [8]. Sodium levels should preferably be measured with flame photometry or indirect ion-selective electrode measurements, as direct ion-selective electrode measurements may give different results [9].

The Peritoneal Equilibration Test

The most common test to evaluate peritoneal transport characteristics in individual patients is to measure the dialysate to plasma solute concentration ratio (D/P) for different solutes during a dialysis exchange [10]. This procedure has been standardized in the PET by Twardowski et al. [8] and has won wide acceptance as a routine method to assess the peritoneal membrane transport characteristics. Briefly, the overnight dialysate is drained and 2 liters of 2.27% glucose dialysis fluid are infused. The PET may include several dialysate samples [8], but usually, the procedure is simplified and dialysate samples are taken after infusion, at 2 and 4 h, at which time the dialysate is drained and the volume

Table 1. Peritoneal transport groups classified according to D/P creatinine at 4 h using Twardowski's initial classification

Transport group	D/P creatinine	Ao/Δx (cm/1.73 m^2 BSA)	Ultrafiltration capacity
Fast	>0.81	>30,000	–
Fast average	0.65–0.81	23,600–30,000	+
Slow average	0.50–0.65	17,200–23,600	++
Slow	<0.50	<17,200	+++

The patients were classified into high, high average, low average and low transport [11]. However, preferably fast and slow transport should be used instead of high and low, as the net removal of very small solutes, e.g. urea, often is low in 'high' transporters due to the poor ultrafiltration and lower drained volume. Fast and fast average transporters have more rapid equilibration of creatinine and poorer net ultrafiltration due to more rapid glucose absorption, whereas slow average and slow transporters will have slower solute transport, resulting in slow glucose absorption and high net ultrafiltration but low peritoneal clearances for creatinine and larger solutes.

recorded. A blood sample is drawn at 2 h dwell time. The net ultrafiltration, D/P for creatinine, and D/D_0 (dialysate concentration/initial dialysate concentration) for glucose are compared to standard values. The patients are usually classified according to D/P creatinine at 4 h using Twardowski's initial classification into high transporters (above mean +1 SD), high average transporters (between mean and mean +1 SD), low average transporters (between mean and mean –1 SD), and low transporters (below mean –1 SD) [8, 11] (table 1). However, most studies show an average creatinine D/P equilibration rate that is slightly more rapid than in the study of Twardowski [11].

It has been suggested to use 3.86% glucose solution instead of 2.27% glucose solution for the PET as it will give a better estimate of ultrafiltration capacity because of the higher ultrafiltration rate, and it also makes it possible to use decrease in dialysate sodium as an additional parameter to identify patients with poor ultrafiltration [12]. There is a marked dip in dialysate sodium concentration due to sieving of sodium with hypertonic solution as about half of the ultrafiltered fluid will pass through the aquaporins (see above). In patients with normal transport characteristics, the decrease in dialysate sodium is marked during the first 60 min, it then decreases slightly to its lowest value after approximately 90 min and thereafter the sodium concentration increases due to sodium diffusion from plasma [10].

PET is a simple procedure and easy to perform, the standard values are well established, and it does not require any complicated calculations. On the other hand, the D/P and D/D_0 results are rather sensitive to laboratory errors (only three samples are used), and the net ultrafiltration is sensitive to variation in the intraperitoneal residual dialysate volume. A commercial computer software program (PD-Adequest^{TM}) has been developed using results from the PET and the preceding overnight exchange to allow for calculation of basic transport parameters and to simulate the effects of changes in treatment schedules in individual patients [6, 7]. The PET has also been modified by using more frequent sampling and adding a tracer to the dialysate to allow for more detailed analysis of changes in intraperitoneal volume [13].

The D/P values generated by the PET procedure show an excellent correlation with PS for small solutes [14], and PS as well as D/P and PS for creatinine can be used to identify patients with loss of ultrafiltration capacity due to increased diffusive transport [15]. When using 2 liters of 4.25%/3.86% dextrose/glucose solution for the PET, it was suggested to define loss of ultrafiltration capacity as a net ultrafiltration below 400 ml after 4 h [12]. This definition has won wide acceptance and the use of hypertonic solution for the PET is strongly recommended for the evaluation of UF capacity.

Personal Dialysis Capacity Test

The personal dialysis capacity test involves urine, blood, and dialysate sampling. The patient collects urine and dialysate during a standardized CAPD day using a special exchange schedule, with two short (2–3 h) and two medium length (4–6 h) exchanges, each with two different glucose solutions, and one long overnight exchange. A sample is taken from each bag and the volume of each bag is measured, to give the variation in net ultrafiltration and solute equilibration with time, with the two glucose-based dialysis fluids [4, 5, 16, 17]. The data are put into a special software program, Personal Dialysis Capacity (PDC^{TM}), based on the three-pore model of peritoneal transport and calculates the following transport parameters (in addition to adequacy parameters and residual renal function): (1) area parameter ($Ao/\Delta x$), determining the diffusion capacity of small solutes, and indirectly, the hydraulic conductance of the membrane (LpA); (2) reabsorption rate of fluid from the peritoneal cavity to the blood after peak time, when the glucose gradient has dissipated, and (3) large pore fluid flow, which determines the loss of proteins to the PD-fluid. The $Ao/\Delta x$ is a more general parameter than PS for a specific solute and can be also used to classify the patients into similar transport groups as the PET (table 1). Because the PDC is based on five different determinations of dialysate concentration, it should have a more reliable classification of individual patient's transport rate [5, 16, 17].

Mini-PET

Recently, a short 'mini-PET' with hypertonic 3.86% glucose solution was suggested for the evaluation of ultrafiltration and free water transport [18]. It was also found to give estimates of solute transport (D/P creatinine) in good agreement with a 4-hour PET. A further improvement of this test was done in the double 'mini-PET' by the combination of two consecutive 1-hour PETs with 1.36 and 3.86% glucose solutions, respectively, which gives the possibility to calculate the free water transport, the osmotic conductance for glucose as well as the ultrafiltration through the small pores. The double 'mini-PET' seems to be a simple, fast and useful test to evaluate particularly patients with reduced ultrafiltration capacity [19].

Large Pore Flux

The large pore flux (plasma loss into the dialysate) is calculated in the PDC test, but may also easily be assessed using the plasma to dialysate albumin clearance in the overnight dwell or in a 24-hour dialysate collection. An increased albumin clearance (large pore flux) has been reported to an independent predictor of poor clinical outcome [20]. It may to some extent be related to inflammation and cardiovascular disease and has, furthermore, been suggested as marker of endothelial dysfunction.

Conclusions

There are several different simplified tests available for the evaluation of different aspects of peritoneal transport. The most widely used are the PET and PDC, which may give relatively detailed information about the peritoneal transport characteristics. Furthermore, computer software is available for the calculation of transport coefficients as well as for simulations of therapy changes. However, for some purposes, in particular the evaluation of patients with poor ultrafiltration, other tests such as the double mini-PET will give additional information. A rough estimate of the diffusive transport can also be made using the 24-hour D/P creatinine from a 24-hour dialysate collection in CAPD patients, which also of course gives the possibility to different calculate adequacy parameters. In summary, which test to use is dependent on the question and the clinical situation (table 2).

Table 2. Recommended transport test for different transport parameters

Parameter	Recommended tests
Diffusion	PET, PDC, mini-PET (24-hour D/P creatinine in CAPD)
Ultrafiltration capacity	short dwell with 3.86% glucose solution, PET, PDC, mini-PET
Free water transport	short dwell 3.86% with glucose solution and dialysate-Na after 1–2 h, double mini-PET
Fluid absorption	use a macromolecular tracer
Large pore flux (plasma loss)	PDC, albumin clearance from overnight or 24-hour dialysate
Kt/V, creatinine clearance, residual renal function, protein intake (PNA)	24-hour collection of dialysate and urine
To simulate therapy changes	PDC, PET with PD-Adequest evaluation

References

1 Heimbürger O: Peritoneal physiology; in Pereira BJG, Sayegh MH, Blake P (eds): Chronic Kidney Disease, Dialysis and Transplantation: A Companion to Brenner and Rector's the Kidney, ed 2. Philadelphia, Elsevier Saunders, 2005, pp 491–513.

2 Rippe B, Rosengren BI, Venturoli D: The peritoneal microcirculation in peritoneal dialysis. Microcirculation 2001;8:303–320.

3 Flessner MF: Peritoneal transport physiology: insights from basic research. J Am Soc Nephrol 1991;2:122–135.

4 Haraldsson B: Assessing the peritoneal dialysis capacities of individual patients. Kidney Int 1995;47:1187–1198.

5 Rippe B: Personal dialysis capacity. Perit Dial Int 1997;17(suppl 2):S131–S134.

6 Vonesh EP, Lysaght MJ, Moran J, Farrell P: Kinetic modeling as a prescription aid in peritoneal dialysis. Blood Purif 1991;9:246–270.

7 Vonesh EF, Rippe B: Net fluid absorption under membrane transport models of peritoneal dialysis. Blood Purif 1992;10:209–226.

8 Twardowski ZJ, Nolph KD, Khanna R, Prowant BF, Ryan LP, Moore HL, Nielsen MP: Peritoneal equilibration test. Perit Dial Bull 1987;7:138–147.

9 La Milia V, Di Filippo S, Crepaldi M, Andrulli S, Marai P, Bacchini G, Del Vecchio L, Locatelli F: Spurious estimations of sodium removal during CAPD when $[Na]^+$ is measured by Na electrode methodology. Kidney Int 2000;58:2194–2199.

10 Heimbürger O, Waniewski J, Werynski A, Lindholm B: A quantitative description of solute and fluid transport during peritoneal dialysis. Kidney Int 1992;41:1320–1332.

11 Twardowski ZJ: Clinical value of standardized equilibration tests in CAPD patients. Blood Purif 1989;7:95–108.

12 Mujais S, Nolph KD, Gokal R, Blake P, Burkhart J, Coles G, Kawaguchi Y, Kawanishi H, Korbet S, Krediet RT, Lindholm B, Oreopoulos DG, Rippe B, Selgas R, International Society of Peritoneal Dialysis Ad Hoc Committee on Ultrafiltration Management in Peritoneal Dialysis: Evaluation and management of ultrafiltration problems in peritoneal dialysis. Perit Dial Int 2000;20 (suppl 4):S5–S21.

13 Pannekeet MM, Imholz ALT, Struijk DG, Koomen GCM, Langedijk MJ, Schouten N, de Waart R, Hiralall J, Krediet RT: The standard peritoneal permeability analysis: a tool for the assessment of peritoneal permeability characteristics in CAPD patients. Kidney Int 1995;48:866–875.

14 Heimbürger O, Waniewski J, Werynski A, Park MS, Lindholm B: Dialysate to plasma solute concentrations (D/P) versus peritoneal transport parameters in CAPD. Nephrol Dial Transplant 1994;9:47–59.
15 Heimbürger O, Tranæus A, Park MS, Waniewski J, Werynski A, Lindholm B: Relationships between peritoneal equilibration test (PET) and 24 h clearances in CAPD (abstract). Perit Dial Int 1992;12 (suppl 2):S12.
16 Haraldsson B: Optimazation of peritoneal dialysis prescription using computer models of peritoneal transport. Perit Dial Int 2001;21 (suppl 3):S148–S151.
17 Johnsson E, Johansson AC, Andréasson BI, Haraldsson B: Unrestricted pore area is a better indicator of peritoneal function than PET. Kidney Int 2000;58:1773–1779.
18 La Milia V, Di Filippo S, Crepaldi M, Del Vecchio L, Dell'Oro C, Andrulli S, Locatelli F: Mini-peritoneal equilibration test: a simple and fast method to assess free water and small solute transport across the peritoneal membrane. Kidney Int 2005;68:840–846.
19 La Milia V, Limardo M, Virga G, Crepaldi M, Locatelli F: Simultaneous measurement of peritoneal glucose and free water osmotic conductance. Kidney Int 2007;72:643–650.
20 Heaf JG, Sarac S, Afzal S: A high peritoneal large pore fluid flux causes hypoalbuminaemia and is a risk factor for death in peritoneal dialysis patients. Nephrol Dial Transplant 2005;20:2194–201.

Olof Heimbürger, MD, PhD
Dept. of Renal Medicine, K56, Karolinska University Hospital, Huddinge
SE–141 86 Stockholm (Sweden)
Tel. +46 8 5858 3978, Fax +46 8 711 47 42, E-Mail olof.heimburger@ki.se

Ronco C, Crepaldi C, Cruz DN (eds): Peritoneal Dialysis – From Basic Concepts to Clinical Excellence. Contrib Nephrol. Basel, Karger, 2009, vol 163, pp 90–95

Dry Body Weight and Ultrafiltration Targets in Peritoneal Dialysis

Raymond T. Krediet[a], *Watske Smit*[a,b], *Annemieke M. Coester*[a], *Dirk G. Struijk*[a,b]

[a]Division of Nephrology, Department of Medicine, Academic Medical Center, University of Amsterdam, and [b]Dianet Foundation Utrecht, Amsterdam, The Netherlands

Abstract

A review is given on methods that can be used for the assessment of dry body weight in peritoneal dialysis patients. Besides clinical examination, the use of natriuretic hormone concentrations in plasma, and the value of multifrequency bio-impedance analysis is discussed. Ultrafiltration targets as formulated in various guidelines are reviewed. Finally, it is concluded that the ultrafiltration target is the amount required to keep patients euvolemic with an exposure to glucose that is as low as possible.

The concept of dry body weight is extensively used in hemodialysis (HD) patients. It can be defined as 'the weight at the end of a hemodialysis, at and below which the patient is free of edema, is normotensive until the next dialysis and has no symptomatic complaints during hemodialysis'. At dry weight, the extracellular volume is at or near normal, but not less than normal [1]. Because of the continuous dialysis treatment as used in peritoneal dialysis (PD), the determination of dry weight is much more difficult. It is dependent on methods to assess the extracellular volume. These methods will be reviewed in the following sections, followed by a discussion on ultrafiltration targets.

Assessment of Volume Status in Peritoneal Dialysis Patients

A careful clinical examination is required. Special attention should be given to blood pressure, postural hypotension, jugular venous pressure, the presence of a gallop rhythm on heart auscultation, and the presence or absence of edema. Clinical examination can be extended by a chest X-ray and measurement of the diameter of the inferior caval vein by ultrasound. The latter method is very accurate in experienced hands [2], but has never become very popular.

Some biochemical parameters may be useful, such as serum concentrations of albumin and natriuretic peptides. The interpretation of serum albumin is, however, difficult because not only is it influenced by hydration status [3], but also by peritoneal losses, nutritional status and inflammation. This explains the association between serum albumin and mortality, mainly due to the presence of a systemic disease [4]. In a more recent analysis of 246 PD patients from the cohort of the Netherlands Study on the Adequacy of Dialysis (NECOSAD), the presence of inflammation was the most important determinant of serum albumin [5]. It can therefore be concluded that serum albumin is not a very good marker for the assessment of volume status in PD patients.

An increase of extracellular volume leads to an increase of serum natriuretic peptides produced by the myocardium. These have been used as biochemical markers for the clinical diagnosis of heart failure [6] and as prognostic markers for mortality in end-stage renal disease patients [7]. In a random sample from the NECOSAD cohort, we found that atrial natriuretic peptides levels above the median of 1,112 pmol/l, and B-type natriuretic peptide concentrations above the median of 7.5 pmol/l, were associated with an 8-fold increase in the mortality hazard ratio [8]. This is illustrated in table 1. BNP is secreted as a prohormone that is cleaved into an active form and the inactive N-terminal propeptide BNP (N-BNP). N-BNP has a longer half-life and therefore more stable plasma concentrations. A recent study showed that serum N-BNP was more than 10 times higher in PD patients than in the normal population. These concentrations were strongly related to the residual GFR. The lower the GFR, the higher the concentrations. In addition, serum N-BNP was associated with cardiovascular congestion, mortality, and cardiovascular death and events [9]. A cross-sectional pilot study in PD patients showed that plasma N-BNP was associated with various parameters of hypervolemia [10]. The relationship of N-BNP concentrations with kidney function makes it difficult to define normal values in dialysis patients. This may limit the usefulness of N-BNP in clinical practice.

Multifrequency bio-impedance analysis (MF-BIA) is an inexpensive and elegant noninvasive technique for the assessment of total body water (TBW) and the extracellular volume (ECV). At low frequencies, the cell membrane acts as a condensator and blocks the flow of the current through the cell, whereas at higher frequencies the current flows both through the intra- and extracellular space. From these measurements, TBW and ECV are calculated using specifically designed software. In these calculations a constant relationship between

Table 1. Mortality hazard ratios (HR) and confidence intervals (CI) for PD patients with N-ANP or BNP above the median value (from Rutten et al. [8])

	Unadjusted HR (95% CI)	p value	Adjusted HR* (95% CI)	p value
N-ANP <1,112 pmol/l	1.0		1.0	
N-ANP >1,112 pmol/l	11.3 (1.4–91.9)	0.02	7.9 (0.9–72.1)	0.07
BNP <7.5 pmol/l	1.0		1.0	
BNP >7.5 pmol/l	11.3 (1.4–91.4)	0.02	8.5 (1.0–73.8)	0.05

* Adjusted for age, comorbidity and residual glomerular filtration rate.

ECV and TBW is assumed, derived from the situation in healthy people. However, in dialysis patients this relationship may be different. This could have been caused by the different type of overhydration compared to normals or the accumulation of 'uremic' waste products.

Evaluation of MF-BIA in PD patients with gold standards has been done in the study of Konings et al. [11]. Compared with deuterium, which is the gold standard, MF-BIA underestimated TBW with on average 2 liters, but with a very wide interindividual range. Assessment of ECV with the gold standard, that is bromide dilution, yielded a mean value 18.4 liters. Using the MF-BIA a mean value of 21 liters was found. In contrast, using segmental MF-BIA the mean value was only 14 liters. Despite this, the values were associated with each other. It appeared that the best correlation was found between TBW measured with deuterium and an estimate made by dual energy X-ray absorptiometry (Dexa). In this approach, lean body mass is multiplied by 0.73 to obtain TBW. Normalization of ECV as obtained by MF-BIA, showed that correction of this parameter for height had a sensitivity of 86% and a specificity of 80% for the presence of hypervolemia [12]. An ECV/height ratio of 10.94 liters/m in males and 9.13 liters/m in females were the cut-off points. It can be concluded from these data that MF-BIA is not suitable for precise measurements of TBW and ECV, but that it may be used for the assessment of overhydration. Although no good data are available, it is likely that MF-BIA can be used for longitudinal follow-up of volume status in individual patients.

Ultrafiltration Targets

According to the definition of the International Society for Peritoneal Dialysis (ISPD) the presence of ultrafiltration failure should be based on the 3

× 4 definition: less net ultrafiltration than 400 ml after a dwell of 4 h with a 4% (3.86%/4.25%) glucose-based dialysis solution [13]. This is a useful definition, because it distinguishes peritoneal ultrafiltration failure from fluid overload, which can be due to various causes. These include loss of residual urine production, a large dietary intake of salt and fluids, and an inappropriate dialysis prescription. Also, patients who drink just a very limited amount of fluids may be euvolemic despite having ultrafiltration failure according to the ISPD definition. The possible discrepancy between ultrafiltration failure and fluid overload makes it very difficult to formulate ultrafiltration targets in guidelines for adequacy of PD. Nevertheless, this has been tried.

In the European Best Practice Guidelines on Peritoneal Dialysis a minimum peritoneal target for net ultrafiltration in anuric patients of 1.0 liters/day was formulated [14]. This target was based on results from a study in Turkey in which the patient quartile with the lowest fluid removal (<1,265 ml/24 h/1.73 m^2) had the lowest survival [15]. An analysis in the NECOSAD cohort showed that a lower ultrafiltration in anuric patients was associated with a higher mortality. However, a cut-off point of <1.0 liters/day only showed a tendency ($p = 0.1$) [16]. In anuric APD patients, a predefined ultrafiltration target of <750 ml/day at baseline was associated with a decreased patients' survival [17]. The more recently published ISPD guidelines/recommendations [18] state: 'Attention should be paid to both urine volume and the amount of ultrafiltration, with the goal of maintaining euvolemia. A small ultrafiltered volume despite the use of dialysis solutions with a high glucose concentration should be regarded as a warning sign for the presence of ultrafiltration failure. This should be investigated further with a peritoneal equilibration test according to the ISPD recommendations on evaluation and management of ultrafiltration problems.'

The considerations discussed above imply that the ultrafiltration target in individual patients is a necessary quantity to keep the patient euvolemic. This is preferably accompanied by the lowest exposure to glucose and glucose degradation products to prevent peritoneal membrane damage. Ways to accomplish these two targets include advises on salt and water consumption, the use of loop diuretics in high dosages, preservation of residual renal function and a PD prescription in which icodextrin is included for the long dwell.

Conclusions

In addition to clinical examination and simple investigations, some new parameters for assessment of hydration status in PD patients have emerged. These include natriuretic peptides and multifrequency bioimpedance analysis. However, all these methods have their pitfalls and the results must be interpreted cautiously. The ultrafiltration target should be determined individually and aimed at keeping patients euvolemic. Dietary advice, preservation of residual renal function and high-dose loop diuretics are important to maintain a euvolemic state. When the achieved ultrafiltration decreases, a standardized test should be done to confirm or reject the presence of peritoneal ultrafiltration failure. To monitor the trends in individual patients, it is most practical to perform such tests on a regular basis.

References

1 Charra B, Laurent G, Chazot C, et al: Clinical assessment of dry weight. Nephrol Dial Transplant 1996;11(suppl 2):16–19.
2 Kouw PM, Kooman JP, Cheriex EC, Olthof GC, de Vries PM, Leunissen KM: Assessment of postdialysis dry weight: a comparison of techniques. J Am Soc Nephrol 1993;4:98–104.
3 Jones CH, Smye SW, Newstead CG, Will EJ, Davison AM: Extracellular fluid volume is determined by bioelectric impedance and serum albumin in CAPD patients. Nephrol Dial Transplant 1998;13:393–397.
4 Struijk DG, Krediet RT, Koomen GCM, Boeschoten EW, Arisz L: The effect of serum albumin at the start of continuous ambulatory peritoneal dialysis treatment on patient survival. Perit Dial Int 1994;14:121–126.
5 De Mutsert R, Grootendorst DC, Indemans F, Boeschoten EW, Krediet RT, Dekker FW, Netherlands Cooperative Study on the Adequacy of Dialysis-II Study Group: Association between serum albumin and mortality in dialysis patients is partly explained by inflammation, and not by malnutrition. J Renal Nutr 2009;19:127–135.
6 Maisel AS, Krishnaswamy P, Nowak RM, et al: Rapid measurement of B-type natriuretic peptide in the emergency diagnosis of heart failure. N Engl J Med 2002;347:161–167.
7 Mallamuci F, Tripeli G, Cutrupi S, Malatino LS, Zoccali C: Prognostic value of combined use of biomarkers of inflammation, endothelial dysfunction, and myocardiopathy in patients with ESRD. Kidney Int 2005;67:2330–2337.
8 Rutten JHW, Korevaar JC, Boeschoten EW, et al: B-Type natriuretic peptide and aminoterminal atrial natriuretic peptide predict survival in peritoneal dialysis. Perit Dialysis Int 2006;26:598–602.
9 Wang AY-M, Lam CW-K, Yu C-M, et al: N-terminal pro-brain natriuretic peptide: an independent risk predictor of cardiovascular outcomes in chronic peritoneal dialysis patients. J Am Soc Nephrol 2007;18:321–330.
10 Gangji AS, Al Helal B, Churchill DN, Brimble KS, Margetts PJ: Association between N-terminal propetide B-type natriuretic peptides and markers of hypervolemia. Perit Dialysis Int 2008;28:308–311.
11 Konings CJAM, Kooman JP, Schonk M, et al: Assessment of fluid status in peritoneal dialysis patients. Perit Dial Int 2002;22:683–692.
12 Van de Kerkhof J, Hermans M, Beerenhout C, Konings C, van der Sande FM, Kooman JP: Reference values for multifrequency bio-impedance analysis in dialysis patients. Blood Purif 2004;22:301–306.
13 Mujais S, Nolph KD, Gokal R, et al: Evaluation and management of ultrafiltration problems in peritoneal dialysis. Perit Dial Int 2000;20(suppl 4):S5–S21.
14 EBPG Expert Group on Peritoneal Dialysis: European Best Practice Guidelines for Peritoneal Dialysis. Nephrol Dial Transplant 2005;20(suppl 9):IX24–IX27.
15 Ates K, Nergizoglu G, Keven K, et al: Effect of fluid and sodium removal on mortality in peritoneal dialysis patients. Kidney Int 2001;60:767–776.

16 Jansen MAM, Termorshuizen F, Korevaar JC, Dekker FW, Boeschoten EW, Krediet RT, NECOSAD Study Group: Predictors of survival in anuric peritoneal dialysis patients. Kidney Int 2005;68:1199–1205.
17 Brown EA, Davies SJ, Rutherford P, et al: Survival of functionally anuric patients on automated peritoneal dialysis: the European APD outcome study. J Am Soc Nephrol 2003;14:2948–2957.
18 Lo W-K, Bargman JM, Burkart J, et al: Guidelines on targets for solute and fluid removal in adult patients on chronic peritoneal dialysis. Perit Dial Int 2006;26:520–522.

R.T. Krediet, MD, PhD
Room F4–215, Academic Medical Center
Division of Nephrology, Department of Medicine
PO Box 22700
NL–1100 DE Amsterdam (The Netherlands)
Tel. +31 20 5665990, Fax +31 20 6914904, E-Mail C.N.deboer@amc.uva.nl

Ronco C, Crepaldi C, Cruz DN (eds): Peritoneal Dialysis – From Basic Concepts to Clinical Excellence. Contrib Nephrol. Basel, Karger, 2009, vol 163, pp 96–101

Acute Central Hemodynamic Effects of Peritoneal Dialysis

W. Van Biesen, A. Pletinck, F. Verbeke, R. Vanholder

Renal Division, Department of Internal Medicine, Ghent University Hospital, Ghent, Belgium

Abstract

Background: The supposed lack of a hemodynamic impact of peritoneal dialysis (PD) has been challenged recently in different studies, although the observed effects are still far below those seen on hemodialysis (HD), and the underlying mechanisms are unclear. **Methods:** Literature overview based on Pubmed search with key words 'peritoneal dialysis, acute dwell, hemodialysis'. **Discussion:** Hemodynamic effects of an acute PD dwell seem to be consistent, but rather limited. Increasing peritoneal pressure, causing enhanced preload and thus better cardiac output, and vasoactive reactions induced by incompatibility of the dialysis fluid seem to be the most prominent causes. The role of hyperglycemia is a matter of debate. In view of the repetitive character of the insults, especially during APD, more in depth investigation of this phenomenon is warranted.

Mortality from cardiovascular disease is higher in patients with renal impairment than in the general population [12, 13] and is the major cause of death in patients on renal replacement therapy. Hemodynamic alterations such as volume overload and hypertension contribute to the accelerated progression of cardiovascular disease in dialysis patients. Acute hemodialysis changes, mostly episodes of hypotension, during hemodialysis (HD) are well known to be related to worse outcome. Most physicians presume that an acute peritoneal dialysis (PD) dwell does not produce acute hemodialysis changes, which might explain why the topic has been investigated only to a limited extent. The few studies evaluating changes in hemodialysis parameters indicate however that some mild changes in blood pressure and vascular tone do occur during an acute dwell [1, 11]. Some potential underlying mechanisms can be forwarded: (1) a change in intra-abdominal volume and/or pressure, leading to enhanced pre-load, and thus, according to Starling's law, an increased output, (2) absorption of glucose, leading to hyperglycemia and insulin release, and (3) a pure bioincompatibility of the dialysis solution.

Such repetitive undulations in hemodialysis status can potentially be harmful, as they might increase cardiac workload, lead to shear stress and can negatively impact residual renal function [5]. Therefore, understanding the nature and causes of these changes is of clinical importance.

Role of Intraperitoneal Volume and Pressure

The intraperitoneal instillation of dialysate causes an abrupt increase in the intraperitoneal pressure, which might cause an enhanced venous retour to the right ventricle. This might result in an improved cardiac output, and thus an increase in blood pressure. Of note, if this mechanism would be true, the increase in blood pressure should not be seen as a bad but rather as a positive side effect. However, it could also be that an increased intraperitoneal pressure reduces venous return by compressing the vena cava inferior. In this situation, cardiac output would decrease. Ivarsen et al. [2] analyzed the evolution of cardiac output using foreign gas rebreathing in 15 PD patients, following the instillation of 2 to 3 liters of PD fluid, and this in the supine and upright position. In the supine position, no effect of a 2 liter infusion was observed. In the upright position, a decrease in cardiac output, and an increase in peripheral systemic resistance was observed, resulting in an absence of a net change in blood pressure. Remarkably, the results were not different when either using a 2- or a 3-liter fill, and no change in heart rate was observed. Boon et al. [1], using fingertip continuous pressure monitoring, found an increase in peripheral resistance and brachial blood pressure after instillation of 2 liters of glucose based peritoneal dialysate, an effect they attributed either to compression of the mesenteric resistance vessels or to the low temperature. Unfortunately, in these studies, hypertonic glucose solutions were used, so an eventual impact of hyperosmolarity and/or glucose cannot be excluded. The role of intraperitoneal volume and/or pressure has been elucidated in a study by Verbeke et al. [15], applying volume instillation with a non-glucose containing solution (icodextrin, Baxter Healthcare, Ireland), at neutral pH and warmed to 37.0°C. Verbeke et al. [15] demonstrated that the increase in intraperitoneal pressure was related to a change in hemodialysis parameters, pointing to the impact of intraperitoneal volume per se on hemodialysis parameters. This was confirmed in a study by Selby et al. [10], where in APD patients, blood pressure decreased during the drain, and increased during the instillation of dialysate (unfortunately all glucose based). The decrease in blood pressure was corre-

Table 1. Overview of clinical studies on hemodialysis effects of an acute peritoneal dwell

	Conditions	Measurement	Results	Remarks
Verbeke et al. [15]	pH neutral, nonglucose-based dialysate; cross-over randomized volume changes	central blood pressure with sphygocor	higher initial fill volumes increase central blood pressure	a circadian effect was demonstrated
Pletinck et al. [8]	1.36% vs. 3.86% glucose, low GDP, normal pH; randomized cross over	central blood pressure with sphygocor	despite increase in glycemia and insulin, no increase in blood pressure	
Boone et al. [1]	instillation of a 3.86% glucose, low pH, high GDP	peripheral blood pressure, finger tip method	increase in blood pressure	no reports on evolution of serum glycemia
Selby et al. [10]	2 dwells of 1.36% and 1 dwell of 3.86% in 4 h APD	finopress	blood pressure increased at the beginning of each dwell	no cross-over
Selby et al. [9]	1.36%, 3.86 glucose, biocompatible vs. amino acid-based dialysate	finopress		
Selby et al. [11]	1.36% glucose followed by 3.86% glucose or icodextrin	finopress	hypertonic glucose but not icodextrin induced blood pressure increase	no cross-over
Vychytil et al. [16]	instillation of amino acid vs. 2.27% glucose-based solution	forearm vasodilation reactivity	amino acid solutions impair flow mediated vasodilation	cross-over
John et al. [3]	instillation of 1.36 vs. 3.86% and low GDP vs. high GDP solutions	barosensor reactivity	hypertonic glucose impairs BRS, no difference biocompatible vs. nonbiocompatible	cross-over

lated with a decrease in peripheral resistance and an incomplete rise in cardiac output.

Impact of Glucose

It is important to know whether glucose has a role in these hemodialysis effects, as this would be an additional argument for the further search for nonglucose osmotic agents, and for the avoidance of hypertonic glucose by appropriate dietary salt restriction and use of icodextrin. In healthy volunteers, it appears that intravenous glucose loading does not alter blood pressure, although it can impair circadian variation in heart rate [7]. Selby et al. [11], using hypertonic 3.86% glucose, found an increase in blood pressure and cardiac output during the dwell, and this irrespective of the biocompatibility of the solution used. Pletinck et al. [8], in contrast, were not able to confirm these results. Despite a significant rise in serum glucose and insulin levels over time, which was more pronounced during the high glucose dwell, there was no demonstrable effect on central systolic or diastolic blood pressure in this study. There are several potential explanations for the discrepancies between the two studies: (1) There was a difference in design, as the Pletinck study used cross-over, and the Selby study not. Of note, Verbeke et al. [15] found a circadian effect of hemodialysis parameters, which might thus explain the differences as the higher blood pressure in the Selby study might rather been induced by the timing than by the higher glucose concentration. (2) The differences between 1.36% and 3.86% glucose, as observed by Selby et al. [11] might merely have been the consequence of a more pronounced ultrafiltration in the 3.86% group, creating a larger intraperitoneal volume. However, also in the Pletinck study, a larger intraperitoneal volume at the end of the dwell was observed. It has been demonstrated that the effect of IP volume is mostly induced in the beginning of the filling procedure, where large changes in IP pressure are induced, whereas the gradual and moderate further increase of IP volume due to ultrafiltration during the dwell has less or even no further impact. However, Selby et al. [11] measured left ventricular filling with cardiac ultrasound during the procedure, and did not observe important changes. (3) The methodology used to measure hemodialysis parameters was different: the Finometer®, which reflects peripheral blood pressure, versus SphygmoCor®, reflecting central blood pressure. The differences in approach of these methods have been discussed previously [14]. From the cardiovascular standpoint, however, the central blood pressure is the most important one, as this is the pressure that the left ventricle actually has to produce (afterload) and that impacts on the wall of the large central vessels.

In any case, the use of low glucose tonicity solutions (1.36%) seemed to have no or very limited hemodialysis effect in both studies. Avoidance of hypertonic exchanges is thus to be advocated.

Impact of Bioincompatibility

It is well conceivable that the bioincompatibility related to temperature, the low pH, hyperosmolarity and/or the presence of GDP's in conventional solutions induces by itself some hemodialysis reaction. Zakaria et al. [17] demonstrated in vitro that exposure to PD solutions resulted in vasodilation of intestinal arterioles in an in vitro model. Also in vitro, Kawabe et al. [4] found that exposure to peritoneal dialysis solutions resulted in contraction of large vessels, and relaxation of microvessels. In an in vivo videomicroscopy model, Mortier et al. [6] found that bioincompatible versus biocompatible solutions caused much more disturbances in capillary recruitment and flow. The main factors appeared to be the presence of GDPs and the low pH. In vivo, the hemodialysis response appears to be different when using conventional versus low GDP solutions or Nutrineal [3]. Vychytil et al. [16] measured forearm reactive hyperemia during dwells with amino acid based dialysate or hypertonic glucose. They found that the amino acid solution significantly impaired forearm dilatation, an effect they postulated was attributable to an increase in ADMA.

Conclusion

Although the underlying mechanisms are still unclear, there appear to be hemodialysis effects of an acute instillation of peritoneal fluid (table 1). Changes in intraperitoneal pressure seem to play a role, and should be avoided. The role of hyperglycemia is debated, but use of nonhypertonic glucose-based solutions, especially those with low GDP content, seem to be without or of very limited hemodialysis effect. The use of high volume, fast cycling hypertonic exchanges should be avoided. The role of amino acid based solutions needs further investigation.

References

1 Boon D, Bos WJ, van Montfrans GA, et al: Acute effects of peritoneal dialysis on hemodynamics. Perit Dial Int 2001;21:166–171.

2 Ivarsen P, Povlsen JV, Jensen JD: Increasing fill volume reduces cardiac performance in peritoneal dialysis. Nephrol Dial Transplant 2007;22:2999–3004.

3 John SG, Selby NM, McIntyre CW: Effects of peritoneal dialysis fluid biocompatibility on baroreflex sensitivity. Kidney Int Suppl 2008:S119–S124.

4 Kawabe T, Zakaria el R, Hunt CM, et al: Peritoneal dialysis solutions contract arteries through endothelium-independent prostanoid pathways. Adv Perit Dial 2004;20:177–183.

5 McIntyre CW: Acute cardiovascular functional effects of peritoneal dialysis: what do we know and why might it matter? Perit Dial Int 2008;28:123–125.

6 Mortier S, De Vriese AS, Van de Voorde J, et al: Hemodynamic effects of peritoneal dialysis solutions on the rat peritoneal membrane: role of acidity, buffer choice, glucose concentration, and glucose degradation products. J Am Soc Nephrol 2002;13:480–489.

7 Petrova M, Townsend R, Teff KL: Prolonged (48-hour) modest hyperinsulinemia decreases nocturnal heart rate variability and attenuates the nocturnal decrease in blood pressure in lean, normotensive humans. J Clin Endocrinol Metab 2006;91:851–859.

8 Pletinck A, Verbeke F, Van Bortel L, et al: Acute central haemodynamic effects induced by intraperitoneal glucose instillation. Nephrol Dial Transplant 2008;23:4029–4035.

9 Selby NM, Fialova J, Burton JO, et al: The haemodynamic and metabolic effects of hypertonic-glucose and amino-acid-based peritoneal dialysis fluids. Nephrol Dial Transplant 2007;22:870–879.

10 Selby NM, Fonseca S, Hulme L, et al: Automated peritoneal dialysis has significant effects on systemic hemodynamics. Perit Dial Int 2006;26:328–335.
11 Selby NM, Fonseca S, Hulme L, et al: Hypertonic glucose-based peritoneal dialysate is associated with higher blood pressure and adverse haemodynamics as compared with icodextrin. Nephrol Dial Transplant 2005;20:1848–1853.
12 Van Biesen W, De Bacquer D, Verbeke F, et al: The glomerular filtration rate in an apparently healthy population and its relation with cardiovascular mortality during 10 years. Eur Heart J 2007;28:478–483.
13 Vanholder R, Massy Z, Argiles A, et al: Chronic kidney disease as cause of cardiovascular morbidity and mortality. Nephrol Dial Transplant 2005;20:1048–1056.
14 Verbeke F, Segers P, Heireman S, et al: Noninvasive assessment of local pulse pressure: importance of brachial-to-radial pressure amplification. Hypertension 2005;46:244–248.
15 Verbeke F, Van Biesen W, Pletinck A, et al: Acute central hemodynamic effects of a volume exchange in peritoneal dialysis. Perit Dial Int 2008;28:142–148.
16 Vychytil A, Fodinger M, Pleiner J, et al: Acute effect of amino acid peritoneal dialysis solution on vascular function. Am J Clin Nutr 2003;78:1039–1045.
17 Zakaria el R, Patel AA, Li N, et al: Vasoactive components of dialysis solution. Perit Dial Int 2008;28:283–295.

W. Van Biesen, MD, PhD
Renal Division, Department of Internal Medicine, University Hospital Ghent
De Pintelaan 185
BE–9000 Ghent (Belgium)
E-Mail wim.vanbiesen@ugent.be

Ronco C, Crepaldi C, Cruz DN (eds): Peritoneal Dialysis – From Basic Concepts to Clinical Excellence. Contrib Nephrol. Basel, Karger, 2009, vol 163, pp 102–109

Cardiovascular Complications in Peritoneal Dialysis Patients

Beth Piraino

Renal Electrolyte Division, University of Pittsburgh, Pittsburgh, Pa., USA

Abstract

Patients on peritoneal dialysis (PD) have a high risk of dying from cardiovascular disease (CVD), including myocardial infarction, arrhythmias, valvular disease, and sudden death. The risk of dying on PD, in contrast to HD, is rather low initially but steadily increases over time. Risk factors for CVD on PD need to be closely examined and research on possible interventions designed. PD patients typically gain weight, often to excess. They develop hyperlipidemia, particularly low-density lipoprotein and triglyceride levels, in part from the glucose load. PD patients are often hypertensive and sometimes volume overloaded. These are all traditional risk factors for CVD. However, in addition, there are some poorly examined risk factors for PD patients, including hypokalemia, shown to be associated with an increased risk of both peritonitis and death. PD patients may be inflamed related to non-physiologic PD fluid and peritonitis. PD patients over time often become profoundly 25(OH) vitamin D deficient, due to dietary constraints, lack of sun exposure and effluent losses. Such insufficiency of 25(OH)vitamin D deficiency seems to be a risk factor for CVD. Many of the above-described risk factors are either treatable or preventable. Very few randomized trials targeting treatment of cardiovascular risk factors have been done in PD patients. Whether a multiple pronged approach to CVD risk in PD patients would lower the later mortality remains to be tested.

Patients on peritoneal dialysis (PD) have a changing pattern of mortality over time, relatively low initially but increasing over the years [1]. The majority of the deaths are due to cardiovascular causes and sudden death [2]. In the United States, the current 1-year survival on PD is 86.8% compared to 79% for HD patients, adjusted for age, gender, race, ethnicity and primary renal diagnosis (excluding the first 90 days) [2]. However, by 5 years, these percentages are 35.2% alive on PD, and 34.5% on HD. As shown in figure 1, for patients ages 65–74 years, the spectrum of cardiovascular disease

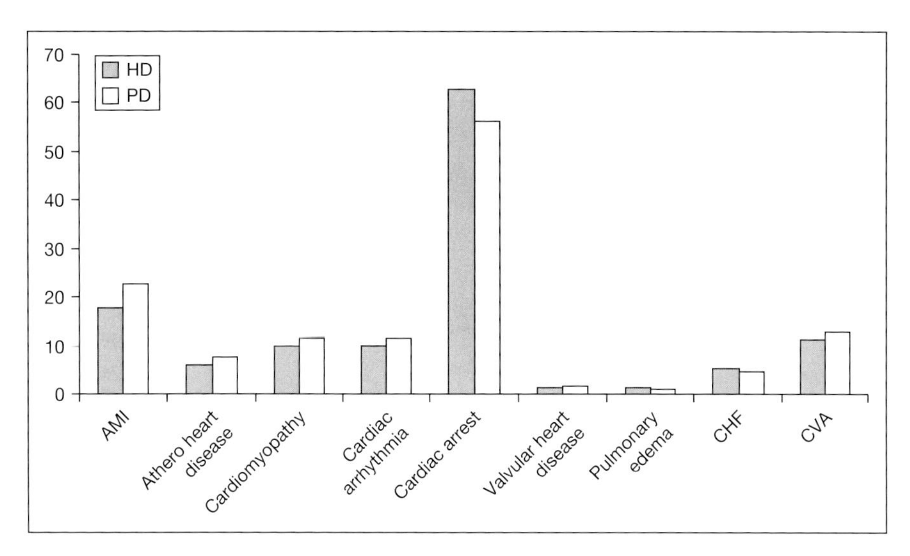

Fig. 1. Cardiovascular causes of death, as deaths per 1,000 patient years [data from ref. 2].

(CVD) is broad for both HD and PD patients. Therefore, risk factors for CVD on PD need to be closely examined and research on possible interventions designed.

In this paper, I explore some of the risk factors associated with CVD risk in the PD population that may account for the striking increase over time of CVD deaths. While there are no interventional studies in PD patients demonstrated to lower CVD risk, I will suggest possible studies that could be done.

Hyperlipidemia

Dyslipidemia is markedly prevalent in PD patients, even more so than hemodialysis (HD) patients [3]. PD patients have high total and low-density lipoprotein (LDL), cholesterol, high apolipoprotein B, and triglyceride levels and reduced high-density lipoprotein (HDL) cholesterol levels [3]. PD patients have small, dense LDL, higher Lp(a) levels and low HDL cholesterol compared to the normal (nondialysis) population. The proposed mechanisms for these findings are the glucose loading from dialysis fluid and peritoneal protein losses, leading to a picture similar to the nephrotic syndrome.

There are few data on the effect of treating hyperlipidemia in PD patients on mortality, but we know in the general population that lowering cholesterol is protective. HMG co-reductase inhibitors reduce the risk of cardiovascular events in the nondialysis population, likely from a combination of lowering cholesterol and decreasing inflammation. The 4D study was done in diabetic patients on HD, and the results should not be imputed to apply to PD patients [4].

The drugs commonly used to treat these abnormalities, the HMG CoA reductase inhibitors, are safe and effective in end-stage renal disease patients [3]. Unfortunately, there are no data regarding impact on mortality and CV outcomes. In the absence of such data, my approach has been to treat with HMC co-reductase inhibitors to an LDL cholesterol level of <2.6 mmol/l, based on the high risk of CVD in this population, even in the absence of clinically obvious disease.

Glucose and the Metabolic Syndrome

Conventional PD solutions are glucose-based which is readily absorbed across the peritoneum leading to a carbohydrate caloric load and weight gain in many patients. In some cases, this leads to the development of new-onset diabetes mellitus. In addition, the glucose absorption contributes to hypertriglyceridemia. Insulin resistance is common, and this is yet another risk factor for CV disease. Because of the caloric loading from glucose, weight gain can occur. Wu et al. [5] performed a retrospective analysis of patients grouped by high, medium and low glucose load over the first 6 months, with follow-up for a mean of 40 months. The glucose load during the first 6 months was an independent predictor of technique survival in multivariate analysis, along with patient age, and pre-existing diabetes mellitus.

Studies examining different PD prescriptions and the effect on weight gain and outcomes are scant. CCPD does not seem to offer any advantage or disadvantage over CAPD, although the latter does result in better sodium removal. Patients randomized to icodextrin for the long exchange in fast transporters leads to a lower weight at 1 year compared to those on dextrose dialysis solution for the long exchange; it was unclear whether this lower weight was due to less edema or a decreased in tissue weight [6].

One can speculate that starting a de novo patient with residual kidney function on an abbreviated cycler prescription with a dry day (knowing that dialysis clearance does not add much to survival) might result in a lesser effect on the peritoneal membrane, and less effect on weight gain. Such an approach minimizes early exposure of the peritoneum to dialysis fluid. The clearance due to residual renal function always exceeds 2 Kt/V weekly early in end stage renal disease but must be followed very carefully as well as symptoms, to adjust the prescription upwards as time passes and residual kidney function deteriorates.

Hyperphosphatemia and Vascular Calcification

We know with increasing time on dialysis, dystrophic calcification of vessels-including the coronary vessels, and cardiac valves can appear and result in significant heart disease [7, 8]. Valvular calcification is present in approximately 1/3 of patients at the start of PD; the 1-year survival is 70% for those with valvular calcification compared to 93% for those without valvular calcification ($p < 0.001$) [7]. The hazards ratio for CV death if valvular calcification is present at the start of PD is 5.39 ($p = 0.0003$), independent of age, male gender, dialysis duration, C-reactive protein, diabetes, and atherosclerotic vascular disease [7]. Cardiac valvular calcification can also be used as an indicator of carotid artery calcification [8]. The pathogenesis of vascular and valvular calcification is complex, but hyperphosphatemia and particularly an elevated calcium-phosphate product is a significant part of the pathology, as is inflammation [9]. Fetuin-A is an inhibitor of the calcification process and is down-regulated in the presence of inflammation [9].

Phosphorus control is difficult in both PD patients as well as HD patients. A low phosphorus diet appears to be difficult for the patient as many are cheese lovers, phosphorus is in many high protein foods, and phosphate binders are disliked by most patients. A diet restricted in phosphorus is critical in combination with binders to control hyperphosphatemia in PD patients as peritoneal clearance alone is inadequate in controlling the levels. Because sevelamer reduces both total and LDL cholesterol, this may be a preferred phosphate binder in PD patients [3].

Inflammation

Inflammation is a risk factor for poor outcomes in PD patients [10]. For each increase of 10 mg/l of serum CRP, the relative risk of death increases 2.69, independent of CVD, age, serum albumin, diabetes and residual kidney function [10]. A high serum C-reactive protein is associated with lower 24-hour fluid removal, as effect that may relate to nutrition [10]. A high CRP should cause one to look for possible obvious causes, such as poor dental hygiene, a potentially correctable problem. Use of HMG co-reductase inhibitors can lower serum TNF-α levels as well as LDL and total cholesterol [11]. Whether this will have any impact on outcome in PD patients is unclear.

Table 1. Areas for possible RCT to prevent CVD in PD patients

Use of dry day with APD to begin dialysis: possible impact on change in D/Pcr over time, weight gain, inflammation, insulin resistance and serum albumin levels
Use of HMG co-reductase inhibitors, perhaps in combination with gemfibrozil: impact on CV events in PD patients
Multi-prolonged approach to patient education regarding assessment of volume status, choice of dialysis dextrose solutions, stricter sodium intake, and closer regulation of blood pressure: impact on CV events
Trial of closer monitoring of cardiac status with routine stress tests and echocardiograms, with planned interventions: impact on CV deaths
Maintenance of normal 25(OH)vitamin D levels: impact on LVH

Hypokalemia

Hypokalemia, seen in about one third of PD patients, has now been recognized as associated with an increase risk of death, as well as a risk of enteric peritonitis [12]. The exact relationship between hypokalemia and death is unclear, but one might speculate that this could be through a cardiovascular cause such as arrhythmias. It might also be a marker of inadequate nutrition. This requires further research. However, hypokalemia is readily treated by a small dose of oral potassium.

Intravascular Volume Expansion and Hypertension

In the ADEMEX trial there was a strong and independent association of baseline NT-proBNP and survival and cardiovascular mortality [13]. The investigators also showed that volume removal was also associated with patient survival [13]. This studied excluded patients with known heart disease, which makes this result even more striking. We know that PD patients often have elevated intravascular volume from studies of patients with central venous pressure measured at the time of transplantation. Hypertension is associated with late mortality especially in patients with less obvious co-morbidity [14]. The relationship among blood pressure, volume overload (related to sodium status) and cardiac status is explored in a recent editorial by Van Biesen et al. [15]. These authors emphasize the importance of matching the PD prescription to transport category to maximize sodium removal (low transporters with longer dwells), and the importance of dietary salt restriction. The balance is delicate as excessive salt and water removal can lead to a decrease in urine volume [16].

All of this suggests that attention needs to be made to the patient's volume status during visits, and that home dialysis patients need to be taught more about volume status and adjustment of the dextrose dialysis. At each visit the treatment records should be examined. The development of new smart technology may make this easier. The prescription should be adjusted to avoid negative ultrafiltration with any part of the prescription (for example, the long dwell). ARB and ace inhibitors, due to their protective effect on residual kidney function and possible protective effect on the peritoneal membrane, should be preferentially used in PD patients [17].

25(OH)Vitamin D Deficiency

PD patients become markedly 25(OH)vitamin D deficient with increasing time on PD due to losses of 25(OH)vitamin D in the effluent, poor intake of vitamin D and lack of exposure to sunshine in some climates. Over time almost all patients on PD become 25(OH)vitamin D insufficient and many have profound deficiency [18, 19]. Wang et al. [19] found that for each one unit increase in log-transformed 25(OH)vitamin D, the hazard of a cardiovascular event decreased (95% CIs of 0.35–0.91, $p = 0.018$). The mechanism may well be through an effect on left ventricular mass, as controlling for such resulted in a loss in the effect. 25(OH)vitamin D lack is readily treated Likely a replacement dose of 1,200 IU each day or 50,000 units each month (once depletion restored by giving 50,000 IU per week for 12 weeks) is sufficient to maintain adequate levels [18].

Conclusions

There is a paucity of interventional trials examining the effectiveness of interventions on CV outcomes in PD patients. Trials that might be studied include an initial prescription that minimizes peritoneal membrane to dialysis fluid by utilizing a dry day, a multi-pronged intervention that would include strict volume control by use of social cognitive theory to enhance adherence with a low-salt diet, strict attention to the prescription and UF as well as sodium removal, control of blood pressure using an ace inhibitor or ARB, repletion of 25(OH)vitamin D, and control of LDL by use of HMG co-reductase inhibitor, as well as control of trigylcerides with niacin or gemfibrozil (table 1). Such interventions need to be studied in rigorous RCTs compared to usual care. Only with such data will we be able to move forward to improve long-term outcomes in PD patients.

References

1 McDonald SP, Marshall MR, Johnson DW, Polkinghorne KR: Relationship between dialysis modality and mortality J Am Soc Nephrol 2009;Doi: 10. 1681/ASN.2007111188.

2 US Renal Data System, USRDS 2008 Annual Data Report: Atlas of Chronic Kidney Disease and End-Stage Renal Disease in the United States. Bethesda, National Institutes of Health, National Institute of Diabetes and Digestive and Kidney Diseases, 2008.

3 Tsimihodimos V, Dounousi E, Siamopoulos KC: Dyslipidemia in chronic kidney disease: an approach to pathogenesis and treatment Am J Nephrol 2008;28:958–973.

4 Wanner C, Krane V, Marz W, Olschewski M, Mann JF, Ruf G, Ritz E, German Diabetes and Dialysis Study Investigators: Atrovastatin in patients with type 2 diabetes mellitus undergoing hemodialysis. N Engl J Med 2005;353:238–248.

5 Wu HY, Hung KY, Huang JW, Chen YM, Tsai TJ, Wu KD: Initial glucose load predicts technique survival in patient on chronic peritoneal dialysis. Am J Nephrol 2008:28:765–771.

6 Wolfson M, Piraino B, Hamburger R, Morton R, Icodextrin Study Group: A randomized controlled trial to evaluate the efficacy and safety of icodextrin in peritoneal dialysis Am J Kidney Dis 2002;40:1055–1065.

7 Wang AY, Wang M, Woo J, Lam C, Li PK, Lui SF, Sanderson JE: Cardiac valve calcification as an important predictor for all-cause morality and cardiovascular mortality in long-term peritoneal dialysis patients: a prospective study. J Am Soc Nephrol 2003;13:159–168.

8 Wang AY, Ho S, Wang M, Liu E, Ho S, Li PKT, Lui S, Sanderson JE: Cardiac valvular calcification as a marker of atheroscleroiss and arterial calcification in end-stage renal disease. Arch Intern Med 2005;165:327–332.

9 Wang A, Woo J, Lam C, Wang M, Chan I, Gao P, Lui S, Li PKT, Sanderson JE: Associations of serum fetuin-A with malnutrition, inflammation, atheroscleroiss and valvular calcification syndrome and outcome in peritoneal dialysis patients. Nephrol Dial Transplant 2005;20:1676–1685.

10 Chung SH, Heimburger O, Stenvinkel P, Wang T, Lindholm B: Influence of peritoneal transport rate, inflammation, and fluid removal on nutritional status and clinical outcome in prevalent peritoneal dialysis patients. Perit Dial Int 2003;23:174–183.

11 Sezer MT, Katirci S, Demir M, Erturk J, Adana S, Kaya S: Short-term effect of simvastatin treatment on inflammatory parameters in peritoneal dialysis patients. Scand J Urol Nephrol 2007;41:436–441.

12 Szeto CC, Chow KM, Kwan BC, Leung CB, Chung KY, Law MC, Li PK: Hypokalemia in Chinese peritoneal dialysis patients: prevalence and prognostic implication. Am J Kidney Dis 2005;46:128–135.

13 Paniagua R, Amato D, Mujais S, Vonesh E, Ramos A, Correa-Rotter R, Horl WH: Predictive value of brain natriuretic peptides in patients on peritoenal dialysis: results from the ADEMEX trial. Clin J Am Soc Nephrol 2008;3:407–415.

14 Udayaraj U, Steenkamp R, Caskey FJ, Rogers C, Nitsch D, Ansell D, Tomson C: Blood pressure and mortality risk on peritoneal dialysis. Am J Kidney Dis 2009;53:70–78.

15 Van Biesen W, Verbeke F, Devolder I, Vanholder R: The relation between salt, volume, and hypertension: clinical evidence for forgotten but still valid basic physiology. Perit Dial Int 2008;28:596–600.

16 Davies SJ, Lopez EG, Woodrow G, Donovan K, Plum J, Williams P, Johansson AC, Bosselmann HP, Heimburger O, Simonsen O, Davenport A, Lindholm B, Tranaeus A, Divino Filho JC: Longitudinal relationships between fluid status, inflammation, urine volume and plasma metabolites of icodextrin in patients randomized to glucose or icodextrin for the long exchange. Nephrol Dial Transplant 2008;23:2982–2988.

17 Kolesnyk I, Noordzij M, Dekker FW, Boeschoten EW, Krediet RT: A positive effect of AII inhibitors on peritoneal membrane function in long-term PD patients. Nephrol Dial Transplant 2009;24:272–277.

18 Shah N, Bernardini J, Piraino B: Prevalence and correction of 25(OH)vitamin D deficiency in peritoneal dialysis patients. Perit Dial Intern 2005;25:362–366.
19 Wang AYM, Lam CWK, Sanderson JE, Wang M, Chan I, Lui SF, Sea M, Woo J: Serum 25-hydroxyvitamin D status and cardiovascular outcomes in chronic peritoneal dialysis patients: a 3-year prospective cohort study Am J Clin Nutr 2008;87:1631–1638.

Beth Piraino, MD
Renal Electrolyte Division, University of Pittsburgh, Suite 200
3504 Fifth Avenue
Pittsburgh, PA 15213 (USA)
Tel. +1 412 383 4899, Fax +1 412 383 4898, E-Mail Piraino@pitt.edu

Ronco C, Crepaldi C, Cruz DN (eds): Peritoneal Dialysis – From Basic Concepts to Clinical Excellence. Contrib Nephrol. Basel, Karger, 2009, vol 163, pp 110–116

Brain Natriuretic Peptide in Peritoneal Dialysis Patients

Mikko Haapio[a], *Eero Honkanen*[a], *Claudio Ronco*[b]

[a]Division of Nephrology, Helsinki University Central Hospital, Helsinki, Finland; [b]Department of Nephrology, St. Bortolo Hospital, Vicenza, Italy

Abstract

Background: The expanding use of brain natriuretic peptides (BNP and NT-proBNP) testing in patients with end-stage kidney disease has increased our knowledge of relating heart-kidney interactions, but also brought concern of the reliability and usefulness of BNPs in dialysis patients. **Methods:** The review highlights the most important recent results of BNP research and discusses applicability of BNPs in the evaluation of peritoneal dialysis (PD) patients. **Results:** Relevant physiological background of BNP with relating aspects of PD treatment are reviewed, along with analysis of BNP measurement limitations and suggestions for rational use in the PD population. **Conclusion:** To date, interpretation of BNP levels in PD patients is limited in areas of hydration status assessment and optimal dry weight determination. However, elevated levels of BNPs exhibit validated correlation and prognostic value with pathological cardiac structure, cardiovascular adverse events and survival, thus assisting in individual patient risk profiling and deeper understanding of cardiorenal factors involved.

Natriuretic peptides are a group of hormones with similar chemical structure and biological actions. They constitute an important part of the vasodilatatory system in human physiology with diversified effects on heart, kidneys, vascular tree and central nervous system. A-type natriuretic peptide (ANP) was the first one discovered in the beginning of the 1980s, followed by characterization of B-type (BNP) and C-type (CNP) natriuretic peptides approximately 20 years ago. Later, similar peptides with effects on blood pressure and fluid homeostasis have been identified, namely Dendroaspis (D-type) natriuretic peptide, urodilatin and guanylin, and uroguanylin [1].

The common physiological effects of natriuretic peptides include enhanced excretion of sodium and water in the kidneys, relaxation of vascular smooth muscle cells and counteraction of the human vasoconstrictor system (sympathetic nervous system, renin-angiotensin-aldosterone system, endothelin and vasopressin) resulting in natriuresis, diuresis and vasodilation. Renal effects are achieved by dilatation of afferent and constriction of efferent arterioles with subsequent increase in renal glomerular filtration and filtered sodium fraction. Effects in heart include lowering of cardiac preload and afterload and improved myocardial perfusion with coronary vasodilation [1–3].

BNP is derived from its precursor pre-proBNP (134 amino acids), which is cleaved by proteases to a signal peptide and proBNP (108 amino acids), continued by formation of the biologically active BNP (32 amino acids, with MW 3.4 kDa, and serum half-life 15–20 min) and inactive N-terminal-proBNP (NT-proBNP, 76 amino acids, MW 8.5 kDa, and half-life 60–120 min). BNP was originally found in porcine brain but has since been recognized to be secreted mostly from myocytes in cardiac ventricles in response to excessive cardiac ventricular wall distension, shear stress, various pathological states (for instance, myocardial ischemia or cardiomyopathies) and increased release of catecholamines, renin, angiotensin II and endothelin. BNP binds to and activates natriuretic peptide receptor A causing production of cyclic guanosyl monophosphate (cGMP) resulting in the biological effects of BNP. BNP is metabolized by natriuretic peptide receptor C, which is widespread in the human body (liver, lungs, kidney and vascular endothelium). BNP, unlike NT-proBNP, is further degraded by neutral endopeptidases (mainly renal tubular NEP) [1, 3].

Effects of Renal Function on B-Type Natriuretic Peptides

Considering the epidemic growth of incidence and prevalence of patients with heart or kidney insufficiency recognized during the past years, the ability to perform accurate diagnosing and comprehensive management of these patients has become a major challenge for modern health system. The situation is further complicated by the many times simultaneous occurrence of heart and kidney insufficiency, called cardiorenal syndrome [4], as it has been reported that 33–56% of patients with heart failure have diminished renal function [2]. BNP has shown its practicability in diagnostics and management of both acute and chronic heart failure without renal impairment, but the proper interpretation of serum BNP levels with simultaneous heart failure (HF) and chronic kidney disease (CKD) has been a subject of contention. Whereas excretion of BNP has been considered in the earlier studies not to be largely affected by renal function, for NT-proBNP the renal clearance was thought to be the principal excretory mechanism, thus compromising the usefulness of NT-proBNP in patients with CKD. The complexity of interpretation lies in the fact that elevated serum BNP levels can be caused multifactorially: by increased burden on myocardium from various (e.g. ischemic) primary heart pathologies, by fluid overload resulting from either (or both) the cardiac and/or renal insufficiency, and, lastly, by the altered clearance of especially NT-proBNP in the kidneys [2, 3]. Recently, new insight has been brought into the association of BNPs with glomerular filtration rate by studies in which fractional renal excretion of both BNP and NT-proBNP remained equivalent across the range of renal function. Furthermore, correlation coefficients between BNP and estimated glomerular filtration rate (eGFR) vs. NT-proBNP and eGFR were found to be similar, giving additional support to consider changes in the level of BNP not to be significantly different from changes of NT-proBNP in patients with renal insufficiency. In patients with diminished eGFR (CKD 1 to CKD 4) the age-dependent cut points for NT-proBNP in assessment of cardiac status need no further adjustment for renal function compared to cut points designed for general population [2]. However, for patients on dialysis, the question of proper cut-off levels for BNPs still awaits a final answer [2, 3].

B-Type Natriuretic Peptide as a Diagnostic Tool

The extent of BNPs secretion from ventricular myocytes is in relation to the severity of the pathological process with pronounced serum levels in more extensive cardiac distress. This had led to the successful use of BNP as a marker of cardiac volume overload in decompensated HF, especially in the differential diagnosing of acute dyspnea. In patients without renal impairment the cut-off levels of under 100 pg/ml (BNP) and under 300 pg/ml (NT-proBNP) give 99% negative predictive value to rule out acute congestive HF [3, 5]. In CKD patients, the confounding issue of renal effects on the level of serum BNPs has been under investigation. In predialysis patients with left ventricular hypertrophy and declining eGFR, both BNPs show increase with a more considerable rise in NT-proBNP, so that reduction in eGFR of every 10 ml/min/1.73 m^2 is associated with an increase of BNP by about 20% and NT-proBNP by about 38% [6, 7]. In the Breathing Not Properly Study, the level of BNP was markedly elevated in patients with severer stage CKD and optimum cut points to diagnose congestive HF were suggested accordingly, rising from approximately 71 pg/ml (CKD 1) to 225 pg/ml (CKD 4) [5]. Although no clear cut-off levels of BNPs in dialysis patients have been established, seemingly higher decision limits are most probably needed. Given the lack of solid evidence to date, BNP testing for heart failure has been suggested to be discouraged in dialysis patients [3].

B-Type Natriuretic Peptide and Cardiovascular Morbidity and Outcomes

A great number of studies have demonstrated the practicability and importance of BNPs (BNP and NT-proBNP) testing in various cardiac pathologies including prediction of left ventricular abnormalities in the general population, recognizing cardiac ischemia, and diagnosing, predicting prognosis and monitoring therapy of heart failure [8]. In end-stage kidney disease, the risk of cardiovascular (CV) morbidity and mortality are grossly increased and CV events are the leading cause for death in this population [9]. Elevated levels of BNPs have been shown to be associated with systolic dysfunction, left ventricular hypertrophy and mortality also in the CKD patient population [6, 8]. In dialysis patients, raised level of BNP or NT-proBNP is strongly and independently associated with left ventricular ejection fraction (negative correlation) and mass index (positive correlation), is predictor of cardiovascular adverse events and risk of cardiovascular and overall mortality [10]. Wang et al. [8] prospectively studied 230 peritoneal dialysis (PD) patients with 3 years' follow-up and found quartiles of NT-proBNP to significantly and independently predict CV congestion, mortality and adverse CV outcomes. Rutten et al. [11] found PD patients with BNP above the median (7.5 pmol/l) to have significantly increased mortality compared to patients with BNP under the median, with a hazard ratio of 8.5 after adjustment for age, comorbidity and residual glomerular filtration rate. ADEMEX Trial comparing CAPD patients on standard treatment (n = 484) with patients in the intervention group (n = 481) resulted in NT-proBNP showing independent and high predictability of overall survival and CV mortality [12].

B-Type Natriuretic Peptide and Volume Status of Peritoneal Dialysis Patients

Hypertension is common in PD patients, and is mostly caused by chronic fluid overload, which is further associated with heightened risk of cardiovascular mortality. Hypervolemia sufficient to cause cardiac stress results in elevated levels of BNPs, setting a basis to look for a connection between BNPs and extracellular fluid measured by bioimpedance analysis (BIA). Results, however, have not been congruent. Further, establishing dry weight of PD patients on solid information provided by BNPs has not been successful. Lee et al. [13] did not find a correlation with NT-proBNP levels and percentage of extracellular water (ECW) in 30 stable CAPD patients examined by a multifrequency bioimpedance analyzer. Haapio and colleagues [unpubl. data] performed a study with 63 PD (45 CAPD, 18 APD) patients divided into tertiles based on BIA measurements of fluid volume and discovered BIA tertiles able to distinquish BNP in a dose-dependent fashion. BNP was significantly higher in patients with most fluid compared to patients with least fluid (364 vs. 87 pg/ml, $p = 0.044$). When considering effects of various PD fluids, the long-term use of icodextrin has been shown to cause significant BNP decrease during a 2-year follow-up, possibly attributable to increased ultrafiltration volume [14].

Influence of Dialysis Modality on B-Type Natriuretic Peptide Levels

Whereas successful renal transplant normalizes BNP levels for almost all, for patients in chronic dialysis the levels remain over normal cut-off values irrespective of the dialysis modality. In PD, however, the levels are usually lower than in hemodialysis, even though use of PD fluids has been reported not to influence BNP levels by clearance [1]. Of note, BNPs are removed from serum especially by high-flux hemodialysis membranes [15]. In concordance with several other studies, Nakatani et al. [16] demonstrated the plasma BNP concentration to be significantly lower in 32 CAPD patients than in 63 hemodialysis patients (114.8 ± 142.7 vs. 296.8 ± 430.4 pg/ml, $p < 0.0001$), suggesting that the cardiac load in CAPD patients may be lower. Treatment with APD is reported to be associated with higher plasma BNP levels compared to CAPD [17]. In summary, BNP shows a fair ability to separate overhydrated from underhydrated dialysis patients and may assist in giving additional information to other measures of volume status, but is unsatisfactory for use as the sole method.

Discussion

Elevated levels of serum BNPs are ubiquitous in chronic PD patients, yet exhibit strong and independent association with cardiovascular morbidity and overall and CV mortality. Though the results of hydration status evaluation with BNPs in PD patients have been variable and in many cases disappointing, BNPs could be useful in combination with other parameters of fluid volume. BNPs are capable of separating overhydrated patients from underhydrated, but lack capacity to detect underhydration as such or to determine dry weight reliably. BNP levels are lower in PD treatment compared to hemodialysis by most reports published, possibly reflecting lower cardiac load in PD. BNPs may be used cautiously and to some extent for assessment of acute cardiovascular congestion, but until now their best benefit for the PD population is in identifying patients with heightened cardiorenal risk profiles.

References

1 Woodard GE, Rosado JA: Recent advances in natriuretic peptide research. J Cell Mol Med 2007;11:1263–1271.
2 DeFilippi C, van Kimmenade RRJ, Pinto YM: Amino-terminal pro-B-type natriuretic peptide testing in renal disease. Am J Cardiol 2008;101(suppl):82A–88A.
3 Maisel A, Mueller C, Adams K Jr, Anker SD, Aspromonte N, Cleland JGF, Cohen-Solal A, Dahlström U, DeMaria A, Di Somma S, Filippatos GS, Ronarow GC, Jourdain P, Komajda M, Liu PP, McDonagh T, McDonald K, Mebazaa A, Nieminen MS, Peacock WF, Tubaro M, Valle R, Vanderhyden M, Yancy CW, Zannad F, Braunwald E: State of the art: using natriuretic peptide levels in clinical practice. Eur J Heart Fail 2008;10:824–839.
4 Ronco C, Haapio M, House AA, Anavekar N, Bellomo R: Cardiorenal syndrome. J Am Coll Cardiol 2008;52:1527–1539.
5 McCullough PA, Nowak RM, McCord J, Hollander JE, Herrmann HC, Steg PG, Duc P, Westheim A, Omland T, Knudsen CW, Storrow AB, Abraham WT, Lamba S, Wu AHB, Perez A, Clopton P, Krishnaswamy P, Kazanegra R, Maisel AS, BNP Multinational Study Investigators: B-type natriuretic peptide and clinical judgment in emergency diagnosis of heart failure: analysis from Breathing Not Properly (BNP) multinational study. Circulation 2002;106:411–422.
6 Austin WJ, Bhalla V, Hernandez-Arce I, Isakson SR, Beede J, Clopton P, Maisel AS, Fitzgerald RL: Correlation and prognostic utility of B-type natriuretic peptide and its amino-terminal fragment in patients with chronic kidney disease. Am J Clin Pathol 2006;126:506–512.
7 Vickery S, Price CP, John RI, Abbas NA, Webb MC, Kempson ME, Lamb EJ: B-type natriuretic peptide (BNP) and amino-terminal proBNP in patients with CKD: relationship to renal function and left ventricular hypertrophy. Am J Kidney Dis 2005;46:610–620.
8 Wang AYM, Lam CWK, Yu CM, Wang M, Chan HIS, Zhang Y, Lui SF, Sanderson JE: N-terminal pro-brain natriuretic peptide: an independent risk predictor of cardiovascular congestion, mortality, and adverse cardiovascular outcomes in chronic peritoneal dialysis patients. J Am Soc Nephrol 2007;18:321–330.
9 Sarnak MJ, Levey AS, Schoolwerth AC, Coresh J, Culleton B, Hamm LL, McCullough PA, Kasiske BL, Kelepouris E, Klag MJ, Parfrey P, Pfeffer M, Raij L, Spinosa DJ, Wilson PW: Kidney disease as a risk factor for development of cardiovascular disease: a statement from the American heart association councils on kidney in cardiovascular disease, high blood pressure research, clinical cardiology, and epidemiology and prevention. Hypertension 2003;42:1050–1065.
10 Zoccali C, Mallamaci F, Benedetto FA, Tripepi G, Parlongo S, Cataliotti A, Cutrupi S, Giacone G, Bellanuova I, Cottini E, Malatino LS: Cardiac natriuretic peptides are related to left ventricular mass and function and predict mortality in dialysis patients. J Am Soc Nephrol 2001;12:1508–1515.
11 Rutten JHW, Korevaar JC, Boeschoten EW, Dekker FW, Krediet RT, Boomsma F, van den Meiracker AH, NECOSAD Study Group: B-type natriuretic peptide and amino-terminal atrial natriuretic peptide predict survival in peritoneal dialysis. Perit Dial Int 2006;26:598–608.
12 Paniagua R, Amato D, Mujais S, Vonesh E, Ramos A, Correa-Rotter R, Horl WH: Predictive value of brain natriuretic peptides in patients on peritoneal dialysis: results from the ADEMEX Trial. Clin J Am Soc Nephrol 2008;3:407–415.
13 Lee JA, Kim DH, Yoo SJ, Oh DJ, Yu SH, Kang ET: Association between serum N-terminal pro-brain natriuretic peptide concentration and left ventricular dysfunction and extracellular water in continuous ambulatory peritoneal dialysis patients. Perit Dial Int 2006;26:360–365.
14 Hiramatsu T, Furuta S, Kakuta H: Favorable changes in lipid metabolism and cardiovascular parameters after icodextrin use in peritoneal dialysis patients. Adv Perit Dial 2007;23:58–61.
15 Haapio M, Ronco C: BNP and a renal patient: emphasis on the unique characteristics of B-type natriuretic peptide in end-stage kidney disease. Contrib Nephrol. Basel, Karger, 2008, vol 161, pp 68–75.
16 Nakatani T, Naganuma T, Masuda C, Uchida J, Sugimura T, Sugimura K: Significance of brain natriuretic peptides in patients on continuous ambulatory peritoneal dialysis. Int J Mol Med 2002;10:457–461.

17 Bavbek N, Akay H, Altay M, Uz E, Turgut F, Uyar ME, Karanfil A, Selcoki Y, Akcay A, Duranay M: Serum BNP concentration and left ventricular mass in CAPD and automated peritoneal dialysis patients. Perit Dial Int 2007;27:663–668.

Dr. Mikko Haapio
Consulting Nephrologist, HUCH Meilahti Hospital
Haartmaninkatu 4
FI–00029 Helsinki (Finland)
Tel. +358 9 4711, Fax +358 9 47177246, E-Mail mikko.haapio@helsinki.fi

Ronco C, Crepaldi C, Cruz DN (eds): Peritoneal Dialysis – From Basic Concepts to Clinical Excellence. Contrib Nephrol. Basel, Karger, 2009, vol 163, pp 117–123

Metabolic Impact of Peritoneal Dialysis

Thyago Proença de Moraes, Roberto Pecoits-Filho

Center for Health and Biological Sciences, Pontifícia Universidade Católica do Paraná, Curitiba, Brazil

Abstract

Peritoneal dialysis (PD) patients are at high risk of developing cardiovascular disease, resulting from both traditional and nontraditional cardiovascular risk factors, including factors related to uremia and dialysis. Glucose metabolism is altered in chronic kidney disease patients even in the earlier stages and the most common manifestations are insulin resistance and dyslipidemia, known factors in the pathogenesis of hypertension and atherosclerosis. Exposure to high glucose concentration solutions during PD intensifies these metabolic abnormalities, which are extremely common in PD patients and potentially have an impact on cardiovascular outcome. Life style and dietary modification and pharmacological interventions may be particularly important to revert these risk factors and glucose sparing solutions may represent an additional strategy to change the PD-induced metabolic profile. Together, these approaches may have a positive impact in reducing the risk of mortality in PD patients.

The complexity of mechanisms, many of them still unknown, that are involved in the genesis and development of cardiovascular (CV) complications in chronic kidney disease (CKD) limits the application of strategies aiming to reduce mortality in this group of patients. Current understanding of CKD-related CV disease points to a combination of Framingham, non-traditional and CKD-specific risk factors involved determining the physiopathology basis of this high mortality disease. An emblematic (since involves all areas of risk factors described above) and potentially important area particularly in patients treated with peritoneal dialysis (PD) is related to carbohydrate metabolism disorders.

These complex metabolic disturbances in glucose/lipid metabolism are extremely important among these risk factors as they permeate all stages of the kidney-cardiovascular interaction, from the generation of diabetic nephropathy through the genesis of uremic insulin resistance and metabolic disorders related to glucose absorption from peritoneal dialysate during renal replacement therapy. This mini-review aims to provide readers with the most recent insights on carbohydrate metabolism disturbances in CKD, with a special emphasis on the metabolic consequences of PD treatment.

Glucose and Insulin Homeostasis in Chronic Kidney Disease

Glucose metabolism is altered in CKD patients even in the early stages of kidney dysfunction. Normally, insulin is freely filtered at the glomerulus, and extensively reabsorbed in the proximal tubule, after enzyme degradation into smaller peptides. With the progression of renal disease, peritubular insulin uptake increases, compensating for the decline in the breakdown of filtered insulin. When renal function is severely impaired, insulin clearance decreases leading to an increase in the half-life of insulin [1].

Insulin resistance also occurs and worsens with the progression of CKD, and it is characterized by sub-normal glucose serum levels in response to a given insulin concentration. Tissue insensitivity to insulin is of primary importance, but alterations in insulin degradation and insulin secretion may also contribute [2]. Possible involved mechanisms in uremic insulin resistance include increased hepatic gluconeogenesis, hepatic and/or skeletal muscle glucose uptake and impaired intracellular glucose metabolism. Since adequate dialysis partially reverses the problem, uremic toxins are most likely involved and a potential candidate is asymmetric dimethylarginine (ADMA), an endogenous nitric oxide synthase inhibitor [3].

The hyperinsulinemia observed in CKD patients contributes to the development of hypertriglyceridemia, which is intensified in PD by the exposure to high glucose concentration solutions. Insulin enhances hepatic very-low-density lipoproteins (VLDL) triglyceride synthesis and may indirectly (via decreased sensitivity of lipoprotein lipase to insulin) reduce the rate of metabolism of VLDL. Because of their adverse implications, particularly endothelial dysfunction, metabolic syndrome and insulin resistance are considered important predictors of vascular complications in dialysis patients and are even more important for PD patients due to the daily glucose load from PD solutions (fig. 1).

Impact of Glucose Absorption on Metabolic Disorders and the Role of Peritoneal Solute Transport

PD patients are constantly exposed to fluids that are bioincompatible as a result of the high glucose concentration, the presence of glucose degradation

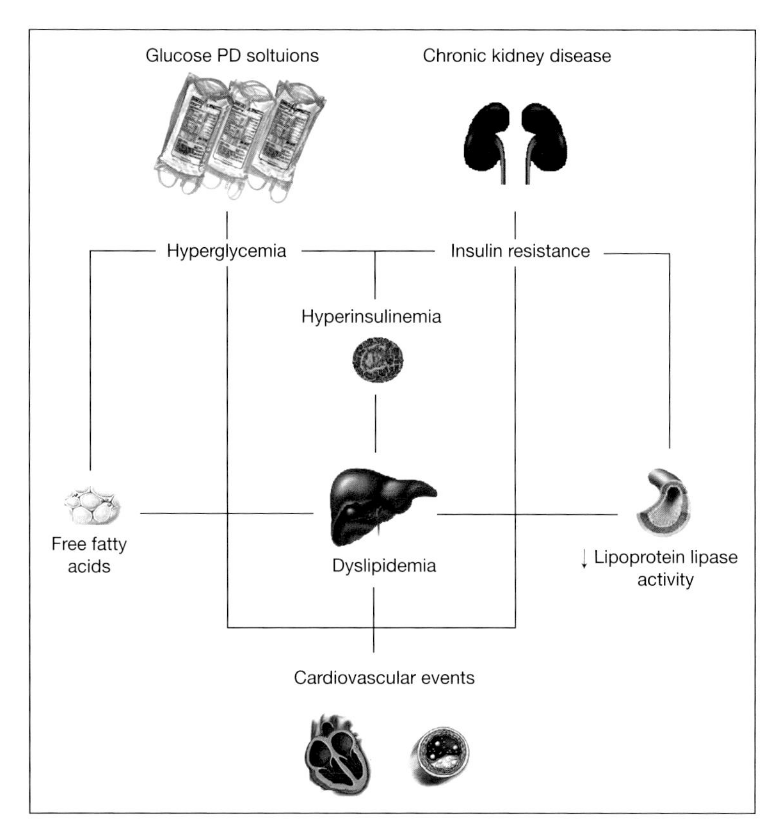

Fig. 1. Combined impact of CKD and PD on glucose and metabolism leading to cardiovascular events.

products (GDP) that enhance the formation of advanced glycosylation end products (AGE), and the low pH of lactate buffers. Conventional PD solutions have long been suspected to be involved in peritoneal membrane injury. The identified mechanisms of injury rely mostly on secretion of cytokines, growth factors, and proteases, complement activation, and increased coagulation. Usually, about 60% of glucose in the dialysate is reabsorbed from the peritoneal cavity during the dwell, creating an imbalance in the carbohydrate metabolism even in nondiabetic patients and despite the partial reversion of insulin resistance after the initiation of dialysis. The resulting glucose absorption may worsen the hyperlipidemia, hyperglycemia, insulin resistance, oxidative stress and inflammation that are normally associated with the uremic milieu. Moreover, new-onset hyperglycemia was described in CKD patients after the start of PD. In fact, upon comparing equivalent doses of oral and intraperitoneal glucose, the

Table 1. Impact of high peritoneal transport status on metabolic and cardiovascular status in PD patients

Impact of increasing potential solute transport rate	Clinical consequences
Increased glucose absorption	obesity dyslipidemia insulin resistance
Increased protein loss	hypoalbuminemia loss of antioxidant capacity edema
Loss of ultrafiltration capacity	fluid overload hypertension inflammatory activation

latter provides significantly higher glucose serum levels [4]. Insulin resistance can be safely assessed by HOMA-index in CKD patients, and a study showed that a high HOMA index predicted mortality in CKD patients.

Glucose in PD solutions is degraded during the heat sterilization procedure and continues during storage. The high GDP levels present in PD solutions are associated with peritoneal membrane injury and can lead to ultrafiltration failure. Over the last decade, an increasing number of reports have identified and quantified glucose degradation products. The mechanisms involved in the biological reactivity of GDP are cytotoxicity, modulation of cellular signal pathways, and protein modification by cross-linking. Accumulation of advanced glycosylation end-products and glucose degradation products over peritoneal vessels due to hypertonic glucose dialysis solution exposure and sequentially induced inflammatory cascades, such as vascular endothelial growth factor (VEGF), interleukin-1β or tumor necrosis factor-α contribute to changes in peritoneal membrane function. Together, these inflammatory changes will lead to the creation of a vicious cycle in the membrane that will intensify the need for hypertonic glucose prescription and metabolic consequences.

Peritoneal solute transport increases with time on treatment in PD patients, contributing to ultrafiltration failure and increasing glucose reabsorption from the abdominal cavity. Fast transporters present a large effective peritoneal surface area and/or higher intrinsic membrane permeability and a consequent large absorption of glucose into the circulation. It has been reported that an increased peritoneal transport rate is associated with lower patient and technique survival. One possible explanation may be that the prolonged exposure of a peritoneal membrane prone to lose the osmotic gradient required for sustained ultrafiltration of hypertonic glucose solutions damages peritoneum tissue, leading to lower drained volume, small solute removal, inadequate fluid balance, and malnutrition [5]. One interesting and still unexplored potential cause for the high mortality in fast transporters may be related to the metabolic consequences of high glucose absorption. This hypothesis should addressed in future studies, including glucose sparing strategies as interventions.

Consequences of Hyperglycemic and Lipid Disorders in Peritoneal Dialysis Patients

Hyperglycemia is associated with poor prognosis in the general population, even in nondiabetics. Several prospective, randomized controlled clinical trials have demonstrated that intensive therapy aimed at lower levels of glycemia results in decreased rates of retinopathy, nephropathy, and neuropathy, thus improving outcome [6]. These benefits are also observed in pre-dialysis patients and patients on renal replacement therapy. Glycated hemoglobin is a sensitive and reliable marker of hyperglycemia recently suggested as an independent predictor of all cause mortality in nondiabetic CKD patients.

Several measures and therapeutic options have been proposed to prevent CV disease and type 2 diabetes in the general population such as: lifestyle modification including diet and exercises, oral hypoglycemic agents such as thiazolidinediones, lipid-lowering agents, antihypertensive therapy, and most recently endocannabinoid inhibition [7]. However, there is not enough evidence to validate these interventions in PD patients. In PD, glucose-sparing solutions could present a potential benefit on the disturbances of carbohydrate metabolism.

Lipid disorders as well as accelerated general atherosclerosis are frequently observed in CKD patients. In addition, lipid abnormalities may cause progressive renal vascular insufficiency deteriorating the already impaired renal function [8]. The pathophysiology of uremic dyslipidemia is not clearly understood, but it is thought to be related to decreased catabolism of apoprotein-B-containing lipoproteins, to decreased activity of lipolytic enzymes, and altered lipoprotein composition [9]. Hypertriglyceridemia is the most common manifestation of the altered lipid profile in PD patients and is sustained primarily by the glucose load reabsorbed daily from the peritoneal cavity.

Role of Inflammation and Oxidative Stress

The mechanisms underlying CV disease have recently shifted to include inflammation and oxidative stress as pivotal factors in the initiation and progression of atherosclerosis in CKD patients. The understanding of the role of oxidative stress and chronic inflammation in membrane permeability, structure, and function has stimulated the search for new strategies that could reduce oxidation and inflammation and drugs that could reverse the structural changes associated with these [10]. Conventional PD fluids can cause neoangiogenesis, accumulation of advanced glycation end-products (AGEs), and other toxins resulting from oxidative stress. Numerous insulting factors associated with the initiation and perpetuation of the cascade of inflammatory events has been identified in PD: glucose load, glucose degradation products, acid pH, hyperosmolality, plasticizers and peritonitis [11]. New measures aiming to reduce the GDPs concentration in PD solutions could reduce cytotoxicity, improve the cellular function, and prolong membrane survival, independently of the glucose concentration contributing to membrane preservation.

Therapeutic Options and Future Interventions

Therapeutic options for carbohydrate metabolism disorders are limited in PD patients. Rosiglitazone (RSG) is an oral agent of a new class of antidiabetic drugs known as peroxisome proliferator-activated receptor modulators (PPAR) which improves insulin resistance in type II diabetic patients [12]. RSG also decreases CRP level. However, more studies are needed to show improvement in patient and technique survival.

Glucose sparing solutions (icodextrin and amino acid based) are an interesting and promising therapy to reduce chronic exposure to glucose and avoid systemic problems related to glucose absorption. Studies using surrogate markers of metabolic control as endpoints are ongoing, and the results may stimulate the realization of large studies testing the hypothesis that glucose sparing solutions may reduce mortality in PD patients through the reduction of metabolic consequences of glucose absorption.

Conclusions

In addition to the disturbances of glucose and lipid metabolism observed in CKD, PD-related risk factors (mainly induced by high glucose exposure) characterize a complex metabolic disorder that potentially harms the patient. The prolonged use of PD solutions with high glucose concentration generates a series of consequences not only in the peritoneal membrane, but also at the systemic level, that vary from hyperglycemia and insulin resistance to dyslipidemia and central obesity. These metabolic abnormalities are common findings in PD patients and may directly influence patient outcome. Life style and dietary modification and pharmacological interventions may be particularly important to revert these risk factors and glucose sparing solutions may represent an additional strategy to change the PD-induced metabolic profile. Together, these approaches may have a positive impact in reducing the risk of mortality in PD patients.

References

1 Mak RH, DeFronzo RA: Glucose and insulin metabolism in uremia. Nephron 1992;61:377–382.
2 Adrogue HJ: Glucose homeostasis and the kidney. Kidney Int 1992,42:1266–1282.
3 Stuhlinger MC, Abbasi F, Chu JW, et al: Relationship between insulin resistance and an endogenous nitric oxide synthase inhibitor. JAMA 2002;287:1420–1426.
4 Delarue J, Maingourd C, Lamisse F, et al: Glucose oxidation after a peritoneal and an oral glucose load in dialyzed patients. Kidney Int 1994;45:1147–1152.
5 Chen HY, Kao TW, Huang JW, et al: Correlation of metabolic syndrome with residual renal function, solute transport rate and peritoneal solute clearance in chronic peritoneal dialysis patients. Blood Purif 2008;26:138–144.
6 UK Prospective Diabetes Study (UKPDS) Group: Intensive blood-glucose control with sulphonylureas or insulin compared with conventional treatment and risk of complications in patients with type 2 diabetes (UKPDS 33). Lancet 1998;352:837–853.
7 Haffner SM, Greenberg AS, Weston WM, et al: Effect of rosiglitazone treatment on nontraditional markers of cardiovascular disease in patients with type 2 diabetes mellitus. Circulation 2002;106:679–684.
8 Avram MM, Fein PA, Antignani A, et al: Cholesterol and lipid disturbances in renal disease: the natural history of uremic dyslipidemia and the impact of hemodialysis and continuous ambulatory peritoneal dialysis. Am J Med 1989;87:55N–60N.
9 Attman PO, Samuelsson O, Alaupovic P: Lipoprotein metabolism and renal failure. Am J Kidney Dis 1993;21:573–592.
10 Pecoits-Filho R, Stenvinkel P, Wang AY, et al: Chronic inflammation in peritoneal dialysis: the search for the holy grail? Perit Dial Int 2004;24:327–339.
11 Yeun JY, Kaysen GA: Acute phase proteins and peritoneal dialysate albumin loss are the main determinants of serum albumin in peritoneal dialysis patients. Am J Kidney Dis 1997;30:923–927.
12 Wong TY, Szeto CC, Chow KM, et al: Rosiglitazone reduces insulin requirement and C-reactive protein levels in type 2 diabetic patients receiving peritoneal dialysis. Am J Kidney Dis 2005;46:713–719.

Roberto Pecoits-Filho
Center for Health and Biological Sciences
Imaculada Conceição, 1155
Curitiba, PR 80215–901 (Brazil)
Tel./Fax +55 41 3271 1657, E-Mail r.pecoits@pucpr.br

Ronco C, Crepaldi C, Cruz DN (eds): Peritoneal Dialysis – From Basic Concepts to Clinical Excellence. Contrib Nephrol. Basel, Karger, 2009, vol 163, pp 124–131

Is Obesity Associated with a Survival Advantage in Patients Starting Peritoneal Dialysis?

Renée de Mutsert, Diana C. Grootendorst, Elisabeth W. Boeschoten, Friedo W. Dekker, Raymond T. Krediet

Leiden University Medical Center, Department of Clinical Epidemiology, C7-P, Leiden, The Netherlands

Abstract

Background: Obesity has been found to be associated with a survival advantage in hemodialysis patients. Results from studies in peritoneal dialysis (PD) patients are inconsistent. The aim of this paper was to study the association between obesity and mortality in the PD population in the Netherlands Co-operative Study on the Adequacy of Dialysis-2 (NECOSAD) cohort and critically discuss the observational data from an epidemiological perspective. **Methods:** Patients starting PD were selected from the Netherlands Co-operative Study on the Adequacy of Dialysis-2 (NECOSAD), a prospective cohort study in incident dialysis patients in The Netherlands and followed for 5 years. Cox regression analysis was used to calculate relative risk of mortality (hazard ratios (HR) with 95% CIs) of baseline and time-dependent BMI, with a BMI of 18.5–25 as the reference. **Results:** In total, 688 patients with end-stage renal disease starting with PD were included (66% men, age: 53 ± 15 years, BMI: 24.6 ± 3.8 kg/m^2). At the start of dialysis, 8.4% of the patients were obese (BMI ≥30). Compared with a normal BMI, obesity at the start of PD (BMI ≥30) was associated with a HR of 0.8 (0.5, 1.3). Time-dependently, this was 0.7 (0.4, 1.2). The HR of BMI <18.5 at the start of PD was 1.3 (95% CI: 0.4, 3.2), and time-dependently this was 2.3 (1.0, 5.3). **Conclusion:** Observational data suggest that PD patients who are obese at the start of dialysis do not have a worse survival compared with PD patients with a normal BMI. PD patients with a low BMI during dialysis have a twofold increased mortality risk. However, it can be argued to what extent the observed association between BMI and mortality in the dialysis population can be causally interpreted.

Whereas obesity is one of the established risk factors for increased morbidity and mortality in the general population, many survival studies in hemodialysis

(HD) patients have indicated opposite associations of obesity [1]. Low values for body mass index (BMI) are associated with increased mortality, and higher values for BMI, even morbid obesity, were found to be protective and associated with improved survival in dialysis patients. Results from studies in peritoneal dialysis (PD) patients, however, are inconsistent. Some studies found a survival advantage of obesity [2–4], whereas others reported an increased mortality risk due to obesity [5, 6], or no association at all [7]. The aim of this paper was to study the association between obesity and mortality in the PD population in the Netherlands Co-operative Study on the Adequacy of Dialysis-2 (NECOSAD) cohort and critically discuss the observational data from an epidemiological perspective.

Methods

Study Design and Patients

The Netherlands Co-operative Study on the Adequacy of Dialysis-II (NECOSAD-II) is an observational prospective cohort study in patients with end-stage renal disease patients starting with their first renal replacement therapy in 38 dialysis centers in The Netherlands. Demographic data and clinical data were collected between four weeks prior to and two weeks after the start of chronic dialysis treatment. Dialysis characteristics and measures of health were furthermore determined at study visits at 3 months and at 6 months after the start of dialysis and subsequently at every 6 months until the end of follow-up. Three months after the start of dialysis was considered as the baseline of the study. Dates and causes of mortality were immediately reported during follow-up. Survival time was defined as the number of days between 3 months after the start of the dialysis treatment (baseline) and the date of death, the date of censoring due to loss to follow-up (kidney transplantation or transfer to a nonparticipating dialysis center), or at a set maximum of 5 years after the start of dialysis. The Medical Ethical Committees of all participating dialysis centers approved NECOSAD-II and all participants gave their written informed consent before inclusion.

Data Collection

Baseline demographic data and clinical data such as age, sex, body mass index, ethnicity, primary kidney disease and comorbidity were recorded in the patient files. Primary kidney diseases and causes of death were classified according to the coding system of the European Renal Association – European Dialysis and Transplantation Association [8]. Routine blood laboratory investigations in the dialysis centers included serum cholesterol and serum albumin concentrations, plasma urea and plasma creatinine. In a corresponding 24-hour urine sample, urea, creatinine, and protein were assessed. Renal function was calculated from the mean of creatinine and urea clearance, adjusted for body surface area (ml/min/1.73 m^2) and expressed as the residual glomerular filtration rate. The daily protein intake was estimated in the peritioneal dialysis patients from the urea excretion in urine and dialysate according to Bergström et al. [9] and expressed as normalized protein equivalent of nitrogen appearance. Comorbid conditions were reported by the patients' nephrologists. The comorbidity index of Khan et al. [10] was calculated classifying patients to have a low, medium or high mortality risk.

Statistical Analysis

Mean values with standard deviations (SD) were calculated for continuous variables at baseline, categorical variables were expressed as proportions. BMI at baseline and the serial measurements of BMI during follow-up were divided into four categories according to the World Health Organization classification for obesity: BMI <18.5, 18.5–25, 25–30 and ≥30 [25]. Absolute mortality rates were calculated within each BMI category per 100 person-years of follow-up. Cox regression analysis was used to calculate hazard ratios (HR, equivalent to relative risks of mortality) with 95% CIs for 5-year all-cause mortality, using a BMI of 18.5–25 as the reference category. Since BMI may vary over time on dialysis, we furthermore performed time-dependent Cox regression analysis to calculate HRs associated with the serial measurements of BMI on subsequent mortality. These relative risks can be considered as short-term mortality risks. Analyses were adjusted for age, sex, smoking, primary kidney disease and comorbidity. We used SPSS 16.0 for Windows (SPSS, Chicago, Ill., USA) for all analyses.

Results

Patient Characteristics

Out of the 1,940 patients with end-stage renal disease who were included in NECOSAD, 689 patients started PD treatment and were still on dialysis and participating in the study after the first 3 months of dialysis, which is considered as the baseline of the study. BMI was missing in one patient, thus 688 patients (457 men and 231 women) were included in the present analysis. Mean age of the patients was 53 ± 15 years, mean BMI was 24.6 ± 3.8 kg/m^2. The main causes of chronic kidney disease were glomerulonephritis (in 20% of the patients), diabetes (15%), and renal vascular disease (13%). Obese patients more often had diabetes as primary kidney disease, whereas patients with a low BMI more often had interstitial nephritis as primary kidney disease (table 1). Patients with overweight and obesity were more often former smokers and had a higher score on the Khan comorbidity index. With a lower BMI, more patients were scored as malnourished (table 1).

Mortality

The median follow-up of patients from 3 months until a maximum of 5 years after the start of dialysis was 2.2 years (25th and 75th percentiles: 1.2, 3.9).

Table 1. Baseline characteristics of 688 incident PD patients per BMI category

Variable	BMI			
	<18.5	18.5–25	25–30	≥30
Number at risk	189 (3)	390 (57)	222 (32)	58 (8)
Sex, % men	39	67	70	57
Age, years	46±12.2	51±16	57±13	54±13
BMI	17.7±0.5	22.4±1.6	27.1±1.4	33.0±2.2
Ethnicity, % white	83	91	95	88
Primary kidney disease, %				
Glomerulonephritis	6	20	20	19
Interstitial nephritis	22	11	12	9
Renal vascular disease	0	14	11	12
Diabetes mellitus	17	11	18	36
Polycystic kidneys	6	10	13	3
Other/unknown	50	33	27	21
rGFR, ml/min/1.73 m^2	3.8±2.9	4.1±3.4	4.7±2.8	4.9±3.4
Comorbidity				
CVD, %	24	22	31	34
Diabetes mellitus, %	18	15	24	39
Malignancy, %	6	4	8	5
Khan index, % high	6	13	22	21
SGA, % malnourished	41	20	17	10
nPNA, g/kg/day	1.2±0.2	1.6±5.5	1.4±5.5	0.9±0.2
Plasma cholesterol, mmol/l	5.9±1.7	5.6±1.3	5.5±1.2	5.7±1.6
Current smoker, %	18	28	19	18
Former smoker, %	24	37	50	54

Values expressed as n (%) or mean ± SD.
BMI = Body mass index; rGFR = residual glomerular filtration rate corrected for body surface area; CVD = cardiovascular diseases; SGA = subjective global assessment of nutritional status; nPNA = normalized protein nitrogen appearance.

During follow-up, 187 patients died and 261 patients left the study because they underwent a kidney transplantation. Other reasons for censoring during follow-up included recovery of renal function (n = 7), transfer to a nonparticipating dialysis center (n = 21), refusal of further participation (n = 73) or other (n = 12).

The absolute mortality in 5 years after the start of dialysis in the total population was 10.9 deaths per 100 person-years. The absolute mortality rates ranged from 6.8 per 100 person-years in the lowest BMI category to 12.9 per 100 person-

Table 2. Absolute mortality rates and relative mortality risks (HR, 95%-CI) of baseline and time-dependent BMI on all-cause mortality in 688 incident PD patients who were followed from 3 months until 5 years after the start of dialysis.

Variable	BMI			
	<18.5	18.5–25	25–30	≥30
Mortality rate (per 100 person-years)	6.76	10.67	11.26	12.86
Baseline BMI (crude model)	HR (95% CI)			
	0.6 (0.2, 1.7)	1	1.0 (0.8, 1.4)	1.2 (0.7, 1.9)
Model 1	1.2 (0.4, 3.3)	1	0.8 (0.6, 1.1)	1.3 (0.8, 2.2)
Model 2	1.2 (0.4, 3.4)	1	0.9 (0.6, 1.2)	1.3 (0.8, 2.2)
Model 3	1.0 (0.4, 2.7)	1	0.8 (0.6, 1.1)	1.0 (0.6, 1.7)
Model 4	1.3 (0.4, 3.2)	1	0.8 (0.6, 1.1)	0.8 (0.5, 1.3)
Time-dependent BMI (crude)	1.4 (0.6, 3.2)	1	1.1 (0.8, 1.5)	1.0 (0.6, 1.5)
Model 1	2.0 (0.8, 4.6)	1	0.9 (0.7, 1.2)	0.9 (0.6, 1.5)
Model 2	2.0 (0.8, 4.5)	1	1.0 (0.7, 1.3)	1.0 (0.6,1.5)
Model 3	2.1 (0.9, 4.8)	1	1.0 (0.7, 1.4)	0.9 (0.5,1.4)
Model 4	2.3 (1.0, 5.3)	1	1.0 (0.7, 1.4)	0.7 (0.4,1.2)

Model 1 = Adjusted for age and sex; model 2 = additionally adjusted for smoking habits; model 3 = additionally adjusted for comorbid conditions; model 4 = additionally adjusted for primary kidney diseases.

years in the highest BMI category (table 2). HRs for all-cause mortality associated with baseline and time-dependent BMI are shown in table 2. Compared with the reference category and adjusted for age, sex, smoking, primary kidney disease and comorbidity the HR (95% CI) associated with overweight at baseline was 0.8 (0.6, 1.1) and the HR associated with obesity at baseline was 0.8 (0.5, 1.3). Also the time-dependent analyses showed no association of overweight and obesity with mortality (table 2). The HR of a low BMI (<18.5) at baseline was 1.3 (95% CI: 0.4, 3.2), and time-dependently it was 2.3 (1.0, 5.3).

Discussion

This prospective longitudinal study in incident PD patients who were followed for 5 years after the start of dialysis showed that overweight and obesity were not associated with increased mortality risks. Time-dependently, a low BMI was associated with a twofold increased mortality risk compared with a normal BMI.

The sample size of 688 PD patients is markedly smaller than those reported in large registry analyses [3, 5, 6] and may be a weakness of the present analysis. As a result of the small sample size, the point estimates (HRs) of the BMI categories have wide confidence intervals. Because the point estimates are close to one, it is not expected that larger sample sizes would result in larger deviations from the reference category. The crude mortality rates in table 2 suggest a positive trend of increasing mortality rates with a higher BMI. However, after adjustment for confounders the HR in obese patients was 0.7 (95% CI: 0.4, 1.2) compared with a normal BMI. Whether this relative risk is interpreted as a survival advantage because of the wide confidence intervals or as the absence of an association of overweight and obesity with mortality, there seems no evidence for an increased mortality risk in PD patients who are obese at the start of dialysis. A recent review of the BMI–mortality relation in PD patients also concluded that survival in obese PD patients and patients with a normal BMI was equivalent [11].

Strengths of the present study include the 6-monthly measurements and the follow-up of 5 years. Another important strength of the study design is that only incident dialysis patients were included. Many large outcome studies in the dialysis population have been performed in prevalent populations and may therefore lead to inconsistent results. The reason for this is that dialysis patients who have a better health status may live longer and may represent a relatively large proportion in a prevalent dialysis cohort.

Another explanation for the inconsistencies reported in the literature may be differences in clinical condition and prognosis of the patients with a normal BMI included in the reference group of the different studies. Because the mortality risk of obese patients is calculated relative to the mortality risk of the reference group within each study, differences in prognosis of the reference groups may result in different relative risks associated with obesity.

The time-dependent risk associated with a low BMI can be interpreted as follows: during time on dialysis patients with a BMI <18.5 had a twofold increased risk to die within the coming half year. Higher mortality in patients with a low BMI is a common finding in the dialysis population [1, 12]. Increased short-term mortality that is associated with thinness is, however, most likely due to illness at baseline [13]. Underweight may be a consequence of disease and thus an indicator of early mortality. This is known as reverse causation. Especially studies with a short follow-up and in elderly people may suffer from reverse causation [14]. Likewise, we showed earlier that age and duration of follow-up may influence the association between BMI and mortality in the dialysis population [12]. In the presence of reverse causation the mortality risk that is associated with a low BMI can not be causally interpreted.

The obesity-survival paradox has resulted in confusion and uncertainty about whether weight loss should be advised in morbidly obese dialysis patients who are awaiting kidney transplantation. Therefore, it is important to know how valid the observational comparisons are, and if the observations can be translated into causal interpretations that eventually may lead to interventions. In other words, should we advise PD patients to lose or to gain weight during dialysis in order to improve their survival?

Association can be interpreted as causation when the exposed group and unexposed group are exchangeable. In clinical trials, randomization is the favorite method to obtain exchangeable groups. When contrasting obese dialysis patients with dialysis patients with a low BMI in observational studies, these patients are considered completely alike, except for their BMI. However, it can be argued to what extent dialysis patients with a high BMI can be considered exchangeable with dialysis patients with a low BMI. In chronic dialysis patients, the underlying reasons for having a low BMI may be fundamentally different from the underlying reasons for having a high BMI. For example, patients with a low BMI may have lost weight due to illness or wasting that is associated with mortality. Furthermore, the main primary kidney disease in obese PD patients was diabetic nephropathy (36%), whereas patients in the lower BMI categories more often had interstitial nephritis as primary kidney diseases (table 1). Since obesity is a risk factor for chronic kidney disease, either directly or through the development of diabetes, a proportion of obese dialysis patients may have developed chronic kidney disease because of their obesity [15]. Differences in disease history are likely to be related with a different health status and a different probability of mortality, irrespective of BMI. As a consequence, dialysis patients with a high BMI may not be exchangeable with dialysis patients with a low BMI and a direct comparison between these two groups may not be valid. Hence, causal interpretations of the effects of BMI on the basis of the observed associations of 'reverse epidemiology' in dialysis patients remain uncertain.

In conclusion, observational data suggest that PD patients who are obese at the start of dialysis do not have a worse survival compared with PD patients with a normal BMI. PD patients with a low BMI during dialysis have a twofold increased risk. However, it can be argued to what extent the observed association between BMI and mortality in the dialysis population can be causally interpreted.

References

1 Kalantar-Zadeh K, Abbott KC, Salahudeen AK, Kilpatrick RD, Horwich TB: Survival advantages of obesity in dialysis patients. Am J Clin Nutr 2005;81:543–554.
2 Johnson DW, Herzig KA, Purdie DM, et al: Is obesity a favorable prognostic factor in peritoneal dialysis patients? Perit Dial Int 2000;20:715–721.
3 Snyder JJ, Foley RN, Gilbertson DT, Vonesh EF, Collins AJ: Body size and outcomes on peritoneal dialysis in the United States. Kidney Int 2003;64:1838–1844.
4 Abbott KC, Glanton CW, Trespalacios FC, et al: Body mass index, dialysis modality, and survival: analysis of the United States Renal Data System Dialysis Morbidity and Mortality Wave II Study. Kidney Int 2004;65:597–605.
5 McDonald SP, Collins JF, Johnson DW: Obesity is associated with worse peritoneal dialysis outcomes in the Australia and New Zealand patient populations. J Am Soc Nephrol 2003;14:2894–2901.
6 Stack AG, Murthy BV, Molony DA: Survival differences between peritoneal dialysis and hemodialysis among 'large' ESRD patients in the United States. Kidney Int 2004;65:2398–2408.
7 Aslam N, Bernardini J, Fried L, Piraino B: Large body mass index does not predict short-term survival in peritoneal dialysis patients. Perit Dial Int 2002;22:191–196.
8 van Dijk PC, Jager KJ, de Charro F, et al. Renal replacement therapy in Europe: the results of a collaborative effort by the ERA-EDTA registry and six national or regional registries. Nephrol Dial Transplant 2001;16:1120–1129.
9 Bergstrom J, Heimburger O, Lindholm B: Calculation of the protein equivalent of total nitrogen appearance from urea appearance: which formulas should be used? Perit Dial Int 1998;18:467–473.
10 Khan IH, Catto GR, Edward N, Fleming LW, Henderson IS, MacLeod AM: Influence of coexisting disease on survival on renal-replacement therapy. Lancet 1993;341:415–418.
11 Abbott KC, Oliver DK, Hurst FP, et al: Body mass index and peritoneal dialysis: 'exceptions to the exception' in reverse epidemiology? Semin Dial 2007;20:561–565.
12 de Mutsert R, Snijder MB, van der Sman-de Beer F, et al: Association between body mass index and mortality is similar in the hemodialysis population and the general population at high age and equal duration of follow-up. J Am Soc Nephrol 2007;18:967–974.
13 Manson JE, Stampfer MJ, Hennekens CH, Willett WC: Body weight and longevity: a reassessment. JAMA 1987;257:353–358.
14 Stevens J, Cai J, Pamuk ER, et al: The effect of age on the association between body-mass index and mortality. N Engl J Med 1998;338:1–7.
15 van Dijk PC, Jager KJ, Stengel B, et al: Renal replacement therapy for diabetic end-stage renal disease: data from 10 registries in Europe (1991–2000). Kidney Int 2005;67:1489–1499.

Renée de Mutsert, PhD
Leiden University Medical Center, Department of Clinical Epidemiology, C7-P
PO Box 9600
NL–2300 RC Leiden (The Netherlands)
Tel. +31 71 526 6534, Fax +31 71 526 6994, E-Mail R.de_Mutsert@lumc.nl

Ronco C, Crepaldi C, Cruz DN (eds): Peritoneal Dialysis – From Basic Concepts to Clinical Excellence. Contrib Nephrol. Basel, Karger, 2009, vol 163, pp 132–139

Systemic and Local Inflammation in Peritoneal Dialysis: Mechanisms, Biomarkers and Effects on Outcome

Antonio Carlos Cordeiro[a,b], *Juan Jesús Carrero*[a], *Hugo Abensur*[c], *Bengt Lindholm*[a], *Peter Stenvinkel*[a]

[a]Divisions of Renal Medicine and Baxter Novum; Department of Clinical Science, Intervention and Technology; Karolinska Institutet, Stockholm, Sweden; [b]Department of Hypertension and Nephrology, Dante Pazzanese Institute of Cardiology, and [c]Department of Nephrology, Hospital das Clínicas, University of Sao Paulo, Sao Paulo, Brazil

Abstract

Thanks to the technological development in peritoneal dialysis (PD) during the last three decades, the most important problem nowadays for the nephrologists is the maintenance of the long-term function of the peritoneal membrane. Although PD may exert an early survival benefit as compared with hemodialysis (HD), long-term PD is often associated with histopathological alterations in the peritoneal membrane that are linked to peritoneal ultrafiltration deficit and increased mortality risk. These alterations are closely related to the presence of a chronic activated (local and systemic) inflammatory response. PD itself may have other factors associated that could further modulate the inflammatory response, such as the bioincompatibility of dialysis solutions, fluid overload and changes in the body composition. Understanding the pathophysiology of inflammation in PD is essential for the adoption of adequate strategies to improve both membrane and patient survival.

Peritoneal dialysis (PD) has undergone a considerable technologic development since the introduction of continuous ambulatory PD (CAPD) more than three decades ago. Currently, PD is the dominant modality for home dialysis [1]. The improvements that have been achieved have clearly changed the nature of the problems faced by nephrologists in the conduction of PD: while peritonitis was one of the most important problems at the beginning of this therapy, nowadays preservation of peritoneal membrane function during long-term PD is the main challenge. The putative merits of PD compared with hemodialysis (HD) have been much debated. Although a recent study reported that treatment with PD may be advantageous initially compared with HD [2], PD induces histopathological alterations in the peritoneal membrane that with time may lead to peritoneal ultrafiltration failure and increased mortality [1, 2]. Peritoneal ultrafiltration failure is closely related to the presence of a chronic activated systemic inflammatory response [3], which will be the focus of the present review.

Inflammation is an adaptive response targeted to the elimination of noxious stimuli and restoration of the homeostasis. Although inflammation has a crucial physiological role when directed against classic instigators, a wide variety of diseases (including type 2 diabetes, cardiovascular disease (CVD), obesity and chronic kidney disease (CKD)) are characterized by the presence of a persistent inflammatory state associated with tissue malfunction. Inflammation is a strong predictor of all-cause and cardiovascular mortality in the general population as well as in CKD patients [3, 4]. Although it has been argued that inflammation is merely a reflection of vascular disease, an inflamed milieu has been proposed to promote endothelial dysfunction, vascular calcification, insulin resistance, oxidative stress and protein energy wasting (PEW) [3, 4].

Inflammation in Chronic Kidney Disease: Causal Mechanisms and General Peritoneal Dialysis-Related Causes

Sustained low-grade systemic inflammation is highly prevalent in CKD already at early stages [5]. CKD per se is accompanied by several factors not related to dialysis (i.e. reduced renal clearance of cytokines, fluid overload, immunological dysfunction and a high prevalence of inflammatory and infectious co-morbidity) that all contribute to inflammation. On the other hand, renal replacement therapy may also promote inflammation such as in cases of dialysis access infections, bio-incompatibility of membranes or solutions and dialysate contamination [6]. In addition, genetic variations may also predispose some dialysis patients to an exaggerated inflammatory response [7]. Of interest, the different techniques may have different impact on the inflammatory outcome, and while initiation of HD may partially reduce CRP levels this was not observed during PD initiation [8]. The reasons for this are not clear, but multiple heparinizations or less increase in fat mass in HD as compared with PD patients may probably contribute. As short daily HD was associated with a reduction in inflammatory factors compared with conventional HD [9], dialysis adequacy may also have an impact on inflammation status. Regardless of the differences in techniques, inflammation in PD patients is associated with poor prognosis [10], and several systemic and local inflammation-related biomarkers have been associated with membrane failure, atherosclerotic disease, CVD, and mortality in PD patients (table 1).

Mechanisms by which Peritoneal Dialysis May Contribute to Inflammation

Bioincompatibility of PD Solutions

Despite its multifactorial genesis, one of the most important causal factors of the inflammatory activation in PD patients is the continued toxicity of the peritoneal membrane by standard PD solution's bioincompatibility. Although this is mainly due to the high concentration of glucose, other factors such as low pH, hyperosmolarity and the presence of lactate and glucose degradation products (GDPs) may contribute. Local inflammation and oxidative stress, which results from the continuous peritoneal injury, accelerate the epithelial-to-mesenchymal transition (EMT) of peritoneal mesothelial cells (MC) resulting in peritoneal fibrosis and ultrafiltration failure [3, 11, 12]. EMT is a process by which the peritoneal MC undergoes a progressive loss of epithelial phenotype and acquire fibroblast-like characteristics, which allows these cells to invade the mesothelial stroma contributing to angiogenesis, fibrosis and ultrafiltration failure [11]. It is not completely clear if uremia per se may initiate EMT; while peritoneal fibrosis and vasculopathy is already observed in uremic pre-PD patients [12], this process seems to be accelerated during PD initiation [11, 12].

Transforming growth factor-β (TGF-β), more specifically TGF-β1, is one of the main mediators of the PD solutions' profibrotic effects through the Smads 2 and 3 pathways. These TGF-β alterations have been associated with fibroblast activation, collagen deposition, inhibition of fibrinolysis, maintenance of fibrosis and neoangiogenesis [11, 13]. Interestingly, angiotensin II inhibitors (which are TGF-β activity suppressors) have recently been shown to reduce peritoneal fibrosis and neoangiogenesis, as well as to prevent the increase of small solute transport in long-term PD patients [13]. Bone morphogenic protein-7 (BMP-7) is a member of the TGF-β superfamily that has opposing effects to TGF-β1, i.e. it tends to preserve the epithelial phenotype of the peritoneal membrane. It is still not yet clear whether the effects of BMP-7 may be beneficial in PD patients, since its high levels in PD effluent are associated with a gradual increase in peritoneal solute transport rate (PSTR) [14]. Another possible therapeutic target is endoglin (CD105), a TGF-β type III receptor predominantly expressed in vascular activated endothelial cells and an important marker of neovascularization and endothelial dysfunction [15]. Whether endoglin has a role in peritoneal transport pathophysiology remains to be investigated.

Table 1. Systemic and local inflammation-related markers already studied in PD patients and their reported role as predictors of outcomes

Marker	n	Outcome
Systemic markers		
IL-6	40	high peritoneal solute transport rate [Pecoits-Filho et al: NDT 2002]
	99	mortality [Pecoits-Filho et al: NDT 2002]
TNF-α	42	anorexia/malnutrition [Aguilera et al: NDT 1998]
CRP	59	carotid atherosclerosis [Ohkuma et al: AJKD 2003]
	73	ischemic heart disease [Kim et al: AJKD 2002]
	82	mortality [Chung et al: Perit Dial Int: 2003]
Fibrinogen	63	carotid plaque score [Papagianni et al: NDT 2004]
	95	cardiovascular and all-cause mortality [Koch et al: NDT 1997]
VEGF	40	high peritoneal solute transport rate [Pecoits-Filho et al: NDT 2002]
ICAM-1	63	carotid intima-media thickness [Papagianni et al: NDT 2004]
VCAM-1	160	cardiac hypertrophy [Wang et al: AJKD 2005]
Albumin	63	carotid intima-media thickness [Papagianni et al: NDT 2004]
	71	mortality [Spiegel et al: AJKD 1994]
Fetuin-A	238	atherosclerosis and valvular calcification [Wang et al: NDT 2005]
	323	mortality [Hermans et al: Kidney Int 2007]
Local (dialysate) markers		
IL-6	40	high peritoneal solute transport rate [Pecoits-Filho et al: NDT 2002]
Albumin	43	cardiovascular disease [Szeto et al Perit Dial Int: 2005]
VEGF	40	high peritoneal solute transport rate [Pecoits-Filho et al: NDT 2002]
Hyaluronan	116	mortality [Szeto et al: AJKD 2000]

n = Number of peritoneal dialysis patients included in each study; IL-6 = interleukin-6; TNF-α = tumor necrosis factor-α; VEGF = vascular endothelial growth factor; ICAM-1 = intercellular adhesion molecule 1; VCAM-1 = vascular cell adhesion molecule 1.

Fluid Overload

With increasing dialysis vintage, PD patients become volume overloaded by peritoneal membrane ultrafiltration failure as well as by the progressive loss of the residual renal function (RRF). The decline of RRF is an important determinant of fluid overload and, consequently, inflammation in PD patients. Furthermore, the decline of RRF per se is associated with inflammation, left ventricular hypertrophy and increased risk of peritonitis in PD patients, thus contributing to increase their mortality risk [16]. Davies et al. [17] showed that hypertonic glucose may play a causative role in alterations in peritoneal membrane function. Fluid overload has an important role in the activation and perpetuation of the systemic inflammatory response. The edema in gut mucosa, which results from the increased fluid retention, compromises its barrier function allowing the translocation of bacteria and their toxins. The most described toxin is lipopolysaccharide (LPS), a potent stimulator of proinflammatory cytokine release by immune cells. LPS stimulus induces the release of TNF- and nitrous oxide, which act as cardiosuppressors. By reducing cardiac output it promotes gut's mucosa hypoxia, further aggravating its function [3, 18].

In PD, the main determinant of fluid overload is PSTR, which has been linked to both inflammation, increased risk for PD dropout and death [19]. Recently, this concept has been expanded by the hypothesis of the existence of four different types of high PSTR [20]: type 1 are those patients that, in addition to a high solute transport rate from the very start of PD, have signals of inflammation (peritoneal and systemic), high prevalence of co-morbidities and high mortality despite the dialysis method. Conversely, type 2 patients are those with no inflammation and with good prognosis; their high PSTR phenotype is related to a large peritoneal surface area, as suggested by a high effluent level of CA125 [3, 20]. Type 3 is present in those patients in whom the high PSTR appears with the increase in dialysis vintage, as a consequence of continuous exposure to bioincompatible PD solutions. Typically, the type 3 patients do not have a higher prevalence of inflammation or co-morbidities, and, thus, generally have a good prognosis provided that they are adequately treated as regards fluid overload using automated peritoneal dialysis (APD) for the short dwells, icodextrin for the long dwell and dietary sodium and fluid restrictions as appropriate [20]. Finally, patients with peritonitis represent a type 4 category with transitory elevation of PSTR [20]. Although the underlying mechanisms for these different PSTR types do need further studies, it could be speculated that the more detailed PSTR identification may contribute in the future to a better personalized therapy and possibly a more effective prevention of fluid overload.

Increased Fat Mass and Protein Energy Wasting

Continuous 24-hour exposure to glucose-based PD solutions leads to a unique metabolic situation (glucose absorption estimated to 100–200 g/day).

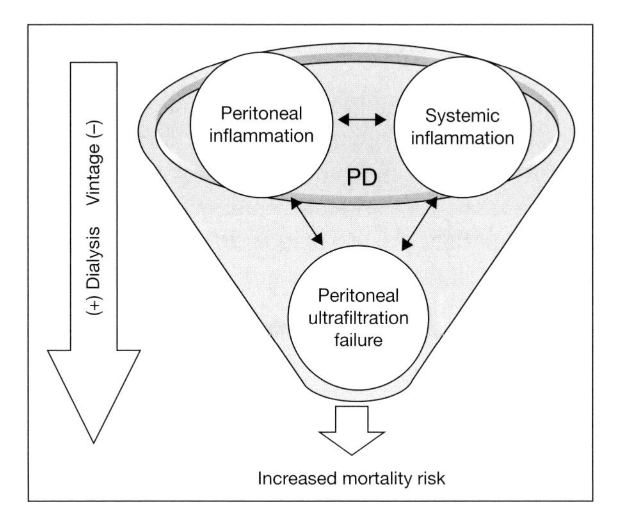

Fig. 1. Interrelations between peritoneal and systemic inflammation, and peritoneal membrane ultrafiltration failure; resulting in an increased mortality risk with the increase in PD vintage.

In consequence, many PD patients will have a considerable increase in body fat mass – especially in the abdominal intraperitoneal region – when starting on long-term PD therapy [21]. As fat mass accumulation does not occur in all PD patients, genetic factors may be operating [22]. As adipose tissue, especially visceral fat, acts like an endocrine organ by secreting several proinflammatory cytokines and adipokines, such as IL-6, TNF-α or adiponectin, the abdominal fat mass gain in PD is potentially related to the systemic inflammatory status [23]. A pro-inflammatory milieu may in turn contribute to appetite loss [24], subclinical hypothyroidism [25] and testosterone deficiency [26], all of them directly or indirectly contributing in a vicious circle to the burden of PEW.

Despite a gain in fat mass, PEW has been reported to be prevalent in up to 60% of PD patients, being a clear risk factor for poor outcome. Several factors in PD patients may contribute to the high prevalence of PEW, such as low protein intake, protein and amino acid losses in dialysate, imbalance between anorexigen and orexigen factors, abdominal distention, absorption of the osmotic agent and metabolic acidosis [21]. Whereas PEW is, at least in part, a consequence of the inflammatory status it may also contribute to the perpetuation of inflammation by favoring the development of oxidative stress and atherosclerosis [21].

Conclusion

Several dialysis risk factors and certain PD-specific characteristics are associated with the inflammatory burden possibly linking inflammation, increased PSTR and decreased RRF to poor outcome. Both local and systemic inflammation may, at the same time, be the cause and consequence of peritoneal membrane failure, and are important prognosticators of mortality in this population (fig. 1). A better understanding of the causes and pathophysiology of these processes is essential in order to adopt adequate strategies to improve both (membrane and patient) survival in PD.

Acknowledgements

This work was supported by the ERA-EDTA, the Karolinska Institutet Center for Gender-based Research, Karolinska Institutet research funds, the Swedish Heart and Lung Foundation, Scandinavian Clinical Nutrition AB, MEC (EX2006–1670) and the Swedish Medical Research Council. Baxter Novum is the result of an unconditional research grant from Baxter Healthcare Inc. to the Karolinska Institutet.

References

1 Khawar O, Kalantar-Zadeh K, Lo WK, Johnson D, Mehrotra R: Is the declining use of long-term peritoneal dialysis justified by outcome data? Clin J Am Soc Nephrol 2007;2:1317–1328.
2 McDonald SP, Marshall MR, Johnson DW, Polkinghorne KR: Relationship between dialysis modality and mortality. J Am Soc Nephrol 2009;20:155–163.
3 Pecoits-Filho R, Stenvinkel P, Wang AY, Heimburger O, Lindholm B: Chronic inflammation in peritoneal dialysis: the search for the holy grail? Perit Dial Int 2004;24:327–339.
4 Stenvinkel P, Carrero JJ, Axelsson J, Lindholm B, Heimburger O, Massy Z: Emerging biomarkers for evaluating cardiovascular risk in the chronic kidney disease patient: how do new pieces fit into the uremic puzzle? Clin J Am Soc Nephrol 2008;3:505–521.
5 Shlipak MG, Fried LF, Crump C, Bleyer AJ, Manolio TA, Tracy RP, Furberg CD, Psaty BM: Elevations of inflammatory and procoagulant biomarkers in elderly persons with renal insufficiency. Circulation 2003;107:87–92.
6 Stenvinkel P, Alvestrand A: Inflammation in end-stage renal disease: Sources, consequences, and therapy. Semin Dial 2002;15:329–337.
7 Liu Y, Berthier-Schaad Y, Plantinga L, Fink NE, Tracy RP, Kao WH, Klag MJ, Smith MW, Coresh J: Functional variants in the lymphotoxin-alpha gene predict cardiovascular disease in dialysis patients. J Am Soc Nephrol 2006;17:3158–3166.
8 Carrero JJ, Axelsson J, Avesani CM, Heimburger O, Lindholm B, Stenvinkel P: Being an inflamed peritoneal dialysis patient: a Dante's journey. Contrib Nephrol. Basel, Karger, 2006, vol 150, pp 144–151.
9 Ayus JC, Mizani MR, Achinger SG, Thadhani R, Go AS, Lee S: Effects of short daily versus conventional hemodialysis on left ventricular hypertrophy and inflammatory markers: a prospective, controlled study. J Am Soc Nephrol 2005;16:2778–2788.
10 Wang AY, Woo J, Lam CW, Wang M, Sea MM, Lui SF, Li PK, Sanderson J: Is single time point C-reactive protein predictive of outcome in peritoneal dialysis patients? J Am Soc Nephrol 2003; 14:1871–1879.

11 Aroeira LS, Aguilera A, Sanchez-Tomero JA, Bajo MA, del Peso G, Jimenez-Heffernan JA, Selgas R, Lopez-Cabrera M: Epithelial to mesenchymal transition and peritoneal membrane failure in peritoneal dialysis patients: pathologic significance and potential therapeutic interventions. J Am Soc Nephrol 2007;18:2004–2013.
12 Honda K, Hamada C, Nakayama M, Miyazaki M, Sherif AM, Harada T, Hirano H: Impact of uremia, diabetes, and peritoneal dialysis itself on the pathogenesis of peritoneal sclerosis: a quantitative study of peritoneal membrane morphology. Clin J Am Soc Nephrol 2008;3:720–728.
13 Kolesnyk I, Noordzij M, Dekker FW, Boeschoten EW, Krediet RT: A positive effect of ACE inhibitors on peritoneal membrane function in long-term PD patients. Nephrol Dial Transplant 2009; 24:272–277.
14 Szeto CC, Chow KM, Kwan BC, Lai KB, Chung KY, Leung CB, Li PK: The relationship between bone morphogenic protein-7 and peritoneal transport characteristics. Nephrol Dial Transplant 2008;23:2989–2994.
15 ten Dijke P, Goumans MJ, Pardali E: Endoglin in angiogenesis and vascular diseases. Angiogenesis 2008;11:79–89.
16 Wang AY, Wang M, Woo J, Lam CW, Lui SF, Li PK, Sanderson JE: Inflammation, residual kidney function, and cardiac hypertrophy are interrelated and combine adversely to enhance mortality and cardiovascular death risk of peritoneal dialysis patients. J Am Soc Nephrol 2004;15:2186–2194.
17 Davies SJ, Phillips L, Naish PF, Russell GI: Peritoneal glucose exposure and changes in membrane solute transport with time on peritoneal dialysis. J Am Soc Nephrol 2001;12:1046–1051.
18 Charalambous BM, Stephens RC, Feavers IM, Montgomery HE: Role of bacterial endotoxin in chronic heart failure: the gut of the matter. Shock 2007;28:15–23.
19 Churchill DN, Thorpe KE, Nolph KD, Keshaviah PR, Oreopoulos DG, Page D: Increased peritoneal membrane transport is associated with decreased patient and technique survival for continuous peritoneal dialysis patients. The Canada-USA (CANUSA) Peritoneal Dialysis Study Group. J Am Soc Nephrol 1998;9:1285–1292.
20 Chung SH, Heimburger O, Lindholm B: Poor outcomes for fast transporters on PD: The rise and fall of a clinical concern. Semin Dial 2008;21:7–10.
21 Carrero JJ, Heimburger O, Chan M, Axelsson J, Stenvinkel P, Lindholm B: Protein-energy malnutrition/wasting during peritoneal dialysis; in Khanna R, Krediet RT (eds): Nolph and Gokal's Textbook of Peritoneal Dialysis. New York, Springer Science and Business Media, 2009.
22 Nordfors L, Heimburger O, Lonnqvist F, Lindholm B, Helmrich J, Schalling M, Stenvinkel P: Fat tissue accumulation during peritoneal dialysis is associated with a polymorphism in uncoupling protein 2. Kidney Int 2000;57:1713–1719.
23 Axelsson J, Rashid Qureshi A, Suliman ME, Honda H, Pecoits-Filho R, Heimburger O, Lindholm B, Cederholm T, Stenvinkel P: Truncal fat mass as a contributor to inflammation in end-stage renal disease. Am J Clin Nutr 2004;80:1222–1229.
24 Carrero JJ, Qureshi AR, Axelsson J, Avesani CM, Suliman ME, Kato S, Barany P, Snaedal-Jonsdottir S, Alvestrand A, Heimburger O, Lindholm B, Stenvinkel P: Comparison of nutritional and inflammatory markers in dialysis patients with reduced appetite. Am J Clin Nutr 2007;85:695–701.
25 Carrero JJ, Qureshi AR, Axelsson J, Yilmaz MI, Rehnmark S, Witt MR, Barany P, Heimburger O, Suliman ME, Alvestrand A, Lindholm B, Stenvinkel P: Clinical and biochemical implications of low thyroid hormone levels (total and free forms) in euthyroid patients with chronic kidney disease. J Intern Med 2007;262:690–701.
26 Carrero JJ, Qureshi AR, Parini P, Arver S, Lindholm B, Barany P, Heimburger O, Stenvinkel P: Low serum testosterone increases mortality risk among male dialysis patients. J Am Soc Nephrol 2009;20:613–620.

Prof. Peter Stenvinkel
Division of Renal Medicine, K56, Karolinska University Hospital at Huddinge
SE–141 86 Stockholm (Sweden)
Tel. +46 8 5858 2532, Fax +46 8 711 4742, E-Mail peter.stenvinkel@ki.se

Ronco C, Crepaldi C, Cruz DN (eds): Peritoneal Dialysis – From Basic Concepts to Clinical Excellence. Contrib Nephrol. Basel, Karger, 2009, vol 163, pp 140–146

How Should We Measure Peritoneal Dialysis Adequacy in the Clinic

Olof Heimbürger

Divisions of Renal Medicine, Department of Clinical Science, Intervention and Technology, Karolinska Institutet, Karolinska University Hospital, Stockholm, Sweden

Abstract

Several different dialysis adequacy indices have been suggested for peritoneal dialysis (PD) patients, but at present mainly Kt/V urea (urea clearance normalized to total body water) and to some extent weekly creatinine clearance (normalized to body surface area) are used as estimates of PD adequacy. These indices can easily be calculated from a 24-hour collection of dialysate and urine, which may also be used to evaluate several other aspects of the dialysis adequacy and the dialysis process including fluid and sodium removal, protein intake, peritoneal albumin clearance and a rough estimate of the diffusive transport. Today, there is general agreement that the target Kt/V urea in PD patients should be 1.7 or higher. However, PD adequacy should also involve many other aspects of the treatment, such as fluid status, adequate mineral metabolism, control of phosphate levels, anemia and acidosis, treatment of comorbidity, and prevention of cardiovascular and infectious complications.

During the 1980s to 1990s, large efforts were made to quantify the dialysis prescription in both peritoneal dialysis (PD) and hemodialysis (HD) treatment based on kinetic modeling. Several different adequacy indices have been suggested, but at present mainly Kt/V urea (urea clearance normalized to total body water) and weekly creatinine clearance normalized to body surface area are used as estimates of PD adequacy. However, PD adequacy should also involve many other aspects of the treatment, and though this short review is mainly focused on small solute clearances it is important not to forget other aspects of dialysis adequacy.

Adequacy of Dialysis

In spite of the large evolution of the understanding of the peritoneal transport process, the modeling of the dialysis prescription, and the randomized clinical trials performed, dialysis adequacy is still not well understood. This may, at least partly, be due to the fact that most indices of dialysis adequacy have been based on the kinetics and removal of small solutes, in particular urea. Though urea is quantitatively the most important end product of nitrogen metabolism and account for 85% of the nitrogen excretion in the urine [1], several other small, middle and large molecular weight compounds may also make a large contribution to uremic toxicity [1]. In addition, the concept of dialysis adequacy should also include several other aspects of patient treatment (table 1).

Adequacy Indices Used in Peritoneal Dialysis

Though several different adequacy indices have been suggested, almost only Kt/V urea and weekly creatinine clearance are presently used as estimates of PD adequacy. Both these indices, as applied to PD patients, involve both dialytic and residual renal clearance components. In PD patients, the residual renal component is particularly important as it accounts for a larger portion of the overall clearance compared to the typical HD patient due to the better preservation of the residual renal function in PD patients. Though residual renal and peritoneal clearances now have been clearly shown to have different impact on patient survival [2], they are commonly added together into total clearances. Based on cohort studies, it was suggested that the higher the total clearance, the better the clinical outcome and that higher levels of Kt/V urea should be aimed for [3]. The US National Kidney Foundation Dialysis Outcome Quality Initiative (K/DOQI) guidelines published in 1997 recommended a (combined renal and peritoneal) Kt/V urea of at least 2.0 for patients on continuous ambulatory PD (CAPD) [4]. This concept was tested in randomized controlled trials of increased PD dose. The ADEMEX study, which is the largest randomized trial done in PD patients, involved 965 CAPD patients in Mexico randomized to receive either the standard therapy of CAPD 2 liters × 4 or an increased volume of dialysis fluid to achieve a peritoneal creatinine clearance of 60 liters/week, and the patients were followed for at least 2 years [5]. Though not all patients in the intervention group reached the clearance target, the separation in peritoneal clearances between the groups was substantial, and possibly as large as can be achieved with CAPD in clinical practice. The results showed that the survival in the two groups was identical (RR 1.00, $p = 0.9878$) and subgroup analysis demonstrated no benefit in any subgroup ana-

Table 1. Different aspects of PD adequacy

Adequate nutrition and nutritional status
Control of acidosis
Adequate clearances and removal of solutes
Small solutes (Kt/V urea, creatinine clearance)
Middle molecules
Normal fluid status
Fluid and electrolyte balance, ultrafiltration
Blood pressure control
Minimal anemia
Adequate mineral metabolism
Calcium-phosphate-PTH control
Avoid osteodystrophy
Avoid vascular calcification
Control of inflammation
MDt/P (Ronco index)

Kt/V urea = Urea clearance normalized to total body water; MDt/P = the medical doctors time per patient.

lyzed (divided by age, sex, diabetes, body size, and anuric patients). There was a higher rate of pressure-related complications such as hernia in the intervention group (likely due to increased intraperitoneal pressure) whereas there was a higher drop-out due to uremia in the standard treatment group. However, this result needs to be interpreted with caution as it was an open study performed at a time when most clinicians believed that increased peritoneal clearances were beneficial. Though quite unexpected, the results of the ADEMEX study were strongly supported by a study from Hong Kong [6] showing no benefit with higher clearance targets in 320 incident CAPD patients randomized to three different Kt/V prescription groups and followed for 2 years. These results may seem opposite to the previous cohort studies, but the PD dose was not varied much in the previous cohort studies (e.g. the CANUSA study [3]) and reanalysis showed that the better outcome with higher total Kt/V urea in the CANUSA was only related to differences in residual renal function, and that differences in peritoneal clearances did not affect outcome [7]. On the other hand, there are clear indications that the clinical outcome may be worse if Kt/V urea falls below 1.7/week [6] and presently there is a general agreement that Kt/V urea should not be below 1.7/week. Both the International Society of Peritoneal Dialysis [2], K/DOQI [8] as well as the European Renal Association

(ERA-EDTA) [9] recommend that PD patients should achieve a total Kt/V urea of at least 1.7 per week. Peritoneal creatinine clearance tends to be used less in the clinical setting today, and to be slightly less well validated. Also, the justification for having targets for both Kt/V urea and creatinine clearance have no strong scientific support and may clearly be questioned [10]. In CAPD patients, there is a close relation between Kt/V urea and weekly creatinine clearance, but during automated PD (APD) it is possible to achieve a higher Kt/V urea in spite of a lower weekly creatinine clearance compared to CAPD, and for APD patients, a weekly total creatinine clearance above 45 liters/week is recommended [2, 9].

It should be noted that to avoid the effect of tubular secretion of creatinine, the residual renal component of creatinine clearance is usually defined as the estimated residual renal glomerular filtration rate as estimated from the average of renal creatinine and urea clearance [10].

Prescription of Peritoneal Dialysis Based on Kinetic Modeling

One advantage of Kt/V urea in PD patients is that it easy to calculate and understand.

As PD patients have continuous dialysis and the slow dialysis allows for equilibration between the stable plasma and interstitial fluid concentrations of urea, Kt/V urea will represent the fractional solute removal during the week. Clearance is defined as the volume of plasma totally cleared from a substance and V represents the total volume to be cleared. Therefore, a Kt/V of, e.g., 1.7 in a PD patient will represent a solute removal index of 1.7 meaning that the total amount of urea has been removed 1.7 times during the week. (Note that this is not the case for HD as the plasma concentration is not stable because of the intermittent HD treatment schedule.)

For standard CAPD the prescription of the dialytic component of Kt/V is simple. Kt/V urea for 1 week is calculated as:

$$\text{Kt/V urea} = 7 \times (\text{D/P urea}) \times V_D/V_{TBW}, \tag{1}$$

where 7 is the number of days/week, D/P urea is the dialysate to plasma concentration of urea in the 24-hour collection of dialysate, V_D is the 24-hour drained dialysate volume, and V_{TBW} is the distribution volume for urea which equals total body water. As urea is almost equilibrated between plasma and dialysate in a CAPD patient, this equation may be simplified to:

$$\text{Kt/V urea} = 7 \times V_D/V_{TBW}. \tag{2}$$

Table 2. Adequacy and transport parameters that may be estimated using a 24-hour dialysate and urine collection

---

Kt/V urea
Weekly creatinine clearance (note interference with glucose if the Jaffe method is used for creatinine measurement)
Pertioneal albumin clearance
Protein intake: PNA (protein equivalent of nitrogen appearance rate = PCR) calculated from urea and protein excretion in the dialysate and urine
Net ultrafiltration (note the overfill of the bags and the weight of the plastic)
Sodium removal and glucose absorption (calculated from mass balance; note the overfill of the bags and the weight of the plastic)
Rough estimate of diffusive transport characteristics in CAPD patients (D/P creatinine in 24-hour dialysate collection)

If we further assume a total body water content of about 58% of the body weight (BW), the equation may be further simplified to (as 7 divided by 0.58 is approximately 12):

$$\text{Kt/V urea} = 7 \times V_D/(0.58 \times BW) \approx 12 \times V_D/BW. \quad (3)$$

Rearranging this equation yields:

$$\text{Kt/V urea} \times BW/12 = V_D. \quad (4)$$

This equation may be used for a rough estimate of the PD prescription. For example, an anuric patient with a body weight of 60 kg would need a daily drained dialysate volume (V_D) of 10 liters to achieve a Kt/V of about 2.0. This calculation also shows how crucial the residual renal function is to achieve the Kt/V urea target in patients that do not have a small body size.

To evaluate the achieved Kt/V urea as well as creatinine clearance, a 24-hour dialysate collection should be performed which may also be used to evaluate several other aspects of the dialysis process (table 2) including protein intake [11], peritoneal albumin clearance and a rough estimate of the diffusive mass transport coefficient for creatinine (from D/P in the 24-hour dialysate collection in a CAPD patient).

Regarding estimation of total body water (V in the calculation of Kt/V urea), the Watson equation has been widely adopted in PD patients for this purpose [12] and it gives a reasonably good estimate of V in most PD patients even though it has been shown to slightly overestimate total body water in obese subjects and to underestimate total body water in lean subjects.

Other Aspects of Dialysis Adequacy

As the results of the ADEMEX trial as well as the Hong Kong study were quite unexpected, these results have led to a large debate. In general, the studies seem to be well performed, but they have also been criticized. Objections have been that they both represent populations with low prevalence of cardiovascular disease (as cardiac disease is relatively rare in Hong Kong and patients with overt cardiac disease were excluded in the ADEMEX study). However, 60% of the deaths in the ADEMEX study were attributed to cardiovascular causes. In general, most people would agree that other aspects of dialysis adequacy may be more important (table 1). In particular, the importance of middle molecule clearance (supported to some extent by the recent MPO study in HD patients [13]) and fluid status have been discussed. Also, it is possible at once frank uremia is prevented by the baseline amount of dialysis, strategies to improve survival need to be focused on other aspects than clearance (table 1).

There are a few studies suggesting that fluid removal is of importance for the survival of PD patients [14, 15]. However, it reasonable to suggest that fluid status, rather than fluid removal should be the primary goal for an adequate dialysis therapy. Both fluid status and middle molecule clearances are discussed in other papers in this issue of contributions to nephrology.

Finally, it was suggested many years ago [Claudi Ronco pers. commun.] that MDt/P, the medical doctors time per patient, should be a more important index of dialysis adequacy. By careful clinical surveillance, the effort can be focused on different clinical problems as well as on prevention of cardiovascular disease and infections, the two major causes of mortality among dialysis patients [10]. This concept has recently got some scientific support from the DOPPS study, in which more frequent and longer patient-doctor contact was associated with reduced mortality in HD patients [16].

In summary, though most of the discussion about adequacy of PD has been focused on small solute clearances, other aspects of dialysis adequacy will need much more focus in the future. However, we should not neglect the importance of achieving the present goals for small solute clearances as a lower limit of the PD prescription.

References

1 Lindholm B, Heimbürger O, Stenvinkel P, Bergström J: Uremic toxicity; in Kopple JD, Massry SG (eds): Nutritional Management of Renal Disease, ed 2. Philadelphia, Lippincott Williams & Wilkins, 2004, pp 63–98.

2 Lo WK, Bargman JM, Burkart J, Krediet RT, Pollock C, Kawanishi H, Blake PG: Guideline on targets for solute and fluid removal in adult patients on chronic peritoneal dialysis. Perit Dial Int 2006;26:520–522.

3 CANADA-USA (CANUSA) Peritoneal Dialysis Study Group: Adequacy of dialysis and nutrition in continuous peritoneal dialysis: association with clinical outcomes. J Am Soc Nephrol 1996;7:198–207.

4 National Kidney Foundation: NKF-DOQI clinical practice guidelines for peritoneal dialysis adequacy. Am J Kidney Dis 1997;30(3 suppl 2):S67–S136.

5 Paniagua R, Amato D, Vonesh E, Correa-Rotter R, Ramos A, Moran J, Mujais S, Mexican Nephrology Collaborative Study Group: Effects of increased peritoneal clearances on mortality rates in peritoneal dialysis: ADEMEX, a prospective, randomized, controlled trial. J Am Soc Nephrol 2002;13:1307–1320.

6 Lo WK, Ho YW, Li CS, Wong KS, Chan TM, Yu AW, Ng FS, Cheng IK: Effect of Kt/V on survival and clinical outcome in CAPD patients in a randomized prospective study. Kidney Int 2003;64:649–656.

7 Bargman JM, Thorpe KE, Churchill DN, CANUSA Peritoneal Dialysis Study Group: Relative contribution of residual renal function and peritoneal clearance to adequacy of dialysis: a reanalysis of the CANUSA study. J Am Soc Nephrol 2001;12:2158–2162.

8 National Kidney Foundation: NKF-DOQI clinical practice guidelines for peritoneal dialysis adequacy: update 2000. Am J Kidney Dis 2000;37(suppl 1):S65–S136.

9 Dombros N, Dratwa M, Feriani M, Gokal R, Heimbürger O, Krediet R, Plum J, Rodrigues A, Selgas R, Struijk D, Verger C, EBPG Expert Group on Peritoneal Dialysis: European best practice guidelines for peritoneal dialysis. 7 Adequacy of peritoneal dialysis. Nephrol Dial Transplant 2005;20(suppl 9):ix24–ix27.

10 Blake P, Suri R: Peritoneal dialysis prescription and adequacy; in Pereira BJG, Sayegh MH, Blake P (eds): Chronic Kidney Disease, Dialysis and Transplantation: A Companion to Brenner and Rector's the Kidney, ed 2. Philadelphia, Elsevier Saunders, 2005, pp 553–568.

11 Bergström J, Heimbürger O, Lindholm B: Calculation of the protein equivalent of total nitrogen appearance from urea appearance. Which formulas should be used? Perit Dial Int 1998;18:467–473.

12 Johansson A-C, Samuelsson O, Attman P-O, Bosaeus I, Haraldsson B: Limitations in anthropometric calculations of total body water in patients on peritoneal dialysis. J Am Soc Nephrol 2001; 12:568–573.

13 Locatelli F, Martin-Malo A, Hannedouche T, Loureiro A, Papadimitriou M, Wizemann V, Jacobson SH, Czekalski S, Ronco C, Vanholder R, Membrane Permeability Outcome (MPO) Study Group: Effect of membrane permeability on survival of hemodialysis patients. J Am Soc Nephrol 2009; 20:645–654.

14 Ates K, Nergizoglu G, Keven K, Sen A, Kutlay S, Ertürk S, Duman N, Karatan O, Ertug AE: Effect of fluid and sodium removal on mortality in peritoneal dialysis patients. Kidney Int 2001;60:767–776.

15 Brown EA, Davies SJ, Rutherford P, Meeus F, Borras M, Riegel W, Divino Filho JC, Vonesh E, van Bree M, EAPOS Group: Survival of functionally anuric patients on automated peritoneal dialysis: the European APD Outcome Study. J Am Soc Nephrol 2003:2948–2957.

16 Kawaguchi T, Bragg-Gresham JL, Fukuhara S, Rayner H, Andreucci V, Pisoni RL, Port F, Morgenstern H, Akizawa T, Saran R: More frequent and longer patient-doctor contact in hemodialyisis care: associations with reduced mortality in the Dialysis Outcomes and Practice Patterns Study (DOPPS) (abstract). J Am Soc Nephrol 2008;19:499A.

Olof Heimbürger, MD, PhD
Department of Renal Medicine, K56, Karolinska University Hospital, Huddinge
SE–141 86 Stockholm (Sweden)
Tel. +46 8 5858 3978, Fax +46 8 711 47 42, E-Mail olof.heimburger@ki.se

Ronco C, Crepaldi C, Cruz DN (eds): Peritoneal Dialysis – From Basic Concepts to Clinical Excellence. Contrib Nephrol. Basel, Karger, 2009, vol 163, pp 147–154

Adequacy of Peritoneal Dialysis: Beyond Small Solute Clearance

Ryan Goldberg, Rajesh Yalavarthy, Isaac Teitelbaum

Division of Kidney Diseases and Hypertension, University of Colorado Denver, Aurora, Colo., USA

Abstract

Peritoneal dialysis adequacy is monitored primarily by indices of small solute clearance, Kt/V_{urea} and creatinine clearance (C_{cr}). Once a threshold of adequacy has been obtained, however, increasing small solute clearance does not result in improved long-term outcomes of PD patients. There are several other factors that may affect optimal dialysis outcomes. These include, but are not limited to: ultrafiltration, inflammation, malnutrition, and mineral metabolism. In this article, we will briefly review data regarding the relationships between these factors and survival on PD.

When considering how well a patient with end-stage renal disease is doing on his/her dialysis regimen a discussion of dialysis adequacy always ensues. The word 'adequate' originates from the Latin *adæquāt* which means to make or become level or equal. Considering potential parameters of dialysis adequacy (table 1), it becomes immediately apparent that even the best dialysis does not equalize any of these to normal kidney function. For example, a peritoneal dialysis patient would be considered to be doing well with a weekly Kt/V_{urea} of 2.0 while a healthy individual with normal GFR has a weekly Kt/V_{urea} of approximately 20! Rather, when considering dialysis 'adequacy' the pertinent question is this: Are there one or more parameters which, when a certain level is achieved, will predict improved survival compared to that in patients in whom these levels have not been achieved? This paper will briefly review the strides that have been taken in the past decade or so to include more than small solute clearance when determining dialysis adequacy.

Table 1. Potential parameters of dialysis adequacy

Small solute clearance
BP and volume homeostasis
Acid-base homeostasis
Control of lipids and CV risk
Nutrition
Calcium/phosphate/bone homeostasis
Inflammation
Middle molecule clearance

Rise and Fall of Traditional Markers of PD Adequacy: Small Solute Clearance

The urea-centric model for peritoneal dialysis adequacy was originally reinforced by two longitudinal observational studies that supported a weekly Kt/V_{urea} ≥2.0 as the appropriate target for 'adequate' peritoneal dialysis [1, 2]. Maiorca et al. [1] demonstrated that two-year patient survival was better in patients with weekly Kt/V_{urea} >1.96 compared to those with weekly Kt/V_{urea} <1.7. The larger CANUSA study showed that survival improved with weekly Kt/V_{urea} >2.1 [2]. However reanalysis of CANUSA revealed that the improved outcome in patients with higher weekly Kt/V_{urea} was due to residual renal function rather than peritoneal clearance [3].

The small solute dependent model for peritoneal dialysis adequacy was further challenged by the results of two randomized controlled trials [4, 5]. In the ADEMEX (ADEquacy of PD in MEXico) trial a total of 965 patients were randomly assigned to an intervention or control group. Patients in the intervention group were prescribed progressive changes in their dialysis regimen to achieve a peritoneal creatinine clearance ≥60 liters/week/1.73 m^2 while patients in the control group continued to perform 4 daily CAPD exchanges of 2 liters each. The time-averaged total weekly Kt/V_{urea} was 2.27 in the intervention group compared to 1.80 in the control group. At 2 years there was no difference in mortality between the two groups [4]. Similarly, Lo et al. [5] found no difference in mortality between three groups of PD patients with weekly Kt/V_{urea} ranging from 1.5 to > 2.0. In the EAPOS (European APD Outcome Study) trial, there was no relationship between small solute clearance and mortality [6] while in the NECOSAD (the NEtherlands COoperative Study on the ADequacy of dialysis) study concluded that the risk of death in anuric peritoneal dialysis patients increased only when weekly Kt/V_{urea} was extremely low, <1.5 [7]. Most recently, Fried et al. [8] detected an increased risk of mortality and hospitalization only at weekly Kt/V_{urea} ≤1.7 (calculated using actual body weight). The indeterminate and fluctuating nature of these studies suggests that there is more to peritoneal dialysis adequacy than small solute clearance.

Ultrafiltration

After considering Kt/V_{urea} as a marker for adequacy, volume status is likely to be the second variable considered. The effect of ultrafiltration (UF) on mortality of PD patients has been examined in several studies with contradictory results. The EAPOS study was a 2-year prospective multicenter study which enrolled a total of 177 anuric patients on APD. The PD prescription was targeted to achieve a creatinine clearance >60 liters/week/1.73 m^2 and UF of >750 ml/24 h. One of the baseline predictors of poor survival on multivariate analysis was UF <750 ml/24 h ($p = 0.047$). On follow-up, however, – and, admittedly, due perhaps to inadequate sample size – the association between time-averaged UF and mortality failed to reach statistical significance ($p = 0.097$) [6]. In another small prospective multicenter cohort study of 130 anuric PD patients (NECOSAD; 102 CAPD and 28 APD), when daily UF was examined as a continuous variable it was significantly associated with 2-year survival ($p = 0.04$). However, while patients in the lowest quintile of UF (<1.15 liters/day) had a relative mortality risk of 3.41 this failed to achieve statistical significance. Likewise, when examined as a dichotomous variable (i.e. UF above or below a specific value) no significant relationship between UF and 2-year survival could be demonstrated ($p > 0.1$ for all) [7].

In contrast, in another study performed in Turkey, 125 patients were followed for three years after starting peritoneal dialysis. Using a Cox proportional hazards model, it was found that total sodium and fluid removal (urinary + dialysate) were each independent factors affecting survival [9]. Thus, the authors of the most recent International Society for Peritoneal Dialysis recommendations regarding dialysis adequacy and ultrafiltration concluded '... from these data that no numerical target for ultrafiltration can be formulated' [10]. It should be noted, however, that two more recent large studies – one utilizing NT-pro-BNP as a surrogate marker for volume [11] and the other examining UF itself [12] – do support an inverse relationship between ultrafiltration and mortality.

The cause of the potentially increased mortality associated with low ultrafiltration has generally been felt to relate to cardiac consequences of volume overload. However, other possibilities must be considered as well. For example, in a study of 82 PD patients followed for a mean of nearly a year Chung et al. [13] observed that among patients with elevated C-reactive protein (CRP) levels (>10 mg/l) there was a higher proportion of patients with low total fluid removal (<1,000 ml/day) and increased mortality (RR 2.69; $p = 0.01$) compared to patients with low CRP levels. This observation suggests a possible association between inflammation and mortality which warrants further examination.

Inflammation and Peritoneal Dialysis Adequacy

The malnutrition, inflammation and atherosclerosis (MIA) syndrome is increasingly recognized as a major cause of cardiovascular morbidity and mortality in dialysis patients. Peritoneal dialysis appears to be a less inflammatory modality than is hemodialysis. Haubitz et al. [14] compared CRP levels between healthy volunteers, patients with chronic kidney disease not on dialysis, and patients on either hemodialysis or peritoneal dialysis. CRP levels were highest in the HD patients while the PD patients had levels comparable to those of the CKD patients not yet on dialysis. The question remains, however, whether within a population of PD patients, the degree of inflammation (or absence thereof) may be used as a marker for peritoneal dialysis adequacy.

Stenvinkel et al. [15] prospectively studied 246 Chinese PD patients for an average of 20 months. In this study, a single measurement of high-sensitivity CRP (hs-CRP) was found to be predictive of mortality: on multivariate analysis, each 1 mg/dl increase in hs-CRP was associated with a 2% increase in all-cause mortality ($p = 0.002$) and a 3% increase in cardiovascular mortality ($p = 0.001$). CRP is not the only inflammatory marker associated with mortality. In another observational study, Stenvinkel et al. [16] demonstrated that mortality increased as serum interleukin-6 (IL-6) levels rose. Furthermore, both serum IL-6 levels and mortality increased as the number of components of the MIA syndrome present in the patients increased.

These studies all suggest that absence of inflammation may be considered a component of 'adequate' renal replacement therapy. However, no clear criteria emerge whereby to ascertain that inflammation is indeed absent. Furthermore, there are no specific guidelines regarding the best methods to monitor or treat occult inflammation in these patients. Finally, as exemplified by the Stenvinkel study, inflammation often coexists with malnutrition suggesting that the potential role of nutrition as a marker for PD adequacy warrants closer scrutiny.

Nutritional Status and Peritoneal Dialysis Adequacy

Malnutrition is a very common problem in PD patients and nutritional indices are predictors of patient outcome. Many patients being referred for dialysis initiation are already malnourished in part due to late nephrology referral. In an international study, the prevalence of malnutrition measured by subjective nutritional assessment, which has 21 components, was 40.6% [17].

A number of studies have demonstrated a relationship between one or more nutritional indices and mortality. For example, in the CANUSA study, each 1 gm/l increase in serum albumin was associated with a 6% decrease in mortality and a one unit increase in SGA score was associated with 25% decrease in relative risk of death [2]. Lo et al. [18] analyzed mortality as a function of both small solute clearance and nutritional status as assessed by the CNI (composite nutritional index which is a function of SGA, albumin, and anthropometric measurements). As mentioned previously, they found no relationship between mortality and weekly Kt/V_{urea} in the range of 1.5 to >2.0. In contrast, they found that patients with a better CNI enjoyed significantly greater 12-month survival ($p = 0.0259$) compared to those with worse nutritional status. Similarly, in both the EAPOS and NECOSAD trials nutritional status independently predicted survival of PD patients at 2 years [6, 7].

More recently, Avram et al. [19] reported on the relationship between nutritional status and mortality in an observational study of 177 patients who started PD from 1991 to 2005. Bioimpedence analysis with determination of phase angle as well as measurement of serum prealbumin and other nutritional indices was performed in subsets of these patients who were then followed for up to 15 years. Phase angle may be understood to be a surrogate marker for the mass of cell membranes; healthy well-nourished individuals should have a phase angle ≥6 degrees. Indeed, over a period of 5 years, PD patients with a phase angle ≥6 degrees at the time of entry enjoyed substantially improved survival ($p = 0.036$) compared to those with a phase angle <6 degrees. Similar findings were obtained when patients were stratified by entry prealbumin level: those with a level ≥32 mg/dl enjoyed significantly superior survival ($p = 0.032$) compared to those with lower levels.

Peritonitis is another major risk factor for PD failure and its occurrence adds to the morbidity and mortality of ESRD patients treated with PD. Malnutrition has been implicated as contributing to the development of peritonitis; however, until recently, there have been no studies examining the relationship between nutritional indices and peritonitis. Prasad et al. [20] enrolled 56 randomly selected Indian patients and followed them prospectively from the day PD was started until the end of the study, approximately 60 months. The proportion of patients developing peritonitis was significantly higher in patients with malnutrition at the start of dialysis as measured by SGA ($p = 0.001$, OR 0.08, 95% CI 0.02–0.36).

The etiology of malnutrition in PD patients is multifactorial. To date, there have been no large-scale interventional trials studying the relationship between nutritional status and mortality in PD patients. However, the available observational data suggest that, along with ensuring adequate small solute clearance, efforts must be made to improve the nutritional status of our PD patients if they are to enjoy superior survival and decreased morbidity.

Mineral Metabolism and Peritoneal Dialysis Adequacy

There is growing interest regarding a possible relationship between mineral metabolism – particularly that of phosphate – and mortality in PD patients. Trivedi et al. [21] followed 191 PD patients for an average of 21 months and examined predictors of mortality. On stepwise logistic regression analysis, they found that the weighted time- averaged serum phosphate level was an independent predictor of death ($p = 0.02$). Similarly, NECOSAD data from 586 patients who began PD between 1997 and 2004 showed that phosphorus levels above the K/DOQI upper limit of 5.5 mg/dl were associated with increased cardiovascular mortality (HR 2.4; $p <0.01$) as was an increased calcium × phosphorus product >55 mg^2/dl^2 (HR 2.2; $p < 0.01$) [22].

To date there are no large database studies regarding the effects of mineral metabolism on cardiovascular outcomes in the PD population. There are, however, two recent abstracts from data obtained in a cohort of 7,034 patients performing PD for at least 3 months that do address this issue. Mehrotra et al. [23] reported that, compared to a reference group of patients with serum phosphorus ranging from 4.5 to 5.5 mg/dl, those with serum phosphorus >8.5 mg/dl were associated with increased mortality (HR 1.37; $p = 0.03$) independent of any confounding variables. In another abstract based upon this same database Khawar et al. [24] reported that each 1 mg/dl increase in albumin-adjusted serum calcium above 9 mg/dl was associated with an increased risk of death (HR 1.15; $p < 0.0001$). It must be noted, however, that in the NECOSAD mineral metabolism data cited previously, increased calcium levels were not associated with worsening mortality [22].

Conclusion

Providing adequate peritoneal dialysis for patients necessitates more than just demonstrating a certain degree of small solute clearance. This paper briefly presented other possible parameters that possibly impact the quality of dialysis, mainly ultrafiltration, inflammation, malnutrition, and mineral metabolism. More research is needed to determine which other parameter(s) – or others not considered in this brief review (e.g. middle molecule clearance) – most closely correlate with outcomes. Large randomized trials will then be needed to determine whether interventions targeted at optimizing these parameters will provide our patients with significant survival benefit.

References

1 Maiorca R, et al: Predictive value of dialysis adequacy and nutritional indices for mortality and morbidity in CAPD and HD patients: a longitudinal study. Nephrol Dial Transplant 1995;10:2295–2305.
2 Churchill DN, Taylor DW: Adequacy of dialysis and nutrition in continuous peritoneal dialysis: association with clinical outcomes. Canada-USA (CANUSA) Peritoneal Dialysis Study Group. J Am Soc Nephrol 1996;7:198–207.
3 Bargman JM, Thorpe KE, Churchill DN: Relative contribution of residual renal function and peritoneal clearance to adequacy of dialysis: a reanalysis of the CANUSA Study. J Am Soc Nephrol 2001;12:2158–2162.
4 Paniagua R, et al: Effects of increased peritoneal clearances on mortality rates in peritoneal dialysis: ADEMEX, a prospective, randomized, controlled trial. J Am Soc Nephrol 2002;13:1307–1320.
5 Lo WK, et al: Effect of Kt/V on survival and clinical outcome in CAPD patients in a randomized prospective study. Kidney Int 2003;64:649–656.
6 Brown EA, et al: Survival of functionally anuric patients on automated peritoneal dialysis: the European APD Outcome Study. J Am Soc Nephrol 2003;14:2948–2957.
7 Jansen MA, et al: Predictors of survival in anuric peritoneal dialysis patients. Kidney Int 2005; 68:1199–1205.
8 Fried L, et al: Association of Kt/V and creatinine clearance with outcomes in anuric peritoneal dialysis patients. Am J Kidney Dis 2008;52:1122–1130.
9 Ates K, et al: Effect of fluid and sodium removal on mortality in peritoneal dialysis patients. Kidney Int 2001;60:767–776.
10 Lo W-K, et al: Guideline on targets for solute and fluid removal in adult patients on chronic peritoneal dialysis. Perit Dial Int 2006;26:520–522.
11 Wang AY, et al: N-terminal pro-brain natriuretic peptide: an independent risk predictor of cardiovascular congestion, mortality, and adverse cardiovascular outcomes in chronic peritoneal dialysis patients. J Am Soc Nephrol 2007;18:321–330.
12 Paniagua R, et al: Predictive value of brain natriuretic peptides in patients on peritoneal dialysis: results from the ADEMEX trial. Clin J Am Soc Nephrol 2008;3:407–415.
13 Chung SH, et al: Influence of peritoneal transport rate, inflammation, and fluid removal on nutritional status and clinical outcome in prevalent peritoneal dialysis patients. Perit Dial Int 2003;23: 174–183.
14 Haubitz M, et al: Chronic induction of C-reactive protein by hemodialysis, but not by peritoneal dialysis therapy. Perit Dial Int 1996;16:158–162.
15 Wang AY-M, et al: Is a single time point C-reactive protein predictive of outcome in peritoneal dialysis patients? J Am Soc Nephrol 2003;14:1871–1879.
16 Stenvinkel P, et al: Malnutrition, inflammation, and atherosclerosis in peritoneal dialysis patients. Perit Dial Int 2001;21(suppl 3):S157–S162.
17 Young GA, et al: Nutritional assessment of continuous ambulatory peritoneal dialysis patients: an international study. Am J Kidney Dis 1991;17:462–471.
18 Lo WK, et al: Relationship between adequacy of dialysis and nutritional status, and their impact on patient survival on CAPD in Hong Kong. Perit Dial Int 2001;21:441–447.
19 Avram MM, et al: Malnutrition and inflammation as predictors of mortality in peritoneal dialysis patients. Kidney Int 2006;70:S4–S7.
20 Prasad N, et al: Impact of nutritional status on peritonitis in CAPD patients. Perit Dial Int 2007;27:42–47.

21 Trivedi H, et al: Predictors of death in patients on peritoneal dialysis: the Missouri Peritoneal Dialysis Study. Am J Nephrol 2005;25:466–473.
22 Noordzij M, et al: Mineral metabolism and cardiovascular morbidity and mortality risk: peritoneal dialysis patients compared with haemodialysis patients. Nephrol Dial Transplant 2006;21:2513–2520.
23 Mehrotra R, et al: Serum phosphorus and mortality in chronic peritoneal dialysis (CPD) patients. Perit Dial Int 2007;27(suppl 3):S12.
24 Khawar O, et al: Serum calcium and survival in a large USA cohort of chronic peritoneal dialysis patients. Perit Dial Int 2007;27(suppl 3):S11.

Isaac Teitelbaum, MD
University of Colorado Hospital, AIP
12605 E.16th Ave F774
Aurora, CO 80045 (USA)
Tel. +1 720 848 7601, Fax +1 720 848 3103, E-Mail Isaac.teitelbaum@ucdenver.edu

Ronco C, Crepaldi C, Cruz DN (eds): Peritoneal Dialysis – From Basic Concepts to Clinical Excellence. Contrib Nephrol. Basel, Karger, 2009, vol 163, pp 155–160

Importance of Residual Renal Function and Peritoneal Dialysis in Anuric Patients

Maria João Carvalho, Anabela Rodrigues

Division of Nephrology, Centro Hospitalar do Porto, Hospital de Santo António, Porto, Portugal

Abstract

Residual renal function (RRF) impacts on patient survival and quality of life of dialysis patients. Its longer preservation is a major advantage of peritoneal dialysis (PD) and should be also a target of adequacy, beyond Kt/V. Anuric patients no longer benefit from such PD advantage, depending only on dialysis schedule to achieve adequate small solute and volume control. This challenge can be successfully dealt with by using automated PD, icodextrine, low-glucose degradation products and individualized PD profiles. There is evidence that PD advances allow nowadays satisfactory patient survival while keeping the benefits of home dialysis and preserving vascular network. An integrated care plan should consider both medical indications and patient preference aiming for the longer total patient survival, even if transfer to HD might be later needed as part of individualized renal replacement strategy.

Peritoneal dialysis (PD) in anuric patients raises concerns about the capacity to obtain adequacy. Kidneys are crucial to internal homeostasis and even a small quantity of residual renal function (RRF) gives important advantages to the dialyzed patient with impact on well-being and survival [1].

Anuric dialysis patients are challenging. But it is unavoidable to face with their peculiarities in a PD program. Most of the patients transferred from hemodialysis (HD) have this condition at admission to the treatment and patients new to dialysis become anuric over time. Much research has been done along the 30 years of PD, so that continuous improvement in the quality and individualization of treatment has been attained. Anuric patients are beneficed by the continuous advances in PD.

Adequacy has been confounded with small solute clearance, as measured by Kt/V urea, but recent evidence was achieved that it is only fairly related with clinical outcomes while ultrafiltration and time of dialysis are relevant in both HD and PD. There is no evidence that anuric patients live longer or better on HD than on PD. On the other hand, PD offers the benefits of a continuous removal of uremic non measured toxins, and sustained ultrafiltration, targets that are being aimed for also in HD. Therefore clinicians must be confident that adjusted and updated PD therapy, in each individual choosing PD, even after RRF loss, will be able to offer him if not a better treatment, at least an as good as it would be accomplished if he was treated by HD.

This article supports the importance of RRF and reviews the present position of PD in anuric patients.

Characteristics and Outcome of Anuric Peritoneal Dialysis Patients

When compared to patients with preserved RRF, more severe left ventricular hypertrophy, poorer blood pressure control, greater anemia, lower serum albumin, higher serum phosphate levels, higher inflammatory status and worse nutritional status were reported in anurics [2]. Cardiovascular mortality remains the leading cause of death in end-stage renal disease. The addition of the above described factors can promote a more adverse metabolic and cardiovascular profile in the subset of anuric patients. The physiopathology of these various clinical parameters and the way they orchestrate themselves to create a metabolic risk profile are topics of ongoing research. Questions are pending: are these unfavorable characteristics conferred predominantly by the condition of anuria or are they the main result of cumulative prolonged time on dialysis and older age or higher associated comorbidity? Are they amenable with PD-adjusted schedules?

However, from the available larger studies on PD in anuric patients [2–4], we can conclude that this group is able to do well, concerning both to patient and technique survival. Reported 2-year patient survival range from 65% [2] to 79% [4], comparable to the described survival in the ADEMEX cohort, which averaged 70% [5]. Besides, the excellent outcomes in anuric patients from EAPOs were achieved despite additional patient comorbidity, using real life PD schedules with automated PD (APD) and icodextrine in a significant subgroup. This underlines the impact of such more recent treatment policies that allow better outcomes. Interestingly, in a modern cohort of HD patients using high flux membranes – the HEMO study – equivalent 2-year patient survival (70%) was also found, irrespective of delivered Kt/V [6].

Prevention of Uremia in Anuric Peritoneal Dialysis Patients

Residual renal clearances and peritoneal clearances are not equivalent, nor are they simply additive; one unit of residual renal clearance is superior to one unit of peritoneal clearance. Targets of small solute removal have been previously erroneously recommended not taking into account this quality difference. After excluding residual renal clearances, minimal PD targets are mandatory. In the NECOSAD subgroup of anurics cohort a Kt/V <1.5 per week and a creatinine clearance <40 liters/week/1.73 m^2 were associated with an increase in the relative risk of death [3]. Szeto et al. [7] reported that dialysis adequacy, as measured by Kt/V and creatinine clearance in anuric, was linked to patient and technique survival and that creatinine clearance was associated with shorter hospitalization.

On the other hand, studies pertaining to the benefits of RRF in HD are scarce, but it is presumably an important predictor of outcome as well in this population. A recent investigation among HD patients in the Dialysis Outcomes and Practice Pattern Study (DOPPS) revealed that patients administered diuretics, as a marker of existing RRF, had a 7% lower all-cause mortality risk ($p = 0.12$) and 14% lower cardiac-specific mortality risk ($p = 0.03$) versus patients not administered diuretics. The relevance of RRF in HD was also highlighted in the cohort of NECOSAD where the contribution of RRF to patient survival in patients treated by HD was prospectively investigated: the effect of Kt/V on mortality was strongly dependent on the presence of renal Kt/V, while low values of dialysis Kt/V influenced mortality only in anuric patients [8].

However, previous focus on Kt/V as a measure of small solute clearance proved to be misleading. For example, in the ADEMEX trial, increments of weekly Kt/V over 1.7 did not significantly improve the 2-year patient survival [5]. In the same way, the HEMO trial cast doubts on the advantages of achieving higher than recommended small solute clearance targets [6]. These randomized trials strengthened the requisite to broaden the definition of adequacy in dialysis. Moreover, focus on Kt/V often neglects the pitfalls of the V calculation: targets are difficult to obtain in obese patients and can precipitate incorrect transfer to HD, while opposite high Kt/V usually offers a false reliance on adequacy in lean, many times malnourished, patients and preclude dialysis enhancement.

Uremic toxins include a wide range of non-measured molecular weight solutes; larger molecules are essentially dependent on elimination by RRF. Creatinine clearance is much dependent of peritoneal membrane transport characteristics, particularly in the anuric patient and importantly phosphorus clearance parallels creatinine clearance. Peritoneal clearance of middle molecules depends mainly on the total PD dwell hours and not on number of dialysis cycles and total nocturnal hours; therefore a dry day should only exceptionally and transitorily be prescribed in an anuric patient. Larger body sized anuric patients with slow membrane transport will be difficult or impossible to manage under APD because of the shorter dwells that impede the equilibration of solutes.

To improve solute PD clearance use of higher dwell volumes and the utilization of an additional exchange, as a PD-plus regimen, are important optimization steps, although we have to be aware of the possible implications of these interventions on the quality of life of the PD patient.

Prevention of Volume Abnormalities in Anuric PD Patients

Presently, there can be no doubts that volume status is crucial to outcomes of PD patients [4, 9]. Volume status is a complex problem, influenced by comorbidity, specially inflammation, cardiac function and malnutrition. Euvolemia is difficult to determine and targets of fluid and salt removal are also difficult to propose. In anuric patients, the link between poor ultrafiltration and increased mortality is not clear since in EAPOs, no differences in comorbidity, nutritional state, or other indices of treatment at baseline were found between patients with ultrafiltration lower than 750 ml/day and the other patients, and lower, instead of higher blood pressure seemed to be related with earlier death [10].

Sodium removal is particularly poor in APD, due to sodium sieving. In addition to icodextrin use in the long dwell, both ultrafiltration and sodium removal can be optimized using hybrid schedules with day time exchanges and perhaps new nocturnal ultrafiltration profiles that are under investigation [11]. Combination of crystalloid and colloid osmotic agents enhances fluid removal by twofold and sodium removal by threefold, a new emerging treatment option for anuric patients, especially if they are fast transporters [12]. Increasing the diffusive transport of peritoneal sodium removal with low-sodium PD solutions might influence positively on volume control of anuric patients [13]. In these patients dietary sodium restriction should not be neglected.

Prevention of Abnormalities beyond Kt/V and Volume in Anuric Peritoneal Dialysis Patients

The prescription of adequate PD must also include the replacement of other vital functions of the kidneys together with the maintenance of a good nutritional and metabolic status and the protection of the peritoneal membrane from deterioration imposed by the procedure itself.

Hyperphosphatemia is strongly associated with an increased risk of death among dialysis patients and in PD. It is known that it deteriorates as RRF declines and that it follows peritoneal creatinine clearance. A recent multicentre study on PD patients described achievement of phosphate control in 58% of subjects [14], not inferior to the disappointing low phosphate control reported in HD surveys. In the anuric patient careful attention should be taken to dietary counseling and phosphate binders prescription; additionally peritoneal creatinine clearance assessment should never be neglected.

Concerning nutritional status, PD seems favorable since it allows caloric and amino acid intraperitoneal nutritional support to avoid malnutrition. The opposite side of the coin is the risk of excessive caloric load in sedentary obese patients although the real impact of such condition in patient outcomes is still unclear.

Anuric patients raise the fear of higher exposure to glucose due to the requirement of higher volumes and tonicity prescription. However, glucose-sparing regimens might prove to be useful in the near future to prevent harmful metabolic systemic complications of PD.

Long-term PD imposes changes in peritoneal membrane structure and function and APD has been also associated with a faster decline in membrane capacity to ultrafiltration. Presently, we have reassuring results also specifically addressed in anuric patients: neutral pH and low GDP solutions showed higher ultrafiltration and better membrane preservation [15] and icodextrin use in the EAPOs trial spared the peritoneal membrane from deterioration [16].

In a still broader definition of adequacy we should aim to include the patient social rehabilitation, quality of life and satisfaction with treatment. Dialysis modality itself has a major impact on individual's patient's life and should be matched to the lifestyle and expectations in addition to medical condition. Autodialysis and home dialysis offered by PD allows autonomy, flexibility and motivation for understanding and deal with the chronic kidney disease, benefits that should not be denied merely by the presence of anuria. Education in PD must also include the patient's awareness of the possible need of timely transfer to HD in the course of renal replacement therapy.

Conclusions

Currently PD is a well-established part of the integrated care of end-stage renal disease and an excellent first renal replacement therapy, as it is recognized to preserve RRF better than HD. We underline that RRF protection is desirable in PD patients and all the efforts should be made to it, as part of the global approach to adequacy in PD. Although challenging and subject to caution, updated PD is feasible in most anuric end-stage renal disease patients. A solid understanding of peritoneal physiology coupled with the modern and emerging resources of APD and new PD solutions allows the clinician to do a better and individualized prescription mitigating some of the adverse characteristics of the anuric patients with impact on survival. There is by now no evidence that outcomes in PD anuric differ substantially from HD anuric patients.

On the other hand, PD gives all the benefits of a continuous dialysis procedure and home dialysis.

References

1 Wang AY, The John F: Maher Award Recipient Lecture 2006. The 'heart' of peritoneal dialysis: residual renal function. Perit Dial Int 2007;27:116–124.
2 Wang AY, et al: Important differentiation of factors that predict outcome in peritoneal dialysis patients with different degrees of residual renal function. Nephrol Dial Transplant 2005;20:396–403.
3 Jansen MA, et al: Predictors of survival in anuric peritoneal dialysis patients. Kidney Int 2005;68:1199–1205.
4 Brown EA, et al: Survival of functionally anuric patients on automated peritoneal dialysis: the European APD Outcome Study. J Am Soc Nephrol 2003;14:2948–2957.
5 Paniagua R, et al: Effects of increased peritoneal clearances on mortality rates in peritoneal dialysis: ADEMEX, a prospective, randomized, controlled trial. J Am Soc Nephrol 2002;13:1307–1320.
6 Eknoyan G, et al: Effect of dialysis dose and membrane flux in maintenance hemodialysis. N Engl J Med 2002;347:2010–2019.
7 Szeto CC, et al: Impact of dialysis adequacy on the mortality and morbidity of anuric Chinese patients receiving continuous ambulatory peritoneal dialysis. J Am Soc Nephrol 2001;12:355–360.
8 Termorshuizen F, et al: Relative contribution of residual renal function and different measures of adequacy to survival in hemodialysis patients: an analysis of the Netherlands Cooperative Study on the Adequacy of Dialysis (NECOSAD)-2. J Am Soc Nephrol 2004;15:1061–1070.
9 Ates K, et al: Effect of fluid and sodium removal on mortality in peritoneal dialysis patients. Kidney Int 2001;60:767–776.
10 Davies SJ, et al: What is the link between poor ultrafiltration and increased mortality in anuric patients on automated peritoneal dialysis? Analysis of data from EAPOS. Perit Dial Int 2006;26:458–465.
11 Vega ND, et al: Nocturnal ultrafiltration profiles in patients on APD: impact on fluid and solute transport. Kidney Int Suppl 2008;108:S94–S101.
12 Freida P, et al: The contribution of combined crystalloid and colloid osmosis to fluid and sodium management in peritoneal dialysis. Kidney Int Suppl 2008;108:S102–S111.
13 Davies S, et al: The effects of low-sodium peritoneal dialysis fluids on blood pressure, thirst and volume status. Nephrol Dial Transplant 2009;24:1609–1617.
14 Yavuz A, et al: Phosphorus control in peritoneal dialysis patients. Kidney Int Suppl 2008;108:S152–S158.
15 Choi HY, et al: The clinical usefulness of peritoneal dialysis fluids with neutral pH and low glucose degradation product concentration: an open randomized prospective trial. Perit Dial Int 2008;28:174–182.
16 Davies SJ, et al: Longitudinal membrane function in functionally anuric patients treated with APD: data from EAPOS on the effects of glucose and icodextrin prescription. Kidney Int 2005;67:1609–1615.

Maria João Carvalho, MD
Division of Nephrology, Centro Hospitalar do Porto – Hospital de Santo António
Largo Professor Abel Salazar
PT–4100, Porto (Portugal)
Tel. +351 2220 77 521, Fax +351 2220 77 520, E-mail: mjcarvalho08@gmail.com

Ronco C, Crepaldi C, Cruz DN (eds): Peritoneal Dialysis – From Basic Concepts to Clinical Excellence. Contrib Nephrol. Basel, Karger, 2009, vol 163, pp 161–168

Insights on Peritoneal Dialysis-Related Infections

Beth Piraino

Renal Electrolyte Division of the University of Pittsburgh School of Medicine, Pittsburgh, Pa., USA

Abstract

Peritonitis remains a serious complication of peritoneal dialysis and can lead to death of the patient. Most peritonitis is due to either contamination with the peritoneal dialysis exchange or exit site infection and can be prevented by protocols for appropriate training and exit site care. The micro-organism infecting the peritoneum is an important clue to the etiology: coagulase-negative *Staphylococcus* is generally due to touch contamination while *Pseudomonas aeruginosa* and *Staphylococcus aureus* are most often due to catheter infections. The etiology of other Gram-negative peritonitis is uncertain but appears to relate to bowel problems especially constipation, hypokalemia, perhaps leading to dysmotility, as well as touch contamination at the time of the exchange. The approach to treating peritonitis is always to rapidly resolve the infection, even if this entails removing the peritoneal catheter. Refractory peritonitis is generally defined as peritonitis treated with an appropriate antibiotic for five days without evidence of resolution. The antibiotic to treat peritonitis should be chosen based on the past history of the organisms and sensitivities of the program, should cover Gram-negative and Gram-positive organisms, and tailored once the cause is identified. Relapsing peritonitis requires catheter exchange. Each program should have continuous quality improvement that tracks each episode, performing root cause analysis, examining organisms, rates of organisms, and outcome. In this way, each program can develop initiatives to lower peritonitis rates. Such protocols might include the use of routine exit site antibiotic cream as part of daily care, re-training of patients, prevention of constipation, and removal of infected catheters. With an aggressive approach peritonitis rates can be lowered to very low rates.

Peritonitis is a serious complication for patients on peritoneal dialysis (PD), contributing to hospitalization, technique failure, catheter loss, and even death [1, 2]. Although the rates of peritonitis have decreased dramatically, rates of above 0.5 episodes per year (an episode every 24 months) still occur commonly [3]. The success of this dialysis technique is very much tied to the ability of the dialysis program to reduce the risk of peritonitis, and when it occurs, manage the patient appropriately.

Modifiable Risk Factors for Peritonitis

A few modifiable risk factors have been identified. The most well-recognized modifiable risk factor (for *S. aureus*-related PD infections) is colonization of the patient with *S. aureus* [4]. A low serum albumin and malnutrition increase the risk of subsequent peritonitis [5, 6]. Peritonitis, of course, worsens the poor nutritional status further. The presence of depressive symptoms has also been associated with increased peritonitis risk [7]. The explanation is unclear. Both of these might be modifiable with aggressive interventions but this has not been well studied. APD may result in lower peritonitis risk than CAPD, but data on this are unclear, perhaps because connectology of the different systems has not been carefully analyzed [8].

Presentation and Definition of Peritoneal Dialysis-Related Peritonitis

The usual presentation of peritonitis in the PD patient is abdominal pain, cloudy effluent or, most often, both. The pain can range from extremely severe to non-existent, related to the organism. In the inexperienced patient, the absence of pain may lead him/her to ignore the cloudy effluent initially, leading to a delay in presentation and thus treatment. Furthermore, with the increase in use of the cycler, examination of the effluent by the patient may become perfunctory or even omitted. Continued education of the patient about the variable presentation of peritonitis is always needed.

If either suspicious abdominal pain or cloudy effluent or both are present, an effluent cell count is ordered. Peritonitis is considered to be present if the white blood cell count in the effluent is 100/μl or greater, with at least 50% polymorphonuclear cells, as in most cases this inflammatory response is due to a micro-organism invading the peritoneal space. If the specimen is from an exchange with an abbreviated dwell time (e.g. if the patient was on the cycler), an aspirate from a drained abdomen, or from a patient already on antibiotics, the percentage of polymorphonuclear cells (i.e. more than 50%) is a more reliable marker for peritonitis than the absolute number of white blood cells.

Table 1. Etiologies and root cause analysis of peritonitis

Contamination: predominately skin and environmental organisms
During an exchange
Contamination from the air (wind, fan in a dusty environment)
Contamination of the tubing or dialysis bags by a cat
Re-use of cassettes on the cycler
Contamination from coming on and off the cycler at night
Product defect (pin-hole in the dialysis bag)
Damage to the external portion of the PD catheter or exchange tubing
Exit site and tunnel infections leading to peritonitis: predominately *S. aureus* and *P. aeruginosa*
Transmural migration of enteric organisms
Colonoscopy without antibiotic prophylaxis especially with polypectomy
Constipation
Diarrhea/colitis
Bacteremia seeding the peritoneum
Transient bacteremia from poor dentition
Dental procedure without antibiotic prophylaxis
Bacteremia from another source (such as intravenous line, pacemaker)
Biofilm-related peritonitis: primarily *Staphylococcus* and *Pseudomonas*
From contamination of biofilm from previous episode of peritonitis
Gynecologic causes
Uterine procedure (biopsy, myotomy) without antibiotic prophylaxis
Vaginal leak of dialysis fluid

Micro-Organisms and Etiologies of Peritonitis

The most common micro-organisms responsible for peritonitis are coagulase-negative *Staphylococcus, S. aureus, Streptococcus*, and Gram-negatives. Much less common are mycobacterium and fungal peritonitis. These proportions vary from program to program and are best expressed as a rate rather than a percentage of the overall rate, to permit comparisons from center to center. The organism is an important clue to the etiology of the peritonitis episode (table 1).

The leading cause of peritonitis continues to be contamination, which is most often at the time of the PD exchange. Peritonitis due to skin organisms such as coagulase-negative *Staphylococcus, Corynebacterium*, and *Bacillus* species are generally accepted as caused by contamination and often cause mild or even no abdominal pain. However, PD patients may be colonized with *Streptococcus viridans*, *S. aureus*, *Micrococcus, Proteus* species, *Klebsiella pneumoniae*, *Enterobacter* species, *Escherichia coli* and *Acinetobacter* species. Therefore, GN peritonitis may also be related to contamination. Training and retraining of the patient is key to preventing peritonitis from contamination [9].

Exit site infections may lead to tunnel infections and peritonitis [10]. This is especially common with *S. aureus* but may also occur with *Pseudomonas* and other organisms. Prophylaxis with gentamicin at the exit site resulted in a reduction in exit site infections and peritonitis episodes [11]. This protocol is well accepted by the patients and can almost eliminate peritonitis from this source.

Enteric peritonitis is a less common cause but important due to the severity of the peritonitis. In some cases, this appears to be from transmural migration as frank intra-abdominal pathology (such as cholecystitis, ischemic bowel, diverticulitis) is a less common cause. Constipation, treatment of constipation with enemas, and colitis may predispose to enteric peritonitis. Hypokalemia appears to be a risk factor [12]. Proton pump inhibitors might be a risk factor but this is somewhat controversial.

Other, much less common, causes of peritonitis are bacteremia from another source, gynecologic categories, dental procedures, and colonoscopy. These are important as most of these are preventable causes of peritonitis. Antibiotic prophylaxis prior to any procedure associated with peritonitis risk is warranted. In addition, the abdomen should be drained prior to pelvic and colonic procedures. Aggressive treatment of constipation may result in enteric peritonitis; therefore, every effort should be made to prevent constipation in the PD patient.

Management of Peritonitis

Upon presentation, a rapid assessment of the patient should include questions on breaks in technique, recent procedures that may have led to peritonitis, change in bowel habits, prior history of peritonitis or catheter infection. The exit site and tunnel should be closely examined for evidence of infection. The patient's abdomen is drained and the effluent sent for cell count with differential, Gram stain, and culture. The cell count with differential will confirm the presence of peritonitis. Pain should be assessed promptly and adequate treatment given.

The culture should be obtained by placing 5 ml in each of two tryptic soy broth blood culture bottles (aerobic and anaerobic). In addition to inadequate culture technique, culture-negative peritonitis may also be due to the presence of antibiotics, so the patient should be questioned closely about recent antibiotic use. Hospitalization will depend on the severity of the peritonitis, a need for intravenous analgesia and fluids.

The causative organism is generally not known when antibiotic therapy is ordered. Therefore, the initial therapy should be active against the most commonly occurring organisms, including *Staphylococcus* (both coagulase-negative and coagulase-positive), and Gram-negative bacilli. In the absence of clinical data suggestive of bowel perforation, anaerobic coverage is generally not given initially. Coverage for fungus should be implemented immediately only if the gram stain is positive for yeast. The ISPD treatment guidelines of 2005 recommended a center-specific protocol for empiric treatment that covers both Gram-positive and Gram-negative organisms [13]. A recent meta-analysis confirms that this is an appropriate approach [14]. Suggested dosing can be found in these guidelines on the ISPD web site.

Within 2–3 days, the organism is usually identified and sensitivities are available. Subsequent therapy is chosen to provide narrow coverage with the least toxicity. If the culture is negative, generally the aminoglycoside is stopped and a single drug such as a first-generation cephalosporin or vancomycin continued alone.

Fungal peritonitis accounts for about 4–6% of episodes. Gram stain is often helpful in establishing the diagnosis early. *Candida* is by far the most common organism. Risk factors include frequent peritonitis, immunosuppression, and antibiotic therapy. Mortality is high, and, therefore, rapid catheter removal is a prudent approach. Therapy with fluconazole (200 mg orally each day), flucytosine (1 g orally each day), and if necessary amphotericin, should be continued after catheter removal for at least an additional 10 days. The catheter can be re-inserted but a waiting period of 1–2 months is advisable. Prophylaxis with nystatin, given to the patient taking antibiotics, successfully reduces the risk of *Candida* peritonitis [15].

Refractory peritonitis is defined as an episode in which there is no improvement 5 days after appropriate antibiotic therapy is initiated [13]. There may be apparent resolution of the peritonitis episode with antibiotic therapy, but a cell count will often show persistence of abnormal inflammatory response. Catheter removal is required and should not be delayed as this can result in peritoneal membrane failure and a high risk of patient death.

Relapsing peritonitis is defined as another episode of peritonitis with the same organism (or sterile culture) as the first within 4 weeks of stopping antibiotics [13]. This may be due to either a catheter infection, which may not be clinically apparent, or infected biofilm. If the peritoneal cell count can be suppressed to less than 100 WBC/μl, then the catheter can be removed and a new catheter inserted at one setting. This approach avoids or minimizes the need for hemodialysis in many patients.

Table 2. Methods to evaluate peritonitis frequency within a program

As a rate for the program: Number of episodes per period/time in years of all patients in program for period examined. Example: During 1 month a program has 24 patients on PD (therefore, 24 months at risk or 2 years) and there is one episode of peritonitis. The rate is: one episode divided by 2 years or 0.5 episodes per year at risk.
As a median of all the rates for each patient: Calculate the rate for each patient for the period of interest, and then obtain a median of all the rates. Example: A program has 5 patients on PD for 1 year, and 4 of the patients have no peritonitis, 1 had 1 episode (so 1 episode per year at risk for that one patient). The median rate for all patients (rates of 0, 0, 0, 0, 1 episode per year) is 0. This indicates that most patients in the program were without peritonitis that particular year.
As a percentage of patients in the program who had peritonitis: Calculate by number of patients with peritonitis in a period divided by all patients in the program during that time. Example: A program has 10 patients at the beginning of the year, and 8 patients start during the year. Two had peritonitis during the year resulting in 2 with peritonitis divided by 18 total patients in the program that year = 11% had peritonitis during the year.

Intraperitoneal immunoglobulin appears to hasten recovery from peritonitis and deserves further study. When compared to antibiotic treatment alone, 320 mg of IgG given intraperitoneally in each exchange of CAPD resulted in a more rapid reduction in neutrophils in the effluent and a quicker resolution of pain [16]. This approach should be studied further, particularly in patients at high risk of peritonitis-related death, including older patients, those with longer PD duration and those with persistently elevated CRP levels [17].

Quality Improvement to Decrease Peritonitis Rates

Each program must monitor and periodically review peritonitis episodes to identify the problem areas (table 2). Rates should be examined for each category of organisms as well to investigate potential causes (since certain organisms cause peritonitis via certain pathways). For example, if the rate of CNS peritonitis is increasing, the program should look at enhancing their patient training since contamination is likely the cause. On the other hand, if *S. aureus* peritonitis is increasing, the program should examine the care of the exit site. If a few patients have numerous episodes, these patients deserve special attention. If the same organism is present repetitively, consideration should be given to simultaneous catheter replacement. With each episode of peritonitis, the dialysis team should perform a root cause analysis to determine the most likely cause. This will allow initiation of a plan to prevent further episodes. Initiatives undertaken by a program might include revision or strengthening of training, implementation of routine re-training in all patients, catheter care protocols, recognition and protocols of appropriate management of contamination (which means the patient needs to recognize such and call the dialysis program), and prophylaxis for high-risk procedures.

In conclusion, peritonitis remains one of the most serious problems facing the PD patient and PD health care worker. Reducing rates of peritonitis can be achieved by careful patient training, use of the best connection technology (in particular avoiding manual spiking), and use of exit site antibiotic cream on a daily basis to prevent exit site infections. As has been shown in Japan, peritonitis can be reduced to a very low rate of 0.13 episodes per year with careful attention [18].

References

1 Sipahiolglu MH, Aybal A, Unal A, Tokgoz B, Oymak O, Utas C: Patient and technique survival and factors affecting mortality on peritoneal dialysis in Turkey: 12 years' experience in a single center. Perit Dial Int 2008;28:238–245.
2 Fried LF, Bernardini J, Johnston JR, et al: Peritonitis influences mortality in peritoneal dialysis patients. J Am Soc Nephrol 1996;7:2176–2182.
3 Kavanagh D, Prescott GJ, Mactier RA: Peritoneal dialysis-associated peritonitis in Scotland (1999–2002). Nephrol Dial Transplant 2004;19:2584–2591.
4 Lye WC, Leong SO, van der Straaten J, et al: *Staphylococcus aureus* CAPD-related infections are associated with nasal carriage. Adv Perit Dial 1994;10:163–165.
5 Wang Q, Bernardini J, Piraino B, Fried L: Albumin at the start of peritoneal dialysis predicts the development of peritonitis. Am J Kid Dis 2003;41:664–669.
6 Prasad N, Gupta A, Sharma RK, Sinha A, Kumar R: Impact of nutritional status on peritonitis in CAPD patients. Perit Dial Intern 2007;27:42–47.
7 Troidle L, Watnick S, Wuerth DB, Gorban-Brennan N, Kliger AS, Finkelstein FO: Depression and its association with peritonitis in long-term peritoneal dialysis patients. Am J Kidney Dis 2003;42:350–354.
8 Rabindranath KS, Adams J, Ali TZ, MacLeod AM, Vale L, Cody JD, Wallace SA, Daly C: Continuous ambulatory peritoneal dialysis versus automated peritoneal dialysis for end-stage renal disease. Cochrane Database System Rev 2007;issue 2:CD006515.
9 Russo R, Manill L, Tiraboschi G, Amar K, De Luca M, Alberghini E, Ghiringhelli P, De Vecchi A, Porri MT, Marinangeli G, Rocca R, Paris V, Ballerini L: Patient re-training in peritoneal dialysis: why and when it is needed. Kidney Int Suppl 2006;103:S127–S132.
10 Piraino B, Bernardini J, Bender FH: An analysis of methods to prevent peritoneal dialysis catheter infections. Perit Dial Int 2008;28:437–443.
11 Bernardini J, Bender F, Florio T, Sloand J, PalmMontalbano L, Fried L, Piraino B: Randomized, double-blind trial of antibiotic exit site cream for prevention of exit site infection in peritoneal dialysis patients. J Am Soc Nephrol 2005;16:539–545.
12 Chuang YW, Shu KH, Yu TM, Cheng CH, Chen CH: Hypokalemia: an independent risk factor of *Enterobacteriaceae* peritonitis in CAPD patients. Nephrol Dial Transplant 2009;24:1603–1608.
13 Piraino B, Bailie GR, Bernardini J, Boeschoten E, Gupta A, Holmes C, Kuijper EJ, Li PKT, Lye WC, Mujais S, Paterson DL, Perez Fontan M, Ramos A, Schaefer F, Uttley L: ISPD Guidelines/Recommendations Peritoneal Dialysis-Related Infections Recommendations: 2005 Update. Perit Dial Int 2005;25:107–131.

14 Strippoli GF, Tong A, Johnson D, Schena FP, Craig JC: Antimicrobial agents for preventing peritonitis in peritoneal dialysis patients. Cochrane Database System Rev 2004;4:CD004679.
15 Lo WK, Chan CY, Cheng SW, et al: A prospective randomized control study of oral nystatin prophylaxis for *Candida* peritonitis complicating continuous ambulatory peritoneal dialysis. Am J Kidney Dis 1996;28:549–552.
16 Coban E, Ozdogan M, Tuncer M, Bozcuk H, Ersoy F: The value of low-dose intraperitoneal immunoglobulin administration in the treatment of peritoneal dialysis-related peritonitis. J Nephrol 2004;17:427–430.
17 Moon SJ, Han SH, Kim DK, Lee JE, Kim BS, Kang SW, Choi KH, Choi KH, Lee HY, Han DS: Risk factors for adverse outcomes after peritonitis-related technique failure. Perit Dial Int 2008;28:352–360.
18 Kawaguchi Y: Various obstacles to peritoneal dialysis development in Japan: Too much money? Too much fear? Perit Dial Int 2007;27(suppl 2):S56–S58.

Beth Piraino, MD
Renal Electrolyte Division of the University of Pittsburgh School of Medicine
Suite 200, 3504 Fifth Avenue
Pittsburgh, PA 15213 (USA)
Tel. +1 412 383 4899, Fax +1 412 383 4898, E-Mail Piraino@pitt.edu

Ronco C, Crepaldi C, Cruz DN (eds): Peritoneal Dialysis – From Basic Concepts to Clinical Excellence. Contrib Nephrol. Basel, Karger, 2009, vol 163, pp 169–176

New Treatment Options and Protocols for Peritoneal Dialysis-Related Peritonitis

Seth B. Furgeson, Isaac Teitelbaum

Division of Kidney Diseases and Hypertension, University of Colorado Denver, Aurora, Colo., USA

Abstract

Peritonitis remains a major complication in patients undergoing peritoneal dialysis. The most recent ISPD guidelines for the empiric initial treatment of peritonitis recommend the use of antibiotics that provide coverage against Gram-positive organisms (vancomycin or cefazolin) and Gram-negative organisms (a third-generation cephalosporin or an aminoglycoside). However, there are some situations in which this regimen may not be desirable. Concerns of resistant organisms, changing microbiology, drug toxicity, or difficulties administering therapy may lead a provider to modify the initial regimen. Drug resistant *Staphylococcus aureus* strains and *Enterococcus* strains may require administration of newer agents such as linezolid, quinipristin/dalfopristin, or daptomycin. Many centers have reported that, over time, the microbiology at those institutions has been changing. Some centers have reported a significant decrease in gram positive organisms and increase in extended spectrum beta-lactamase (ESBL) organisms. It is important for each center to examine its microbiology to document such trends. Although the currently recommended therapies have low toxicities, it is possible that concerns for untoward side effects in an individual patient may dictate changing the regimen. Finally, there is evidence from many prospective studies that monotherapy with different agents (oral quinolones or cefepime) is efficacious; if ease of therapy is a consideration, these may also be appropriate agents.

As is true of hemodialysis, infectious complications are a major cause of morbidity for patients on peritoneal dialysis. A significant number of patients switch from peritoneal dialysis to hemodialysis due to infections. In fact, using four cohorts of PD patients in the United States, a recent study showed that infections were the most common, single reason for a modality change to hemodialysis [1]. Importantly, peritonitis also increases mortality risk in patients on peritoneal dialysis. Multiple studies in different patient populations have confirmed this association. Appropriate, prompt treatment of peritonitis is, therefore, paramount.

Historical Context of Antibiotic Regimens for Peritonitis

Once an organism has been identified and sensitivities are known, treatment will be guided by these results. Therefore, this review will focus on the initial selection of empiric therapy for the treatment of peritonitis. The International Society for Peritoneal Dialysis (ISPD) has convened many times to establish recommendations for empiric treatment of peritonitis. In 1993, the ISPD recommended empiric vancomycin and gentamicin. Due mainly to concerns regarding the emergence of vancomycin resistance, later guidelines suggested cefazolin (or equivalent) with gentamicin (1996) or ceftazidime (2000). Gentamicin was not favored in the latter recommendations because of the possibilities of nephrotoxicity and ototoxicity. Most recently, in 2005, the ISPD encouraged flexibility with empiric antibiotic therapy [2]. Empiric Gram-positive therapy may be provided by either a cephalosporin or vancomycin and Gram-negative coverage may be provided by a third generation cephalosporin or an aminoglycoside.

Potential Reasons for Changing the Empiric Treatment of Peritonitis

Although the above recommendations will lead to effective cures in the majority of patients, there are circumstances in which one may wish to consider other regimens (table 1). Certainly, if there is concern about resistant organisms, initial antibiotics may need to be altered. If the microbiology for causative agents of peritonitis or their antimicrobial susceptibility pattern is changing in an individual institution, the initial empiric antibiotic regimen may need to be changed. Concerns about toxicity of current empiric therapies may warrant different treatments. Finally, as always, consideration should be given to ease of therapy and patient-specific characteristics.

Emergence of Resistant Bacteria

There are documented strains of *S. aureus* with intermediate susceptibility to vancomycin (VISA). VISA strains are defined by a minimal inhibitory concentration (MIC) of 8–16 μg/ml of vancomycin. The first case of VISA peritonitis was reported in 1999 [3]. Though there have been no further case reports

Table 1. Potential reasons to change the regimen for empiric therapy

Concern regarding the emergence of resistant organisms with the potential for diffuse, perhaps even global, dissemination
Alterations in the microbiology of PD-related peritonitis
Alterations in susceptibility patterns to antibacterial agents
Toxicity of the existing regimen
Ease of administration

of VISA peritonitis to date, the widespread use of vancomycin certainly raises the risk of further cases. Vancomycin resistant *S. aureus* species (MIC >32 μg/ml) have been described but fortunately no cases of VRSA peritonitis have been reported. Vancomycin-resistant enterococcal (VRE) infections, however, are an increasingly common cause of peritonitis. For over 15 years, there have been many documented cases of VRE peritonitis worldwide.

The last few years have seen the development and approval of several antibiotics with activity against otherwise resistant *S. aureus* and *Enterococcus* species. These include quinupristin/dalfopristin, linezolid, and daptomycin. Though data on the use of these agents for treatment of peritonitis is limited, a brief review of their pharmacology and antimicrobial spectra is in order.

Quinipristin/dalfopristin is a streptogramin antibiotic which is FDA approved for serious or life-threatening VRE infections. Many *S. aureus* isolates are resistant to quinipristin/dalfopristin as are all *Enterococcus faecalis* strains. Quinipristin/dalfopristin has poor penetration into the peritoneal space when given intravenously. Therefore, for treatment of peritonitis it should be given intraperitoneally. Quinipristin/dalfopristin has been used successfully to treat VRE peritonitis in both pediatric and adult patients [4].

Linezolid has also been used successfully in peritonitis. Linezolid is an oxazolidinone antibiotic which received FDA approval in 2000. It has activity against most strains of VRE and MRSA and has virtually 100% bioavailability when administered orally. Although the published data are sparse, there is evidence of good peritoneal penetration of linezolid. In 2001, Bailey et al. [5] reported successful treatment of VRE peritonitis with linezolid. A subsequent paper demonstrated adequate linzeolid levels using a dose of 600 mg twice daily in a patient with VRE peritonitis [6]. When using linezolid, consideration should be given to toxicities. Linezolid is similar to MAO inhibitors and should not be used with tricylic antidepressants; it should be used carefully with selective serotonin reuptake inhibitors (SSRI) or sympathomimetic drugs.

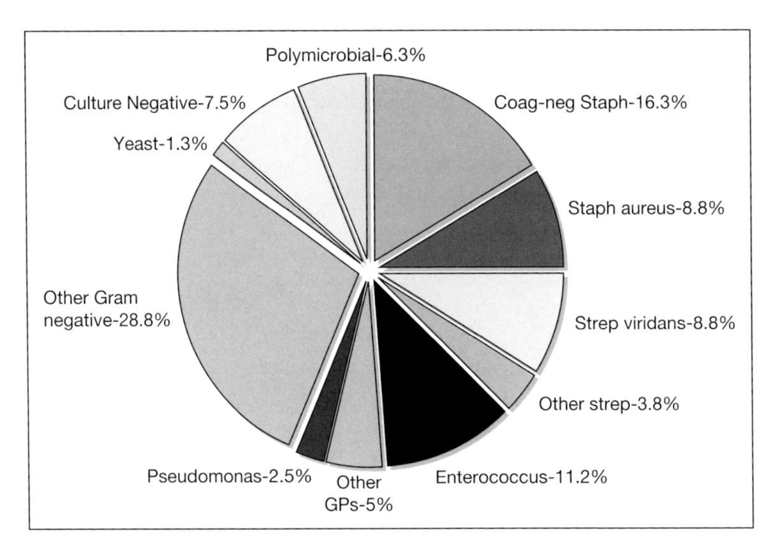

Fig. 1. Distribution of causative organisms for PD-related peritonitis at UCH, 2003–2007.

The most recently approved antibiotic for resistant Gram-positive infections is daptomycin. It is the first member of the cyclic lipopeptide class and received FDA approval in 2003. Until recently, there were no reports of daptomycin resistance; however, two recent articles have documented the existence of daptomycin-resistant Staph aureus [7, 8]. A recent publication described the successful use of daptomycin for a case of non PD-associated peritonitis. Daptomycin was given intravenously resulting in nadir peritoneal concentrations of 5 mg/l (MIC 4 mg/l) [9]. Given those findings, it seems preferable to administer daptomycin intraperitoneally rather than intravenously. However, given the lack of data using daptomycin in PD patients and, more importantly, concerns about developing resistance to a new class of antibiotics, daptomycin should not be considered a first line therapy for resistant *S. aureus* infections.

Alterations in Microbiology of Peritoneal Dialysis-Related Peritonitis

Many centers worldwide are reporting significant evolution in the microbiology of PD-related peritonitis. For example, Kim et al. [10] reported that the microbiology of peritonitis in a large PD center in Korea had changed. Over 10 years, there were fewer infections with Gram-positive organisms and more with Gram-negative organisms. The incidence of *S. aureus* infections had not changed but there were significantly fewer coagulase-negative *Staphylococcus* infections (although higher numbers of methicillin-resistant isolates). Similar results were reported by Perez-Fontan et al. [11] in Spain. The cause of this change in microbiology is not apparent. Changes in PD apparatus, such as the Y set, and improvements in sterile technique have likely contributed to this shift by decreasing the likelihood of infection with skin organisms.

Given the variability of microbiology from center to center, it is important for each provider to evaluate the microbiology of PD-related peritonitis at his or her institution(s). For example, at the University of Colorado Hospital, over the 5-year period of 2003- 2007, 7.5% of episodes were culture negative, 31.3% of causative organisms were Gram negative, and only 53.8% of the organisms were Gram positive (fig. 1). Likewise, providers should be familiar with the antimicrobial susceptibility patterns of the causative organisms at their institution(s) (see below). While the recommended ISPD empiric therapy should be effective if there is an increase in Gram-negative organisms, charting a center's microbiology can help illuminate other potential problems, such as an increase in yeast, which may lead a provider to modify therapy.

Alterations in Susceptibility Patterns to Antibacterial Agents

Arguably more important than changes in bacteriology are changes in the antibiotic susceptibility profile of organisms. At the University of Colorado Hospital, only 38% of coagulase negative staphylococci are sensitive to methicillin as are but 59% of *S. aureus* isolates. 27% of enterococci are resistant to vancomycin. At our institution, over 80% of gram negative organisms retain susceptibility to 3rd generation cephalosporins, such as ceftazidime. However, in some centers, infections with extended spectrum beta-lactamase (ESBL) organisms are increasing. Yip et al. [12] reported the incidence of ESBL *Escherichia coli* at one center in Hong Kong over the period of 1994 to 2003. There were no cases of ESBL *E. coli* peritonitis prior to 1996. This increased progressively over time, exceeding 20% in 2002 and 2003. Patients with ESBL *E. coli* were more likely to have received previous antibiotics or gastric acid inhibitors.

The increasing incidence of ESBL organisms has led to randomized trials examining the efficacy of different regimens. A study in Hong Kong compared imipenem/cilastatin monotherapy to cefazolin plus ceftazidime. Imipenem is structurally similar to β-lactams but is resistant to most β-lactamases. In this trial of 102 patients, imipenem/cilastatin proved to be a satisfactory alternative to cefazolin plus ceftazidime as there was no difference in response rates, cure rates, or catheter removal rates [13].

Toxicity of the Existing Regimen

One of the major concerns regarding the empiric use of aminoglycoside therapy has been the potential for nephrotoxicity. The risk of aminoglycoside toxicity in peritonitis treatment is not clearly defined. Previous observational studies have yielded conflicting results. A more recent prospective, randomized trial has evaluated the risk for aminoglycoside nephrotoxicity [14]. Investigators randomized patients with peritonitis to therapy with either cefazolin and ceftazidime or cefazolin and netilmicin. Netilmicin is an aminoglycoside antibiotic which is sometimes active against gentamicin-resistant Gram negatives and may have a lower incidence of ototoxicity than gentamicin. However, it also has less activity against *Pseudomonas auruginosa*. In this trial, there was no difference in cure rates between the two regimens. Importantly, there was also no difference in loss of urine volume or residual GFR. Based on the current literature, the ISPD sanctions short term use of aminoglycosides (≤3 days) in patients with residual renal function but discourages longer courses thereof unless there is no suitable alternative. In general, the risk of toxicity from current therapies seems to be low; however, in an individual patient, concerns about toxicity or side effects may make existing regimens unacceptable.

Ease of Administration

A final reason for which one might consider a change in the empiric therapy for peritonitis is the ease of administration. Monotherapy using intraperitoneal, intravenous, or oral antibiotics is an attractive option for many providers and patients. Many prospective trials have studied the efficacy of various treatments using a single antibiotic. For example, in 2001, Wong et al. [15] from Hong Kong published a study comparing cefepime monotherapy to vancomycin plus netilmicin. Eighty-one patients were randomized and there was no significant difference in cure rates or relapse rates. It should be noted, however, that the microbiological spectrum of peritonitis at this center was atypical as they had relatively high rates of both culture-negative and fungal peritonitis.

Oral monotherapy with quinolones has been another well-studied treatment. A recent meta-analysis evaluated studies using a quinolone as oral monotherapy [16]. Four trials had compared oral quinolone therapy to vancomycin/aminoglycoside combination therapy. In none of the trials was there a significant difference in treatment failure or any secondary outcome. There is, however, one important limitation to this analysis: all of the studied trials were published before 2000. If an institution has had any change in microbiology of peritonitis or change in sensitivity of organisms (i.e. quinolone resistance), using quinolones empirically without first ascertaining local sensitivity patterns would clearly carry risk.

Conclusion

Treatment of PD-related peritonitis can usually be given effectively and without significant adverse effects using protocols recommended by the ISPD. However, under certain circumstances, the practitioner may need or desire to deviate from established algorithms. Most importantly, emergence of resistant organisms and changing microbiology may lead some to alter initial empiric therapy of PD-related peritonitis. If emergence of resistant Gram-positive organisms (e.g. VRE or VISA) were to become a significant concern in an individual unit, strong consideration would have to be given to agents such as linezolid or daptomycin. While the current therapies recommended by the ISPD are generally well tolerated, it is important to realize that many different regimens have proven efficacious and alterations can be made if concerns of toxicity arise. Knowledge of the microbiology and antimicrobial susceptibilities at an individual institution is imperative and empiric therapies should be guided by this data. Finally, it should be emphasized that one needs to monitor efficacy of the chosen regimen often and should be prepared to change if either efficacy or toxicity becomes unacceptable.

References

1 Mujais S, Story K: Peritoneal dialysis in the US: evaluation of outcomes in contemporary cohorts. Kidney Int Suppl 2006;103:S21–S26.
2 Piraino B, Bailie GR, Bernardini J, Boeschoten E, Gupta A, Holmes C, Kuijper EJ, Li PK, Lye WC, Mujais S, Paterson DL, Fontan MP, Ramos A, Schaefer F, Uttley L, ISPD Ad Hoc Advisory Committee: Peritoneal dialysis-related infections recommendations: 2005 update. Perit Dial Int 2005;25:107–131.
3 Smith TL, Pearson ML, Wilcox KR, Cruz C, Lancaster MV, Robinson-Dunn B, Tenover FC, Zervos MJ, Band JD, White E, Jarvis WR: Emergence of vancomycin resistance in *Staphylococcus aureus*: Glycopeptide-Intermediate *Staphylococcus aureus* Working Group. N Engl J Med 1999;340:493–501.
4 Lynn WA, Clutterbuck E, Want S, Markides V, Lacey S, Rogers TR, Cohen J: Treatment of CAPD-peritonitis due to glycopeptide-resistant *Enterococcus faecium* with quinupristin/dalfopristin. Lancet 1994;344:1025–1026.
5 Bailey EM, Faber MD, Nafziger DA: Linezolid for treatment of vancomycin-resistant enterococcal peritonitis. Am J Kidney Dis 2001;38:E20.
6 DePestel DD, Peloquin CA, Carver PL: Peritoneal dialysis fluid concentrations of linezolid in the treatment of vancomycin-resistant *Enterococcus faecium* peritonitis. Pharmacotherapy 2003; 23:1322–1326.
7 Kirby A, Mohandas K, Broughton C, Neal TJ, Smith GW, Pai P, Nistal de Paz C: In vivo development of heterogeneous glycopeptide-intermediate *Staphylococcus aureus* (hGISA), GISA and

daptomycin resistance in a patient with meticillin-resistant *S. aureus* endocarditis. J Med Microbiol 2009;58:376–380.
8 Cunha BA, Pherez FM: Daptomycin resistance and treatment failure following vancomycin for methicillin-resistant *Staphylococcus aureus* (MRSA) mitral valve acute bacterial endocarditis (ABE). Eur J Clin Microbiol Infect Dis 2009, Epub ahead of print.
9 Burklein D, Kirchhoff C, Ozimek A, Mutschler W, Heyn J, Heindl B, Traunmuller F, Joukhadar C, Rothenburger M: Analysis of plasma and peritoneal fluid concentrations of daptomycin in a patient with *Enterococcus faecium* peritonitis. Int J Antimicrob Agents 2008;32:369–371.
10 Kim DK, Yoo TH, Ryu DR, Xu ZG, Kim HJ, Choi KH, Lee HY, Han DS, Kang SW: Changes in causative organisms and their antimicrobial susceptibilities in CAPD peritonitis: a single center's experience over one decade. Perit Dial Int 2004;24:424–432.
11 Pérez Fontan M, Rodríguez-Carmona A, García-Naveiro R, Rosales M, Villaverde P, Valdés F: Peritonitis-related mortality in patients undergoing chronic peritoneal dialysis. Perit Dial Int 2005;25:274–284.
12 Yip T, Tse KC, Lam MF, Tang S, Li FK, Choy BY, Lui SL, Chan TM, Lai KN, Lo WK: Risk factors and outcomes of extended-spectrum beta-lactamase-producing *E. coli* peritonitis in CAPD patients. Perit Dial Int 2006;26:191–197.
13 Leung CB, Szeto CC, Chow KM, Kwan BC, Wang AY, Lui SF, Li PK: Cefazolin plus ceftazidime versus imipenem/cilastatin monotherapy for treatment of CAPD peritonitis: a randomized controlled trial. Perit Dial Int 2004;24:440–446.
14 Lui SL, Cheng SW, Ng F, Ng SY, Wan KM, Yip T, Tse KC, Lam MF, Lai KN, Lo WK: Cefazolin plus netilmicin versus cefazolin plus ceftazidime for treating CAPD peritonitis: effect on residual renal function. Kidney Int 2005;68:2375–2380.
15 Wong KM, Chan YH, Cheung CY, Chak WL, Choi KS, Leung SH, Leung J, Chau KF, Tsang DN, Li CS: Cefepime versus vancomycin plus netilmicin therapy for continuous ambulatory peritoneal dialysis-associated peritonitis. Am J Kidney Dis 2001;38:127–131.
16 Wiggins KJ, Johnson DW, Craig JC, Strippoli GF: Treatment of peritoneal dialysis-associated peritonitis: a systematic review of randomized controlled trials. Am J Kidney Dis 2007;50:967–988.

Isaac Teitelbaum, MD
University of Colorado Hospital, AIP
12605 E.16th Ave F774
Aurora, CO 80045 (USA)
Tel. +1 720 848 7601, Fax +1 720 848 3103, E-Mail Isaac.teitelbaum@ucdenver.edu

Ronco C, Crepaldi C, Cruz DN (eds): Peritoneal Dialysis – From Basic Concepts to Clinical Excellence. Contrib Nephrol. Basel, Karger, 2009, vol 163, pp 177–182

Lipid Disorders, Statins and the Peritoneal Membrane

Olof Heimbürger

Divisions of Renal Medicine, Department of Clinical Science, Intervention and Technology, Karolinska Institutet, Karolinska University Hospital, Stockholm, Sweden

Abstract

Lipid disturbances are common in patients with chronic renal failure and peritoneal dialysis (PD) patients have in general an even more atherogenic lipid profile. The pathogenesis of the lipid profile in PD patients is not well understood, but both the peritoneal protein loss and the glucose absorption from the dialysate may contributes to these alterations. Hydroxymethylglutaryl coenzyme A reductase inhibitors, more commonly known as statins, are effective to reduce low density cholesterol levels in PD patients, but so far no large studies has been performed concerning the effects on clinical outcome in PD patients. Except from the lipid lowering effects, statins also have pleiotropic effects that are independent of their lipid lowering effect. Several of these pleiotropic effects may inhibit pathogenetic pathways involved in the long-term structural and functional changes in the peritoneal membrane, which also is supported by limited data from animal studies. Further studies in PD patients are needed concerning the potentially beneficial effects of statins on the peritoneal membrane.

Lipid disorders are common in patients treated with peritoneal dialysis (PD) and statin treatment is often used to treat these alterations. However, statins also have other pleiotropic effects, that potentially might prevent some of the pathogenetic pathways behind the long-term changes in the function and structure of the peritoneal membrane.

Lipid Disorders in Peritoneal Dialysis Patients

Lipid disturbances are common in patients with chronic renal failure and PD patients have in general an even more atherogenic lipid profile with high total and low-density lipoprotein (LDL) cholesterol, high triglycerides (TG), low high-density lipoprotein (HDL) cholesterol, high apolipoprotein B (apoB), low apolipoprotein A1 and high lipoprotein(a) levels. Compared to hemodialysis (HD) patients, PD patients usually have higher LDL cholesterol, TG, apoB and lipoprotein(a) levels [1]. The pathogenesis of the lipid profile in PD patients is not well understood, but the peritoneal protein loss may at least partly be involved in the pathogenesis of the increased LDL levels, whereas the glucose absorption from the dialysate contributes to the hypertriglyceridemia [1].

The association between lipid disturbances and cardiovascular outcome in patients with chronic renal failure is also complicated by the so-called reverse epidemiology phenomenon, i.e. the well-known association between established risk factors in the general population, such as hypercholesterolemia, and obesity appear to be reversed in patients with advanced kidney disease [2]. One reason for this could be that different time profiles exist for different risk factors in the different populations as premature deaths in CKD patients preclude the impact of complications, which are more important for long-term mortality. Furthermore, the common occurrence of persistent inflammation and/or protein-energy wasting in advanced CKD seems to a large extent account for the seemingly paradoxical association between hypercholesterolemia and cardiovascular outcome in this patient group [3]. In summary, in spite of the reverse epidemiology between lipid levels and clinical outcome in patients with advanced renal failure, high lipid levels are still considered to be harmful and to contribute to atherosclerosis and cardiovascular disease in the longer perspective.

Use of Statins to Treat Lipid Disorders in Peritoneal Dialysis Patients

In the population without renal disease, there is strong evidence that hydroxymethylglutaryl coenzyme A reductase inhibitors, more commonly known as statins, reduces the progression of coronary atherosclerosis and reduces mortality from cardiovascular disease. No similar studies have been performed in PD patients, but a recent meta-analysis concluded that cardiovascular events were reduced with statins irrespective of stage of renal disease [4], but statins had no significant effect on total mortality. However, the so far only reported large-scale statin study in dialysis patients, the 4D study [5], in which 1,255 prevalent HD patients with diabetes were randomized to receive atorvastatin or placebo, showed no effect of statin treatment on the composite primary end point of cardiovascular death, nonfatal myocardial infarction, and stroke. However, there was a positive effect of atorvastatin on all cardiac events combined (which was a secondary end-point). In PD patients, statins treatment is effective to improve the lipid profile by reducing LDL cholesterol as well as apo B levels [1], and statins may theoretically have a higher impact on clinical outcome in PD patients, compared to HD patients, due to the more atherogenic lipid profile in PD patients.

Long-Term Changes in the Peritoneal Membrane

In patients treated with PD for 4 years or more, there is a tendency toward decreasing ultrafiltration and increasing small solute transport, and the risk of developing loss of ultrafiltration capacity (UFC) increases markedly with time on PD [6, 7]. In contrast, macromolecule transport (as assessed by protein clearances) has been reported to be stable or to decrease with time on PD.

Though there are several pathophysiologic mechanisms behind the permanent loss of UFC, increased transport of small solutes with rapid glucose absorption (likely due to neoangiogenesis of the peritoneal membrane) is the most common mechanism [7]. The rapid glucose absorption results in rapid loss of the osmotic driving force (glucose gradient) and, consequently, a rapid decline in ultrafiltration rate. However, detailed kinetic analyses of patients with UFC due to rapid diffusive transport also show that the remaining osmotic gradient cannot induce water flow as effectively as in patients with normal UFC, indicating a decreased osmotic conductance of the peritoneal membrane [8], which likely is due to interstitial changes.

Parallel with the changes in peritoneal transport, marked changes in peritoneal morphology have been reported in patients treated with long-term PD, including mesothelial denudation, marked thickening of the submesothelial compact zone with fibrosis, and vascular changes with neoangiogenesis, subendothelial hyalinization, and vasculopathy [6, 9]. In addition, accumulation of advanced glycation end products (AGEs) can be found in the peritoneum [6, 10]. Patients with membrane failure had higher submesothelial thickness and also a higher density of blood vessels, which correlated with the degree of fibrosis [9]. In a few patients that had been treated with PD for several years, progressive peritoneal fibrosis with development of encapsulating peritoneal sclerosis (EPS, previously called sclerosing encapsulating peritonitis) have also been reported.

The pathogenetic mechanisms behind these alterations are only partly understood. Devuyst [11] has suggested a model where the increased reactive carbonyls (due to uremia and the carbonyls in the PD fluid) will amplify the AGE formation in the peritoneal membrane. The carbonyls and AGEs will have several effects, including stimulation of transdifferentiation of mesothelial cells to undergo epithelial to mesenchymal transition (EMT) [12], and stimulation of peritoneal cells (particularly EMT derived fibroblasts) to produce vascular endothelial growth factor (VEGF). VEGF will stimulate neoangiogenesis and interact with endothelial cells to produce endothelial nitric oxide synthase (eNOS), which is markedly increased in long-term PD patients [13]. The correlation seen between submesothelial fibrosis and neoangiogenesis suggests that these two processes are related [9], and resident submesothelial myofibroblasts (originating from mesothelial cells undergoing EMT) may participate in the extracellular matrix accumulation and neoangiogenesis [12]. It is in this context of interest that uremia per se and the binding of VEGF to the extracellular matrix will induce the release of basic fibroblast growth factor (bFGF), which has fibrotic as well as angiogenetic effects [11]. Furthermore, several inflammatory cytokines may also stimulate neoangiogenesis and fibrosis.

Pleiotropic Effects of Statins and Potential Effects of Statins to Preserve the Peritoneal Membrane

Except from the lipid-lowering effects, statins also have pleiotropic effects that are independent of their lipid-lowering effect. Through the inhibition of the mevalonic acid pathway, statins decrease the synthesis of other isoprenoids (e.g. farnesyl pyrophosphate and geranyl geranyl pyrophosphate) which inhibits isoprenylation of Ras and Rho GTPases [14]. Also, statins will activate the P13/Akt pathway. The effects of statins on these pathways will result in improvement of many of the cellular dysfunctions found in atherosclerosis [15], such as decreased production of cytokines by inflammatory cells, increased expression of nitric oxide synthase, improved balance between TPA and PAI-1, decreased synthesis of extracellular matrix and inhibition of cellular proliferation triggered by growth factors and cell migration in response to inflammatory factors.

In this context, it is of interest that the pleiotropic effects of statins may prevent several of the pathogenetic pathways suggested to involved in the evolution the long-term changes in the peritoneal membrane structure and function. In particular, statins have been shown to have positive effects on the intraperitoneal fibrinolytic system by suppressing tissue factor expression and to increase the fibrinolytic activity in TNF-α in activated human mesothelial cells [16]. Intraperitoneal injections of statins have been shown to increase tPA mRNA expression as well as to reduce postoperative adhesions in an animal model using the ischemic button model [17]. Even more important, statins have been shown to prevent EMT in tubular epithelial cells in vitro [18].

Unfortunately, no human studies of the effects of statins on the long-term changes of the peritoneal membrane have been performed. However, there are some data from animal studies available. In a four week study in rats, daily intraperitoneal injection of hypertonic glucose-based PD solution resulted in decreased ultrafiltration capacity, increased glucose absorption, increased peritoneal thickness, and increased effluent levels of VEGF and TFG- [19]. Addition of atorvastatin to the drinking water prevented most of these alterations [19]. In another study in rats, atorvastatin in the drinking water partly inhibited the changes in peritoneal structure and function in an animal model of encapsulating peritoneal sclerosis induced by intraperitoneal injection of chlorhexidine gluconate [20].

In summary, several of the pathways involved in the long-term changes in the structure and function of the peritoneal membrane may be prevented by statins. This concept is supported by some animal data, but no clinical studies have so far been performed.

Further animal research as well as clinical studies are needed to evaluate the potential benefits of statins in this respect.

References

1 Prichard SS: Metabolic complications of peritoneal dialysis; in Daugirdas JT, Blake PG, Ing TS (eds): Handbook of Dialysis, ed 4. Philadelphia, Lippincott Williams & Wilkins, 2007, pp 446–452.

2 Kalantar-Zadeh K, Block G, Humphreys MH, Kopple JD: Reverse epidemiology of cardiovascular risk factors in maintenance dialysis patients. Kidney Int 2003;63:793–808.

3 Liu Y, Coresh J, Eustace JA, Longnecker JC, Jaar B, Fink NE, Tracy RP, Powe NR, Klag MJ: Association between cholesterol level and mortality in dialysis patients. Role of inflammation and malnutrition. JAMA 2004;291:451–459.

4 Strippoli GFM, Navaneethan SD, Johnson DW, Percovic V, Pellegrini F, Nicolucchi A, Craig JC: Effects of statins in patients with chronic renal disease: meta-analysis and meta-regression of randomized controlled trials. BMJ 2008;336:645–651.

5 Wanner C, Krane V, März W, Olschewski M, Mann J, Ruf G, Ritz E, German Diabetes and Dialysis Study Investigators: Atorvastatin in patients with type 2 diabetes mellitus undergoing hemodialysis. N Engl J Med 2005;353:238–248.

6 Heimbürger O: Peritoneal physiology; in Pereira BJG, Sayegh MH, Blake P (eds): Chronic Kidney Disease, Dialysis and Transplantation: a Companion to Brenner and Rector's the Kidney, ed 2. Philadelphia, Elsevier Saunders, 2005, pp 491–513.

7 Heimbürger O, Waniewski J, Werynski A, Tranæus A, Lindholm B: Peritoneal transport in CAPD patients with permanent loss of ultrafiltration capacity. Kidney Int 1990;38:495–506.

8 Waniewski J, Heimbürger O, Werynski A, Lindholm B: Osmotic conductance of the peritoneum in CAPD patients with permanent loss of ultrafiltration capacity. Perit Dial Int 1996;16:488–496.

9 Williams JD, Craig KD, Topley N, von Ruhland C, Fallon M, Newman GR, Mackenzie RK, Williams GT: Morphological changes in the peritoneal membrane of patients with renal disease. J Am Soc Nephrol 2002;13:470–479.

10 Nakayama M, Kawaguchi Y, Yamada K, Hasegawa T, Takazoe K, Katoh N, Hayakawa H, Osaka N, Yamamoto H, Ogawa A, Kubo H, Shigematsu T, Sakai O, Horiuchi S: Immunohistchemical detection of advanced glycosylation end products in the peritoneum and its possible pathophysiological role in CAPD. Kidney Int 1997;51:182–186.

11 Devuyst O: New insight in the molecular mechanisms regulating peritoneal permeability. Nephrol Dial Transplant 2002;17:548–551.

12 Aroeira LS, Aguilera A, Sánchez-Tomero JA, Bajo MA, del Peso G, Jiménez-Heffernan JA, Selgas R, López-Cabrera M: Epithelial to mesenchymal transition and peritoneal membrane failure in

peritoneal dialysis patients: pathologic significance and potential therapeutic interventions. J Am Soc Nephrol 2007;18:2004–2013.
13 Combet S, Miyata T, Moulin P, Pouthier D, Goffin E, Devuyst O: Vascular proliferation and enhanced expression of endothelial nitric oxide synthase in human peritoneum exposed to long-term peritoneal dialysis. J Am Soc Nephrol 2000;11:717–728.
14 Fried LF: Effects of HMG-CoA reductase inhibitors (statins) on progression of kidney disease. Kidney Int 2008;74:571–576.
15 Selwyn AP: Antiatherosclerosic effects of statins: LDLD versus non-LDL effects. Current atherosclerosis reports 2007;9:281–285.
16 Haslinger B KR, Toet KH, Kooistra T: Simvastatin suppresses tissue factor expression and increases fibrinolytic activity in tumor necrosis factor-alpha-activated human peritoneal mesothelial cells. Kidney Int 2003;63:2065–2074.
17 Aarons CB, Cohen PA, Gower A, Reed KL, Leeman SE, Stucchi AF, Becker JM: Statins (HMG-CoA reductase inhibitors) decrease postoperative adhesions by increasing peritoneal fibrinolytic activity. Ann Surg 2007;245:176–184.
18 Patel S, Mason RM, Suzuki J, Imaizumi A, Kamimura T, Zhang Z: Inhibitory effect of statins on renal epithelial-to-mesenchymal transition. Am J Nephrol 2006;26:381–387.
19 Duman S, Sen S, Sozmen E, Oreopoulos DG: Atorvastatin improves peritoneal sclerosis induced by hypertonic PD solution in rats. Int J Artif Organs 2005;28:170–176.
20 Sipahi S, Sezak M, Duman S, Ozkan S, Sen S, Ok E: Atorvastatin ameliorates morphological changes in encapsulated peritoneal sclerosis rat model. Nephrol Dial Transplant 2006;21(suppl 4): iv501.

Olof Heimbürger, MD, PhD
Department of Renal Medicine, K56, Karolinska University Hospital, Huddinge
SE–141 86 Stockholm (Sweden)
Tel. +46 8 5858 3978, Fax 46 8 711 47 42, E-Mail olof.heimburger@ki.se

Ronco C, Crepaldi C, Cruz DN (eds): Peritoneal Dialysis – From Basic Concepts to Clinical Excellence. Contrib Nephrol. Basel, Karger, 2009, vol 163, pp 183–197

Complications of the Peritoneal Access and Their Management

Rodrigo Peixoto Campos[a], *Domingos Candiota Chula*[a], *Miguel Carlos Riella*[a,b]

[a]Department of Medicine, Renal Division, Evangelic School of Medicine, Curitiba, and [b]Catholic University of Paraná, Paraná, Brazil

Abstract

Although peritoneal catheter insertion is relatively considered a minimal invasive procedure, it is associated with some complications. These complications are divided into mechanical (bleeding, visceral perforation, dialysate leaks, catheter dysfunction, hernia formation, cuff extrusion) and infectious (early peritonitis, surgical wound, tunnel and exit site infections). It is well recognized that the appearance of these complications can increase morbidity and the chance of peritoneal dialysis treatment failure. Independent of the insertion technique, the operator must be prepared to an immediate recognition and adequate management of complications. Pre-operative evaluation and identification of potential risk factors are essential to prevent them.

The introduction of a permanent and flexible catheter-converted peritoneal dialysis (PD) is a viable chronic treatment for patients with end-stage renal disease. However, the peritoneal access continues to be a barrier for an adequate PD treatment. It is well recognized that since the development of the original Tenckhoff catheter [1], many other models of catheters, made of different materials and with different designs, some associated with recently minimally invasive surgery techniques, have increased catheter and, consequently, PD treatment survival [2, 3]. In a way, the creation of PD access teams has shortened the delays for catheter insertion and even increased the number of patients on PD programs [4–6]. Unfortunately, complications associated with catheter insertion may occur with different degrees of incidence, depending on the technique employed and type of complication [7–24]. Prevention, early recognition, and appropriate management of these complications are important because of the associated patient morbidity and technique failure. PD catheter complications are increasing in relative importance as a cause of technique failure and account for approximately 20% of transfers to hemodialysis [25]. To better outcomes, independent of the operator (nephrologist, interventional radiologist or surgeon), it is obligatory to acquire knowledge of abdominal anatomy, to be experienced and to follow all steps of placement techniques and recognize patient risk factors for complications and catheter dysfunction for rapid detection and management. The present review discusses the most common complications which may occur on PD catheter insertion and their management. These complications may be divided into mechanical and infectious (table 1). The most common mechanical complications are bleeding, bowel perforation, catheter dysfunction by tip migration, obstruction or omental entrapment, dialysate leak, cuff extrusion and hernia formations. Infections related to the procedure are peritonitis, surgical wound, exit site and tunnel infections.

Bleeding

In different studies, the incidence of bleeding may vary from 1 to 23% when was used a surgical open approach [26–29]. However, these studies do not characterize its criteria for a bleeding event. In a retrospective study with 292 catheters inserted by open surgery [30], the authors defined major bleeding as a decrease in the hematocrit by greater than or equal to 3%, or the requirement for a blood transfusion or surgical intervention. An incidence rate of 2% was found. Almost all patients with bleeding had received anticoagulation or had coagulopathy. It is possible that the rates of hemorrhagic complications may be lower using minimally invasive techniques of insertion. Two randomized studies compared surgical versus laparoscopic insertion [16, 31]. One used classical laparoscopic material and the other a peritoneoscope. In both, hemorrhagic complications were not discussed. Again, in a 10-year experience of percutaneous fluoroscopic insertion, no episodes of bleeding were described [32]. We believe this lack of information results from the absence of major episodes of bleeding in these studies. Although the laparoscopic technique is less aggressive, in a report of 362 PD catheters placement by this technique, 22 (6%) patients experienced blood-tinged dialysate on irrigation [33].

A volumous bleeding may occur when the epigastric artery or its branches is inadvertently injured. This artery lies behind the rectus muscle and generally is not visualized by percutaneous techniques (trocar, fluoroscopy or peritoneoscopy). Recently, Maya et al. [34] reported the associated use of ultrasound to a

Table 1. Complications of the PD catheter insertion

Mechanical complications	– bleeding – visceral perforation (bowel, bladder) – leaks (surgical wound, exit site, abdominal wall, pleural, external genitalia) – flow dysfunction (intraluminal clots or fibrin, catheter wrapping by stool-filled bowel, omental wrapping, tip dislocation, tip entrapment in peritoneal pockets – cuff extrusion – hernias
Infectious complications	– early peritonitis – surgical wound infection – exit site or tunnel infection

fluoroscopic technique to visualize the epigastric artery and to avoid its injury. In this study, no perioperative bleeding occurred.

It is mandatory to perform coagulation studies to stop the use of warfarin or heparin and to correct any abnormalities before the procedure. Related to aspirin, a retrospective study of PD catheter insertion evaluated if its therapy augmented the risk of bleeding. Nevertheless, the incidence of bleeding events was not different from patients who did not use it [35]. It looks as if it is not necessary to stop aspirin before catheter insertion. Furthermore, it is our opinion that if patients have not been dialyzed recently, and they appear to be clinically uremic, one could consider prophylactic use of parenteral DDAVP prior to the procedure.

Hemorrhage can occur during the procedure or in the early postoperative phase. Any visible bleeding vessel must be cauterized to avoid posterior hematoma. If immediately after the infusion test the dialysate returns bloody, a washing-out of the cavity with 0.5–1.0 liters of cool dialysate solution should be initiated in the operating room. Generally, it becomes clearer after 2 or 3 liters of dialysate washing-out. If it does not occur, this patient must remain hospitalized and a hematocrit obtained. These patients should be monitored during the first 24 h postoperatively. For patients with hemodynamic instability, a decision about surgical intervention should be rapidly instituted. Blood transfusion must be administrated to maintain a hematocrit around 30% associated with DDAVP and/or cryoprecipitate use, as recommended for uremic bleeding [36]. In these patients with bloody dialysate, after irrigating the abdominal cavity in the operating room, they should be maintained on low-volume automated PD until the effluent becomes clear, and then daily irrigation must be instituted, without leaving the cavity dry. Gadallah et al. [33] demonstrated a restricted relationship between bloody dialysate and dry break-in period with catheter dysfunction and intra-abdominal adherences. Hemorrhage around the exit site or at the surgical wound usually resolves with compressive dressing.

Visceral Perforation

Bowel Perforation

Bowel perforation is an unusual but serious complication of PD catheter insertion. The operator performing this procedure should be well prepared in establishing the diagnosis promptly and treating the patient effectively. The risk of bowel perforation is reduced in a surgical open approach due direct visualization of viscerae. The diagnosis is immediately established by the return of fecal material in the trocar, needle or catheter and the release of foul-smelling gas, during a percutaneous procedure, independently if it was performed by trocar, peritoneoscopy or fluoroscopy. In a retrospective 12-year period 750 catheters were inserted by peritoneoscopic technique [37]. Only 6 patients (0.8%) experienced bowel perforation. Of these, 5 were diagnosed promptly during the procedure. Although the risk of bowel perforation after PD catheter insertion is low, this complication is associated with increased morbidity and mortality [38–40]. Patients with previous abdominal surgery have a greater risk of bowel perforation due to the formation of intra-abdominal adhesions. Nevertheless, in the report of Asif et al. [37], half of the patients had a history of previous abdominal surgery and of these only 2 (33%) had experienced bowel perforation. Another risk factor for bowel perforation is the presence of abdominal distension. That is why some recommend the use of enemas or laxatives on the evening before the procedure, but this is not mandatory [41]. To avoid perforation of the cecum, the catheter should be placed in the left side of the abdominal wall [42].

Some reports documented successful results of conservative treatment of bowel perforation in PD catheter insertion [43–46] and colonoscopic procedures [47–49]. A majority of these perforations are usually small and seal spontaneously. These perforations close within 24–48 h, most likely secondary to omental adherence [49, 50]. Initially, all patients must be maintained NPO and intravenous antibiotics and fluids administered. The antibiotic therapy should coverage Gram-positive, Gram-negative and anaerobic bacteria. On the second day, if clinically stable, oral fluids may be initiated and then slowly advanced to diet. Any signs of clinical deterioration as fever, vomiting or abdominal irritation, should prompt a surgical consult.

Bladder Perforation

Perforation of bladder during placement of PD catheter is a very rare complication. Few cases are reported in the literature [16, 51–56]. Generally, the diagnosis is made after the procedure. During installation of dialysate the patient complained of abdominal discomfort and urinary urgency. With the subsequent increase in input volume of the dialysate, there is a marked increase in urinary volume. Added to that, urinalysis shows positive for glucose. To prevent this complication, every patient must void his bladder completely just before the procedure. It is important to emphasize that diabetic patients and others with previous history of neurological diseases may have neurogenic bladder. In these patients, a Foley catheter should be considered before the procedure [41].

Dialysate Leak

A dialysate leak is defined as any loss of dialysate from the abdominal cavity, with exception of the loss from the PD catheter lumen, due to damage of peritoneal membrane integrity. Dialysate leaks should be classified as early or late, according to how long after the catheter insertion they occur. Late leaks occur 30 days after catheter insertion, usually within 2 years of initiation of continuous ambulatory peritoneal dialysis (CAPD). They are often related to a mechanical or surgical tear in the peritoneal membrane and not directly to the procedure of catheter insertion. These leaks present as internal leakage (pleural cavity, abdominal wall, external genitalia). The diagnosis of late leaks is usually confirmed by computed tomography after infusion of dialysis fluid containing radiocontrast material in the peritoneal cavity. Pleural leaks can also be confirmed by lung scintigraphy after injection of technetium-tagged macroaggregated albumin into the peritoneal space. Thoracocentesis and pleural fluid analysis may be helpful and almost diagnostic, revealing a transudate with high glucose concentration [57].

Differently from late leaks, early leaks are directly related to PD catheter placement and thought to be the result of inadequate healing of the peritoneum around the catheter. They present as a leak in the exit site or in the surgical wound immediately prior to CAPD initiation. In practice, the diagnosis is made clinically by the leakage of dialysis fluid at the exit site or surgical wound. When dialysate leakage occurs into subcutaneous tissues, it is sometimes occult and difficult to diagnose. The incidence reports of dialysate leaks may varies from less than 1–20% [16, 18, 19, 32, 56, 58–60]. This variation may be explained by the nonuniformity of definitions. In the majority of studies which were defined as early leaks, the incidence did not exceed 5%. Two reports with more than 10 years of experience in PD [57, 61] reported an incidence around 9%. Nevertheless, less than half were early leak and directly related to the procedure.

Before the procedure, it is important to recognize the potential risk factors for dialysate leak. These factors may be subdivided into three categories: those related to any weakness of the abdominal wall associated to elevated intra-abdominal pressure, those related to the technique of PD catheter insertion and those related to the way PD is initiated [57]. The first is found in patients with multiple surgeries, chronic ascites, multiple pregnancies, obesity, previous use of steroids, hypothyroidism, polycystic kidney disease and chronic lung disease. Related to the second factor, some studies demonstrated lower incidences (around 1%) of early leaks when the technique performed was open surgery [16, 18, 58]; however, there is a tendency to recognize that mini-invasive procedures such as percutaneous with or without fluoroscope and peritoneoscopic have a lower risk of leakage because of less damage to peritoneum, rectus muscle and aponeurosis. Independently of the technique, a paramedian incision [42] is recommended due to this area having higher resistance and supporting more intra-abdominal pressures than median incisions. In this manner, the risk of leakage is lower. To remove the third category of risk factors, the PD initiation should be delayed for 10–14 days to guarantee complete healing around the internal cuff. Initially, it must be started in the supine position with low volumes to avoid increasing intra-abdominal pressure.

When a patient presents with a dialysate leak, several treatment modalities may be advocated, including surgical repair, temporary transfer to hemodialysis, lower dialysate volumes, and PD with a cycler. In the case of early leaks, a 2-week period of resting should be instituted during which the patient, if in need of dialysis, is maintained on hemodialysis, low-volume supine CAPD or APD. If after this period the leakage resumes, surgical repair is essential and the site of the leak may be localized using computed tomography [62]. In patients with apparent leak at the exit site or surgical wound, the risk of tunnel infection and peritonitis is increased. There is no consensus about prophylactic antibiotic coverage in these patients; however, some recommend its administration [62, 63].

Catheter Dysfunction

Immediate Dysfunction

Catheter dysfunction may be evidenced during the procedure of PD catheter insertion. Just after the positioning of the catheter in the abdominal cavity, a test for flow function must be performed. In our institution, 1 liter of dialysate solution is infused and the outflow time must be equal to or less than 5 min. However, in practice the liquid must drain in a continuous fashion, and not by drops. If an outflow dysfunction is observed in the operating room, the catheter is probably located out of the true pelvis due to insertion in an incorrect location such as the omentum or among the viscerae, or because of the presence of adhesions or even by clots inside the catheter. Sometimes the bowels are so distended that it impedes the catheter positioning into the true pelvis. That is one reason why some patients must be prepared with laxatives before the procedure. If the procedure is executed through open mini-incision surgery or blind techniques (trocar, percutaneous), the operator cannot identify the real cause of poor outflow. In such cases, the use of fluoroscope, peritoneoscope or laparoscope during the procedure would assist the diagnosis and cause of dysfunction [64]. Although laparoscopic insertion is more expensive and laborious, this technique has some advantages to prevent immediate and late causes of poor outflow by selective prophylactic omentopexy to prevent omental entrapment, selective resection of epiploic appendices to prevent catheter obstruction, adhesiolysis to eliminate compartmentalization, and simultaneous repair of previously undiagnosed abdominal wall hernias [65]. After the first 1 liter, a second repositioning attempt with more infusion of dialysate solution can be performed. If the catheter maintains a poor outflow, the procedure must be stopped, the catheter must be removed and another date set to open surgery, peritoneoscopic or laparoscopic insertion. The majority of studies did not report unsuccessful attempts at insertion. In a report of fluoroscopic insertion, it reached a rate of 2.4% of 209 PD catheter placements [32]. Other study using fluoroscopic insertion demonstrated a 5% unsuccessful insertion rate [11]. A randomized trial comparing laparoscopic and open surgery insertions demonstrated that 4 of 25 laparoscopic placements had to be converted to open surgery due to technical difficulties [31].

Inflow dysfunction generally occurs due to a kink in the subcutaneous tunnel or by the presence of intraluminal clots. The first is correct by the creation of a new subcutaneous tunnel and the second by the aspiration of clots.

Late Dysfunction

We defined late dysfunction as a poor inflow or outflow at the moment of initiation of CAPD after the break-in period, or at any moment during the treatment. Depending on the technique used, the incidence may change from 0.5 to 20% [16, 18, 32, 56, 58, 64]. It seems that technique does not influence the appearance of late dysfunction. However, in a laparoscopic report when possible reasons for posterior poor flow were identified during the procedure and immediately corrected, these problems reduced to less than 1% [64]. The causes of late dysfunction are divided between intraluminal, as clots or fibrin, and extraluminal, such a stool-filled bowel wrapping the catheter (constipation), an omental wrapping, a catheter tip dislocation out of the true pelvis and a tip entrapment in peritoneal pockets conditional on adhesions [62].

Some clues may help identify the cause of dysfunction. When there is resistance to infusion of the dialysate fluid, the presence of fibrin or clots is suspected. The patient may also refer to the presence of fibrin in the dialysate bag. For patients with chronic constipation, a wrapping of the catheter by bowel should be suspected. A plain abdominal radiography may show catheter tip migration, but not the cause of migration. A recent study demonstrated no correlation between catheter tip localization out of the true pelvis by radiography and poor outflow [66]. The presence of adequate inflow with total absence of outflow suggests omental entrapment. Nevertheless, in the majority of cases the exact cause of outflow is only diagnosed by direct visualization. The recommendation is to follow a step-by-step approach [62].

Firstly, conservative or noninvasive approaches such as body position change, administration of laxatives and vigorous flushing with heparinized saline should be undertaken. If it does not restore flow, the instillation of fibrinolytic agents may be tried. The use of recombinant plasminogen tissue activator (rt-PA) showed to be effective in restoring catheters flow [67–69]. In a study with 29 cases of obstruction, a mixture of 8 mg of rt-PA in 10 ml of sterile water was injected into the catheter and allowed to dwell for 1 h. A patency was restored in 24 of 29 instances with no adverse effects [69]. A pediatric study using urokinase demonstrated a positive result in 50% of PD catheters [70]. If after these conservative maneuvers the catheter still demonstrates poor outflow, a fluoroscopic guidewire manipulation should be considered. This treatment is usually reserved for catheters with radiographic evidence of migration, although malfunctioning catheters that are properly positioned in the true pelvis may be entrapped in an adhesion and benefit from guidewire manipulation. The rate of initial success with fluoroscopic guidewire manipulation is in the range of 60–85%, but the improvement is often short lived and the patency rate after 1 month is less than 60% in most series [71–75]. Another possible way to restore catheter patency is to use an endoluminal cleaning brush. In a report, this method re-established catheter function in 80% of patients [76]. The manipulation of PD catheter migrations with a Fogarty catheter demonstrated good results and prevented patients from having to have more invasive procedures. Under fluoroscopy, a Fogarty catheter is introduced until the end of the PD catheter, it is then insufflated and the catheter manipulated by tugging movements until reaching the true pelvis. Gadallah et al. [77] reported a successful repositioning in 71% of migrated catheters and only one remigrated after 90 days. If under these circumstances the catheter still demonstrates poor flow, it is almost certain that the omentum is obstructing the catheter.

The actual incidence of catheter obstruction accompanying omentum extending into the deep pelvis is not known. Nevertheless, in our experience it may occur in up to 20% of cases. It is generally suspected after exclusion of other causes. When omental entrapment is suspected, a laparotomy with partial omentectomy and repositioning of the catheter in the true pelvis is an effective technique to solve this problem. However, open revision of the peritoneal cavity increases the risk of dialysis fluid leakage and the potential development of an incisional hernia. Furthermore, PD treatment needs to be stopped for at least 2 weeks and the patient transferred to hemodialysis. The use of laparoscopy to revise the cavity and if necessary to perform omentectomy or adhesiolysis can prevent these complications [78, 79].

The role of preventive omentectomy has been investigated. In a retrospective series of 300 inserted catheters by open surgery, partial omentectomy was performed in 38% of the cases. In a multiple regression model used to further analyze the data, omentectomy was the only factor having an independently beneficial effect on catheter survival (RR = 0.36; 95% CI = 0.22–0.6) [80]. Another case series with 60 PD insertions with prophylactic omentectomy demonstrated only 1 catheter (2%) with late dysfunction [81]. An alternative to omental resection is omentopexy, an omental tack-up procedure. Omentopexy is preferred to omentectomy because it can be performed in a fraction of the time and cost with an outcome that is equal or better [78]. Omentopexy should be performed selectively when it is recognized laparoscopically that the omentum extends into the pelvis juxtaposition of the peritoneal catheter tip [64, 82]. This technique was first described by Ogunc [83] in 1999. In a later report, this author observed no mechanical dysfunction in 44 PD insertions [84]. In a retrospective study with this technique, the rate of catheter failure was less than 1%, while in the group with conventional implantation the rate was around 15% [64]. In summary, the rate of success of the different treatment procedures for catheter dysfunction has not been evaluated in randomized trials. However, it is reasonable to begin with the noninvasive procedures before more drastic steps are employed. Furthermore, preventive measures should be established whenever possible.

Other Mechanical Complications

Cuff extrusion and hernia formation are not directly related to the insertion procedure of the PD catheters; however, these complications can be prevented if the technique is performed correctly and risk factors are avoided.

If the outer cuff is left closer to the exit site and repetitive catheter tractions occur, the cuff may extrude. In such cases, this can be prevented when attention to exit site location is advocated. Published guidelines for optimal peritoneal access have historically supported both downwardly and laterally directed tunnel segment and exit site configurations [41, 62]. A recent comparison of catheter types employing downward and lateral tunnel tract and exit site configurations produced equal outcomes for infectious and mechanical complications [85]. A catheter with a straight intercuff segment should never be bent more than to produce a laterally directed exit site. If a downward facing exit site is required, then a catheter with a preformed bend (swan-neck catheters) should be used [65]. Independently of catheter configuration, the skin must be incised at a point where the exit site lies 2–3 cm beyond the outer cuff. The extruded cuff should be shaved because the continuous contact with exit site skin can irritate and propitiate infection. Another risk for cuff extrusion is the presence of tunnel infection. Generally in this situation, the infection does not respond to antibiotics and the catheter must be removed or replaced.

The presence of hernias do not contraindicate catheter insertion or PD treatment, nevertheless it must be repaired surgically, especially if significant, otherwise the hernia will worsen because of the increased pressure on the abdominal wall created by the intraperitoneal dialysis fluid. The catheter insertion and hernia correction can be performed concomitantly. Incisional hernia through the catheter placement site is more frequent if the implantation of the catheter is made through the midline, due to the fragility of this location. The most frequently occurring hernias during PD treatment are incisional, umbilical, and inguinal. After repairing of these hernias, patients can be maintained on CAPD, using low volumes in the supine position, or transferred temporarily to APD or hemodialysis.

Infections

Peritoneal infections are the leading cause of PD treatment failure and transfer to hemodialysis [25]. The catheter implantation is associated with every source of catheter infection. There is no particular recommendation showing that one catheter has an advantage in reducing peritonitis over the standard Tenckhoff catheter [41, 42, 86]. Although there is a tendency to believe that minimally invasive procedures may reduce infections, only one randomized study comparing peritoneoscopic implantation versus open surgery demonstrated reduction on early peritonitis episodes [16]. If an infection is suspected, a swab from surgical wound or exit site secretion, or peritoneal fluid must be collected to carry out cultures and to identify the specific organism. The decision about which antibiotics should be started must be decided according to the experience and routine of the PD care team. However, empiric treatment should cover *Staphylococcus aureus* and it is rational to follow the PD-related infections recommendations from the International Society for Peritoneal Dialysis [86].

Prophylactic antibiotics should always be administrated before the procedure [86]. Generally, a first-generation cephalosporin is used. A randomized study compared the role of preoperative administration of antibiotics in preventing early peritonitis. One group used 1 g of vancomycin, the other 1 g of cefazolin and the last one did not use antibiotics. The vancomycin group had less evidence of peritonitis than the others (1% compared to 7% in cefazolin and 12% in the group without antibiotics) [87]. Therefore, each program must consider using vancomycin for prophylaxis of catheter placement, carefully weighing the risk of its use in hastening resistant organisms. Postoperative care should consist of aseptic management of the exit site during the healing phase. A dressing should be applied aiming for immobilization of the catheter to avoid trauma and bleeding in the exit site. Sutures in the site are contraindicated because they increase the risk of infection. The dressing should not be changed more than once a week during the first 2 weeks, unless bleeding occurs or infection is suspected [42].

Conclusions

Every operator must be prepared to recognize an eventual complication during the insertion of a PD catheter, especially in patients with established risk factors. During the break-in period or even after the initiation of CAPD, late complications may occur and again the operator is directly responsible for managing these complications. In this manner, independently of the technique to be performed, every operative care must be followed. Firstly, preoperative evaluation should be carried out when possible. Secondly, information about the procedure and possible complications must be explained to the patient and his family. Thirdly, it is obligatory that the operator is familiar with the technique, indumentary and equipment in detail. Fourthly, the operator is sought to obtain knowledge on how to prevent and how to recognize a complication. Lastly, immediate management of the specific complication diagnosed. In this way, we can reduce morbidity and prolong catheter patency and PD treatment.

References

1 Tenckhoff H, Schechter H: A bacteriologically safe peritoneal access device. Trans Am Soc Artif Intern Organs 1968;14:181–187.
2 Ash SR: Chronic peritoneal dialysis catheters: overview of design, placement, and removal procedures. Semin Dial 2003;16:323–334.
3 Ash SR: Chronic peritoneal dialysis catheters: challenges and design solutions. Int J Artif Organs 2006;29:85–94.
4 Asif A, Byers P, Gadalean F, Roth D: Peritoneal dialysis underutilization: the impact of an interventional nephrology peritoneal dialysis access program. Semin Dial 2003;16:266–271.
5 Asif A, Pflederer TA, Vieira CF, Diego J, Roth D, Agarwal A: Does catheter insertion by nephrologists improve peritoneal dialysis utilization? A multicenter analysis. Semin Dial 2005;18:157–160.
6 Goh BL, Ganeshadeva YM, Chew SE, Dalimi MS: Does peritoneal dialysis catheter insertion by interventional nephrologists enhance peritoneal dialysis penetration? Semin Dial 2008;21:561–566.
7 Kelly J, McNamara K, May S: Peritoneoscopic peritoneal dialysis catheter insertion. Nephrology (Carlton) 2003;8:315–317.
8 Cheng YL, Chau KF, Choi KS, Wong FK, Cheng HM, Li CS: Peritoneal catheter-related complications: a comparison between hemodialysis and intermittent peritoneal dialysis in the break-in period. Adv Perit Dial 1996;12:231–234.
9 Garcia Falcon T, Rodriguez-Carmona A, Perez Fontan M, Fernandez Rivera C, Bouza P, Rodriguez Lozano I, Valdes F: Complications of permanent catheter implantation for peritoneal dialysis: incidence and risk factors. Adv Perit Dial 1994;10:206–209.
10 Ates K, Erturk S, Karatan O, Duman N, Nergisoglu G, Ayli D, Erbay B, Ertug AE: A comparison between percutaneous and surgical placement techniques of permanent peritoneal dialysis catheters. Nephron 1997;75:98–99.
11 Savader SJ, Geschwind JF, Lund GB, Scheel PJ: Percutaneous radiologic placement of peritoneal dialysis catheters: long-term results. J Vasc Interv Radiol 2000;11:965–970.
12 Roueff S, Pagniez D, Moranne O, Roumilhac D, Talaszka A, Le Monies De Sagazan H, Dequiedt P, Boulanger E: Simplified percutaneous placement of peritoneal dialysis catheters: comparison with surgical placement. Perit Dial Int 2002;22:267–269.
13 Rosenthal MA, Yang PS, Liu IL, Sim JJ, Kujubu DA, Rasgon SA, Yeoh HH, Abcar AC: Comparison of outcomes of peritoneal dialysis catheters placed by the fluoroscopically guided percutaneous method versus directly visualized surgical method. J Vasc Interv Radiol 2008;19:1202–1207.
14 Weber J, Mettang T, Hubel E, Kiefer T, Kuhlmann U: Survival of 138 surgically placed straight double-cuff Tenckhoff catheters in patients on continuous ambulatory peritoneal dialysis. Perit Dial Int 1993;13:224–227.
15 Asif A, Tawakol J, Khan T, Vieira CF, Byers P, Gadalean F, Hogan R, Merrill D, Roth D: Modification of the peritoneoscopic technique of peritoneal dialysis catheter insertion: experience of an interventional nephrology program. Semin Dial 2004;17:171–173.
16 Gadallah MF, Pervez A, el-Shahawy MA, Sorrells D, Zibari G, McDonald J, Work J: Peritoneoscopic versus surgical placement of peritoneal dialysis catheters: a prospective randomized study on outcome. Am J Kidney Dis 1999;33:118–122.
17 Pastan S, Gassensmith C, Manatunga AK, Copley JB, Smith EJ, Hamburger RJ: Prospective comparison of peritoneoscopic and surgical implantation of CAPD catheters. ASAIO Trans 1991;37:M154–M156.
18 Ozener C, Bihorac A, Akoglu E: Technical survival of CAPD catheters: comparison between percutaneous and conventional surgical placement techniques. Nephrol Dial Transplant 2001;16:1893–1899.
19 Mellotte GJ, Ho CA, Morgan SH, Bending MR, Eisinger AJ: Peritoneal dialysis catheters: a comparison between percutaneous and conventional surgical placement techniques. Nephrol Dial Transplant 1993;8:626–630.
20 Zappacosta AR, Perras ST, Closkey GM: Seldinger technique for Tenckhoff catheter placement. ASAIO Trans 1991;37:13–15.

21 Blessing WD Jr, Ross JM, Kennedy CI, Richardson WS: Laparoscopic-assisted peritoneal dialysis catheter placement, an improvement on the single trocar technique. Am Surg 2005;71:1042–1046.
22 Keshvari A, Najafi I, Jafari-Javid M, Yunesian M, Chaman R, Taromlou MN: Laparoscopic peritoneal dialysis catheter implantation using a Tenckhoff trocar under local anesthesia with nitrous oxide gas insufflation. Am J Surg 2009;197:8–13.
23 Sanderson MC, Swartzendruber DJ, Fenoglio ME, Moore JT, Haun WE: Surgical complications of continuous ambulatory peritoneal dialysis. Am J Surg 1990;160:561–565; discussion 565–566.
24 Dequidt C, Vijt D, Veys N, Van Biesen W: Bed-side blind insertion of peritoneal dialysis catheters. Edtna Erca J 2003;29:137–139.
25 Mujais S, Story K: Peritoneal dialysis in the US: evaluation of outcomes in contemporary cohorts. Kidney Int Suppl 2006:S21–S26.
26 Yeh TJ, Wei CF, Chin TW: Catheter-related complications of continuous ambulatory peritoneal dialysis. Eur J Surg 1992;158:277–279.
27 Rubin J, Didlake R, Raju S, Hsu H: A prospective randomized evaluation of chronic peritoneal catheters: insertion site and intraperitoneal segment. ASAIO Trans 1990;36:M497–M500.
28 Di Paolo N, Manganelli A, Strappaveccia F, De Mia M, Gaggiotti E: A new technique for insertion of the Tenckhoff peritoneal dialysis catheter. Nephron 1985;40:485–487.
29 Cronen PW, Moss JP, Simpson T, Rao M, Cowles L: Tenckhoff catheter placement: surgical aspects. Am Surg 1985;51:627–629.
30 Mital S, Fried LF, Piraino B: Bleeding complications associated with peritoneal dialysis catheter insertion. Perit Dial Int 2004;24:478–480.
31 Wright MJ, Bel'eed K, Johnson BF, Eadington DW, Sellars L, Farr MJ: Randomized prospective comparison of laparoscopic and open peritoneal dialysis catheter insertion. Perit Dial Int 1999;19:372–375.
32 Vaux EC, Torrie PH, Barker LC, Naik RB, Gibson MR: Percutaneous fluoroscopically guided placement of peritoneal dialysis catheters: a 10-year experience. Semin Dial 2008;21:459–465.
33 Gadallah MF, Torres-Rivera C, Ramdeen G, Myrick S, Habashi S, Andrews G: Relationship between intraperitoneal bleeding, adhesions, and peritoneal dialysis catheter failure: a method of prevention. Adv Perit Dial 2001;17:127–129.
34 Maya ID: Ultrasound/fluoroscopy-assisted placement of peritoneal dialysis catheters. Semin Dial 2007;20:611–615.
35 Shpitz B, Plotkin E, Spindel Z, Buklan G, Klein E, Bernheim J, Korzets Z: Should aspirin therapy be withheld before insertion and/or removal of a permanent peritoneal dialysis catheter? Am Surg 2002;68:762–764.
36 Hedges SJ, Dehoney SB, Hooper JS, Amanzadeh J, Busti AJ: Evidence-based treatment recommendations for uremic bleeding. Nat Clin Pract Nephrol 2007;3:138–153.
37 Asif A, Byers P, Vieira CF, Merrill D, Gadalean F, Bourgoignie JJ, Leclercq B, Roth D, Gadallah MF: Peritoneoscopic placement of peritoneal dialysis catheter and bowel perforation: experience of an interventional nephrology program. Am J Kidney Dis 2003;42:1270–1274.
38 Simkin EP, Wright FK: Perforating injuries of the bowel complicating peritoneal catheter insertion. Lancet 1968;i:64–66.
39 Wakeen MJ, Zimmerman SW, Bidwell D: Viscus perforation in peritoneal dialysis patients: diagnosis and outcome. Perit Dial Int 1994;14:371–377.
40 Krebs RA, Burtis BB: Bowel perforation: a complication of peritoneal dialysis using a permanent peritoneal cannula. JAMA 1966;198:486–487.
41 Flanigan M, Gokal R: Peritoneal catheters and exit-site practices toward optimum peritoneal access: a review of current developments. Perit Dial Int 2005;25:132–139.
42 Dombros N, Dratwa M, Feriani M, Gokal R, Heimburger O, Krediet R, Plum J, Rodrigues A, Selgas R, Struijk D, Verger C: European best practice guidelines for peritoneal dialysis. 3 Peritoneal access. Nephrol Dial Transplant 2005;20(suppl 9):ix8–ix12.
43 Ianhez LE, Chocair PR, Sergio L, de Azevedo F, Romao JE Jr, Sabbaga E: Conservative treatment of bowel perforation complicating chronic peritoneal dialysis. Report of six cases. AMB Rev Assoc Med Bras 1980;26:89–90.

44 Kahn SI, Garella S, Chazan JA: Nonsurgical treatment of intestinal perforation due to peritoneal dialysis. Surg Gynecol Obstet 1973;136:40–42.
45 Grzegorzewska A, Deja A: Conservative treatment of perforation of the transverse colon caused by a catheter for continuous peritoneal dialysis: a case report. Pol Arch Med Wewn 1989;81:368–372.
46 Rubin J, Oreopoulos DG, Lio TT, Mathews R, de Veber GA: Management of peritonitis and bowel perforation during chronic peritoneal dialysis. Nephron 1976;16:220–225.
47 Donckier V, Andre R: Treatment of colon endoscopic perforations. Acta Chir Belg 1993;93:60–62.
48 Berry MA, Rangraj M: Conservative treatment of recognized laparoscopic colonic injury. JSLS 1998;2:195–196.
49 Christie JP, Marrazzo J 3rd: 'Mini-perforation' of the colon: not all postpolypectomy perforations require laparotomy. Dis Colon Rectum 1991;34:132–135.
50 Damore LJ 2nd, Rantis PC, Vernava AM 3rd, Longo WE: Colonoscopic perforations: etiology, diagnosis, and management. Dis Colon Rectum 1996;39:1308–1314.
51 Vidaur F, Rentero R, Naranjo P, Torrente J, D'Ocon MT: Perforation of the bladder as a complication of peritoneal dialysis. Rev Clin Esp 1976;140:485–488.
52 Rall KL, Beagle GL: Inadvertent puncture of the urinary bladder by a peritoneal dialysis catheter. South Med J 1993;86:1398–1399.
53 Moreiras M, Cuina L, Rguez Goyanes G, Sobrado JA, Gil P: Inadvertent placement of a Tenckhoff catheter into the urinary bladder. Nephrol Dial Transplant 1997;12:818–820.
54 Bamberger MH, Sullivan B, Padberg FT Jr, Yudd M: Iatrogenic placement of a Tenckhoff catheter in the bladder of a diabetic patient after penectomy. J Urol 1993;150:1238–1240.
55 Ekart R, Horvat M, Hojs R, Pecovnik-Balon B: An accident with Tenckhoff catheter placement: urinary bladder perforation. Nephrol Dial Transplant 2006;21:1738–1739.
56 Allon M, Soucie JM, Macon EJ: Complications with permanent peritoneal dialysis catheters: experience with 154 percutaneously placed catheters. Nephron 1988;48:8–11.
57 Leblanc M, Ouimet D, Pichette V: Dialysate leaks in peritoneal dialysis. Semin Dial 2001;14:50–54.
58 Eklund BH: Surgical implantation of CAPD catheters: presentation of midline incision-lateral placement method and a review of 110 procedures. Nephrol Dial Transplant 1995;10:386–390.
59 Caramori JCT, Lopes AA, Bartoli LD, Redondo AP, Kawano PR, Fellipe MJDB, Barretti P: Sobrevida de 172 cateteres de Tenckhoff implantados cirurgicamente para diálise peritoneal crônica. J Brasileiro Nefrol 1997;19:11–15.
60 Ash SR, Alan EA, Bloch R: Peritoneoscopic placement of the Tenckhoff catheter: further clinical experience. Perit Dial Bull 1983;3:8–12.
61 Tzamaloukas AH, Gibel LJ, Eisenberg B, Goldman RS, Kanig SP, Zager PG, Elledge L, Wood B, Simon D: Early and late peritoneal dialysate leaks in patients on CAPD. Adv Perit Dial 1990;6:64–71.
62 Gokal R, Alexander S, Ash S, Chen TW, Danielson A, Holmes C, Joffe P, Moncrief J, Nichols K, Piraino B, Prowant B, Slingeneyer A, Stegmayr B, Twardowski Z, Vas S: Peritoneal catheters and exit-site practices toward optimum peritoneal access: 1998 update. Perit Dial Int 1998;18:11–33.
63 Holley JL, Bernardini J, Piraino B: Characteristics and outcome of peritoneal dialysate leaks and associated infections. Adv Perit Dial 1993;9:240–243.
64 Crabtree JH, Fishman A: A laparoscopic method for optimal peritoneal dialysis access. Am Surg 2005;71:135–143.
65 Crabtree JH: Selected best demonstrated practices in peritoneal dialysis access. Kidney Int Suppl 2006:S27–S37.
66 Palomar R, Morales P, Dominguez-Diez A, Martin L, de Francisco AL, Arias M: The position of the peritoneal dialysis catheter is not essential for a correct performance. Clin Nephrol 2008;70:554–557.
67 Sahani MM, Mukhtar KN, Boorgu R, Leehey DJ, Popli S, Ing TS: Tissue plasminogen activator can effectively declot peritoneal dialysis catheters. Am J Kidney Dis 2000;36:675.
68 Shea M, Hmiel SP, Beck AM: Use of tissue plasminogen activator for thrombolysis in occluded peritoneal dialysis catheters in children. Adv Perit Dial 2001;17:249–252.

69 Zorzanello MM, Fleming WJ, Prowant BE: Use of tissue plasminogen activator in peritoneal dialysis catheters: a literature review and one center's experience. Nephrol Nurs J 2004;31:534–537.
70 Stadermann MB, Rusthoven E, van de Kar NC, Hendriksen A, Monnens LA, Schroder CH: Local fibrinolytic therapy with urokinase for peritoneal dialysis catheter obstruction in children. Perit Dial Int 2002;22:84–86.
71 Jones B, McLaughlin K, Mactier RA, Porteous C: Tenckhoff catheter salvage by closed stiff-wire manipulation without fluoroscopic control. Perit Dial Int 1998;18:415–418.
72 Moss JS, Minda SA, Newman GE, Dunnick NR, Vernon WB, Schwab SJ: Malpositioned peritoneal dialysis catheters: a critical reappraisal of correction by stiff-wire manipulation. Am J Kidney Dis 1990;15:305–308.
73 McLaughlin K, Jardine AG: Closed stiff-wire manipulation of malpositioned Tenckhoff catheters offers a safe and effective way of prolonging peritoneal dialysis. Int J Artif Organs 2000;23:219–220.
74 Kappel JE, Ferguson GM, Kudel RM, Kudel TA, Lawlor BJ, Pylypchuk GB: Stiff wire manipulation of peritoneal dialysis catheters. Adv Perit Dial 1995;11:202–207.
75 Plaza MM, Rivas MC, Dominguez-Viguera L: Fluoroscopic manipulation is also useful for malfunctioning swan-neck peritoneal catheters. Perit Dial Int 2001;21:193–196.
76 Kumwenda MJ, Wright FK: The use of a channel-cleaning brush for malfunctioning Tenckhoff catheters. Nephrol Dial Transplant 1999;14:1254–1257.
77 Gadallah MF, Arora N, Arumugam R, Moles K: Role of Fogarty catheter manipulation in management of migrated, nonfunctional peritoneal dialysis catheters. Am J Kidney Dis 2000;35:301–305.
78 Crabtree JH, Fishman A: Laparoscopic omentectomy for peritoneal dialysis catheter flow obstruction: a case report and review of the literature. Surg Laparosc Endosc Percutan Tech 1999;9:228–233.
79 Lee M, Donovan JF: Laparoscopic omentectomy for salvage of peritoneal dialysis catheters. J Endourol 2002;16:241–244.
80 Nicholson ML, Burton PR, Donnelly PK, Veitch PS, Walls J: The role of omentectomy in continuous ambulatory peritoneal dialysis. Perit Dial Int 1991;11:330–332.
81 Reissman P, Lyass S, Shiloni E, Rivkind A, Berlatzky Y: Placement of a peritoneal dialysis catheter with routine omentectomy: does it prevent obstruction of the catheter? Eur J Surg 1998;164:703–707.
82 Crabtree JH, Fishman A: Selective performance of prophylactic omentopexy during laparoscopic implantation of peritoneal dialysis catheters. Surg Laparosc Endosc Percutan Tech 2003;13:180–184.
83 Ogunc G: A new laparoscopic technique for CAPD catheter placement. Perit Dial Int 1999;19:493–494.
84 Ogunc G: Minilaparoscopic extraperitoneal tunneling with omentopexy: a new technique for CAPD catheter placement. Perit Dial Int 2005;25:551–555.
85 Crabtree JH, Burchette RJ: Prospective comparison of downward and lateral peritoneal dialysis catheter tunnel-tract and exit-site directions. Perit Dial Int 2006;26:677–683.
86 Piraino B, Bailie GR, Bernardini J, Boeschoten E, Gupta A, Holmes C, Kuijper EJ, Li PK, Lye WC, Mujais S, Paterson DL, Fontan MP, Ramos A, Schaefer F, Uttley L: Peritoneal dialysis-related infections recommendations: 2005 update. Perit Dial Int 2005;25:107–131.
87 Gadallah MF, Ramdeen G, Mignone J, Patel D, Mitchell L, Tatro S: Role of preoperative antibiotic prophylaxis in preventing postoperative peritonitis in newly placed peritoneal dialysis catheters. Am J Kidney Dis 2000;36:1014–1019.

Miguel C. Riella, MD, PhD
Evangelic School of Medicine
Rua Bruno Filgueira 369
Curitiba 80240–220 (Brazil)
Tel. +55 41 3342 5849, Fax +55 41 3244 5539, E-Mail mcriella@pro-renal.org.br

Ronco C, Crepaldi C, Cruz DN (eds): Peritoneal Dialysis – From Basic Concepts to Clinical Excellence. Contrib Nephrol. Basel, Karger, 2009, vol 163, pp 198–205

Phosphate Balance in Peritoneal Dialysis Patients: Role of Ultrafiltration

Carlos Andres Granja, Peter Juergensen, Fredric O. Finkelstein

Hospital of St. Raphael, Yale University, Renal Research Institute, New Haven, Conn., USA

Abstract

Current National Kidney Foundation's Disease Outcome Quality Initiative (K/DOQI) clinical practice guidelines for bone metabolism and disease in chronic kidney disease (CKD) recommend maintenance of serum phosphorus levels below 5.5 mg/dl. About 40% of patients maintained on chronic peritoneal dialysis (CPD) have phosphate levels above 5.5 mg%. The present study was designed to examine the relative contribution of ultrafiltration to phosphate removal in CPD patients. 24-hour dialysate collections were obtained in 28 CPD patients and the diffuse and ultrafiltration (UF) contributions to phosphate removal determined. 11% of phosphate removal was accounted for by UF. There was a highly significant correlation between UF rate and the % of phosphate removed by UF. The results of this study underscore the importance of individualizing the peritoneal dialysis prescription.

Phosphate is essential for normal bone mineralization, and plays a critical role in a number of other biological processes such as signal transduction, nucleotide metabolism and enzyme regulation [1]. Consequently, the maintenance of appropriate phosphorus homeostasis is crucial for the well being of the organism.

Hyperphosphatemia is a common clinical finding in end-stage renal disease (ESRD) patients and has been associated with increased overall and cardiovascular mortality [2–4] and various hemodynamic abnormalities, including systolic dysfunction, left ventricular hypertrophy and elevated coronary artery calcification score [5, 6]. Importantly, in vitro studies with vascular smooth muscle cells exposed to elevated in vitro phosphate levels have demonstrated that an association of elevated phosphate levels with phenotypic changes predisposing to calcification, increased expression of osteogenic markers, and increased mineral deposition in vascular cells exists [7, 8].

Current National Kidney Foundation's Disease Outcome Quality Initiative (K/DOQI) clinical practice guidelines for bone metabolism and disease in chronic kidney disease (CKD) recommend maintenance of serum phosphorus levels below 5.5 mg/dl [9]. However, achieving this goal is a difficult task. About 40% of peritoneal dialysis (PD) patients in the NECOSAD database have phosphate levels to ≥5.5 mg% and amongst anuric CAPD patients in Hong Kong only 56% of their population attained serum phosphate levels below 5.5 mg/dl [10]. Phosphate removal with PD is limited. It has been suggested that phosphate removal with nocturnal cycling regimens is less than with standard CAPD [5, 11]. This is of particular interest since in the United States close to half of chronic PD patients are now maintained on cycler therapy [12].

The present study was undertaken to examine the role of ultrafiltration (UF) in contributing to phosphate removal with automated PD therapy. A cohort of 28 patients receiving automated PD therapy was examined, phosphate clearances measured, and the role of diffusion and ultrafiltration in phosphate removal estimated.

Patients and Methods

The study is a prospective cohort that included 28 ESRD patients, cared for in the New Haven CAPD unit (a free-standing facility located in an urban area), maintained on automated cycling PD therapy. The organization and structure of this unit has been described previously [17, 18].

Patients were eligible to participate if they were medically stable patients who had been on PD for at least 3 months and did not have acute medical problems in the preceding 12 weeks.

All patients received standard nocturnal CCPD during the night for 8–10 h and had a daytime dwell with either dextrose (1.5, 2.5 or 4.25%) or icodextrin-based solutions. The dialysis dose was adjusted per standard urea kinetics to maintain a Kt/V greater than 1.70 per week.

All patients were studied during their routine monthly visit to the CAPD unit. The patients' daily dialysis prescription and 24-hour total dialysate volume were noted and recorded. They were also instructed to bring a well-mixed aliquot of PD drainage to the dialysis unit. Phosphate concentrations in the serum and dialysate were measured using a phosphate fluorometric assay. Drainage volume was recorded by the cycler. The following calculations were made:

(1) The net UF volume was obtained after subtracting the daily dialysis prescription volume from the 24-hour drained dialysate:

net UF = 24-h drained dialysate – dialysis prescription.

(2) Since nearly 40–50% of UF during PD occurs through the aquaporin channels [13], nearly half of the UF volume is solute-free fluid. Consequently, the amount of UF that participates in actual PO_4 clearance is presumed to be 50% of the UF volume. This is called corrected UF for aquaporin transport:

corrected UF (for aquaporin transport) = net UF × 0.5.

(3) The total amount of phosphate drained from the PD fluid was determined by multiplying the 24-hour dialysate drain volume by the total PD fluid phosphate concentration:

total PO_4 drained = 24-hour dialysate drain volume × total PD fluid PO_4 concentration.

(4) The amount of phosphate that was removed by net UF was determined by multiplying the corrected UF volume times the serum phosphate level (the assumption was made that phosphate concentration associated with UF was equivalent to serum levels):

UF PO_4 = corrected UF volume × serum PO_4 level.

(5) Phosphate clearance was calculated by multiplying the phosphate concentration in the dialysate by the volume of dialysate and dividing by the serum phosphate concentration:

PO_4 clearance = (dialysate PO_4 concentration × 24-hour dialysate drain volume)/PO_4 serum concentration.

(6) Finally, the quantity of diffused phosphate that was removed was measured by subtracting the UF phosphate from the total amount of phosphate:

diffused PO_4 = total PO_4 drained – corrected UF PO_4.

Correlation analysis was performed and statistical analysis was carried out using the statistical program Stata™ 8.2 (StataCorp LP, Tex., USA). Results are expressed as mean ± SD.

The protocol conformed to the ethical guidelines of our institution and was approved by the Institutional Review Board.

Results

Demographic characteristics of the study population are shown in table 1; the mean age was 59.3 ± 12.5 years; 61.9% of the patients were males.

PD characteristics are summarized in table 2. The average 24-h PD prescription was 14,642.86 ± 3,019.86 ml. The mean Kt/V was 2.01 ± 0.40 and the average dialysate Kt/V was 1.68 ± 0.4. Sixty-five percent of the patients were high or high-average transporters.

The mean 24-hour UF for the 28 patients was 1,129.96 ± 1,118.33 ml. Mean total PO_4 removed by PD for the entire population was 303.81 ± 135.14 mg; that translates to a PO_4 clearance of 4.9 ± 1.7 liters/24 h/1.73 m^2.

Of the 303.81 ± 135.14 mg PO_4 removed with PD, a mean of 34.21 ± 25.13 mg was removed by UF and 269.61 ± 120.99 mg by diffusion.

Table 1. Demographic characteristics

Age, years	59.3±12.5
Males/females	17/11
Race, %	
Caucasian	72.5 (?)
Black	27.5 (?)
Diabetic patients, %	45 (?)
Nondiabetic patients, %	55 (?)

Data are presented as mean ± SD where appropriate.

Table 2. PD characteristics

PD prescription, ml	14,642.86±30,19.86
24-hour UF, ml	1,129.96±1,118.33
Corrected 24-hour UF for aquaporin transport, ml	564.98±318.54
Kt/V_{urea}	2.01±0.40
Dialysate Kt/V_{urea}	1.68±0.43
BUN clearance, liters/week/1.73 m^2	72.29±17.75
Creatinine clearance, liters/week/1.73 m^2	62.27±24.12

Data are presented as mean ± SD.

Low UF (less than 750 ml daily) has been associated with decreased survival in patients on PD [16]. The patients were then divided in two groups, a high UF (HUF) group (>750 ml/day) that included 18 patients (64.3%) and a low UF (LUF) group (<750 ml/day) with 10 patients (35.7%). Results from the HUF and LUF groups are summarized in table 3.

The mean serum PO_4 for the LUF group was 4.7 ± 1.3 mg/dl and for the HUF group was 6.1 ± 1.2 mg/dl ($p = 0.006$); this difference is a major determinant in the higher net PO_4 removed during PD in the HUF group, 361.8 ± 103.5 vs. 199.4 ± 115.6 mg observed in the LUF group ($p = 0.001$).

A mean of 88.97 ± 5.23% of PO_4 removed by PD in a 24-hour period was diffusive removal and only 11.03 ± 5.23% was cleared by UF. In the HUF group, diffusive removal represented 87.10 ± 5.31% compared to 92.36 ± 2.98% in the LUF group (0.007).

Table 3. Phosphate clearance results in low and high UF groups

	Total study group population (n = 28)	Low UF group (<750 ml) (n = 10)	High UF group (>750 ml) (n = 18)
Serum PO_4 level, mg/dl	5.62±1.38	4.70±1.26	6.13±1.18§
Total PD PO_4 removed, mg†	303.81±135.14	199.37±103.46	361.84±115.57§
UF PO_4 removed, mg†	34.21±25.13	13.10±4.53	45.93±24.1§
Diffusion PO_4 removed, mg†	269.61±120.99	186.26±102.46	315.91±106.47§
PO_4 clearance, liters/24 h/1.73 m^2†	4.85±1.74	4.20±1.80	5.21±1.65
UF PO_4 clearance, liters/24 h/ 1.73 m^2†	0.52±0.33	0.28±0.07	0.65±0.33§
Diffusion PO_4 clearance, liters/ 24 h/1.73 m^2†	4.33±1.65	3.92±1.81	4.56±1.56
% of PO_4 removed by UF†	11. 03±5.23	7.64±2.98	12.90±5.31§
% of PO_4 removed by diffusion†	88.97±5.23	92.36±2.98	87.10±5.31§

Data are presented as mean ± SD.
† 24-hour results.
§ $p < 0.05$ between low and high UF groups.

A total of 45.9 ± 24.1 mg of PO_4 was removed by UF in the HUF and 13.10 ± 4.53 mg in the LUF group ($p < 0.001$). In the HUF group, 12.90 ± 5.31% of PO_4 was removed by UF, as opposed to 7.64 ± 2.98% seen in the LUF group (0.007). In terms of clearances, mean UF PO_4 clearance of 0.65 ± 0.33 l/24 h/1.73 m^2 was observed in the HUF and 0.28 ± 0.07 l/24 h/1.73 m^2 in the LUF group ($p = 0.001$).

A strong correlation ($r = 0.810$) exists between UF rate and percent of phosphate removed by ultrafiltration, as shown in figure 1.

Discussion

Phosphate balance is problematic for many PD patients in whom the absorption of phosphate from the diet exceeds the capacity of elimination thought PD. Thus, 40–50% of patients maintained on PD have elevated serum phosphate levels [10]. These elevated levels are particularly important since over the last few years hyperphosphatemia has emerged as an independent risk factor for increased mortality in ESRD patients.

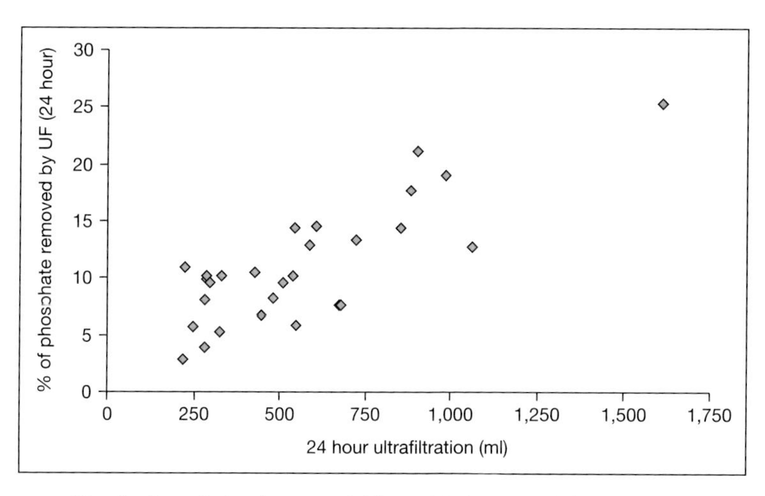

Fig. 1. Association between 24-hour UF (corrected for aquaporin transport) and percent of phosphate removed by UF. r = 0.809739831.

A recent, previous study by our group demonstrated the importance of UF for maintenance of calcium balance in PD patients [17]. Patients with lower UF rates were more likely to have a positive transperitoneal calcium balance (TCB) [17]. Maintenance of high UF helped to achieve a neutral or negative TCB. These findings underscored the importance of UF in maintaining calcium balance and we speculated on the possible links amongst UF rates, calcium balance, and mortality. Thus, the present study was designed to examine the role of UF in phosphate removal in PD patients.

The mean net PO_4 removed by PD during a 24-hour period was 303.81 ± 135.14 mg. Only 34.21 ± 25.13 mg was cleared by UF and 269.61 ± 120.99 by diffusion. There was a statistically significant difference in serum PO_4 in the HUF and LUF group (6.1 ± 1.2 mg/dl *vs.* 4.7 ± 1.3 mg/dl, respectively) that in large part accounted for the significant difference in net PD fluid removed observed between these 2 groups (361.8 ± 103.5 vs. 199.4 ± 115.6 mg, respectively), as noted in table 3. Patients with UF higher than 750 ml had statistically significant higher amounts of PO_4 removed by UF 45.9 ± 24.1 mg compared to 13.10 ± 4.53 mg in the group with UF less than 750 ml. Although this difference can be explained in part by the higher serum PO_4 levels seen in the HUF group, a higher percentage of PO_4 cleared by UF was seen in the HUF group, 12.90 ± 5.31%, compared to 7.64 ± 2.98% in the LUF group, suggesting that higher UF contributes significantly to phosphate removal. In fact, as shown in figure 1, there is a strong correlation between UF rates and the percent of phosphate removed by UF (r = 0.81).

Low UF (less than 750 ml daily) has been associated with decreased survival in patients on PD [16]. The mechanisms linked to this observation are not fully determined and may well be multifactorial. Poor nutrition, use of solutions with higher dextrose concentration, and inflammation have been postulated. The possible contribution of UF in maintaining calcium and phosphate balance needs to be considered.

The effect of increasing dialysis cycles and volume in continuous cycling PD was recently reported [18]. Increasing cycling volume from 15 to 24 liters results in only a 10% phosphate clearance; this increase in phosphate clearance translates into <50 mg net phosphate removal in 9 h, assuming a serum phosphate of 6 mg/dl. Another approach to increase PO_4 removal would be to increase UF in conjunction with increased PD volume.

The results from the present study demonstrate a significant role of UF in PO_4 removal reinforcing the importance of individualizing the PD prescription. Maintaining adequate UF rates may not only contribute to maintenance of fluid balance but also to maintenance of calcium and phosphate balance.

References

1 Berndt TJ, Schiavi S, Kumar R: 'Phosphatonins' and the regulation of phosphorus homeostasis. Am J Physiol Renal Physiol 2005;289:F1170–F1182.

2 Block GA, Hulbert-Shearon TE, Levin NW, Port FK; Association of serum phosphorus and calcium x phosphate product with mortality risk in chronic hemodialysis patients: a national study. Am J Kidney Dis 1998;31:607–617.

3 Kalantar-Zadeh K, Kuwae N, Regidor DL, Kovesdy CP, Kilpatrick RD, Shinaberger CS, McAllister CJ, Budoff MJ, Salusky IB, Kopple JD: Survival predictability of time-varying indicators of bone disease in maintenance hemodialysis patients. Kidney Int 2006;70:771–779.

4 Noordzij M, Korevaar JC, Bos WJ, Boeschoten EW, Dekker FW, Bossuyt PM, Krediet RT: Mineral metabolism and cardiovascular morbidity and mortality risk: peritoneal dialysis patients compared with haemodialysis patients: Nephrol Dial Transplant 2006;21:2513–2520.

5 Badve SV, McCormick BB: Phosphate balance on peritoneal dialysis. Perit Dial Int 2008;28:S25–S32.

6 Stompor TP, Pasowicz M, Sułowicz W, Dembińska-Kieć A, Janda K, Wójcik K, Tracz W, Zdzienicka A, Konieczyńska M, Klimeczek P, Janusz-Grzybowska E: Trends and dynamics of changes in calcification score over the 1-year observation period in patients on peritoneal dialysis. Am J Kidney Dis 2004;44:517–528.

7 Jono S, McKee MD, Murry CE, Shioi A, Nishizawa Y, Mori K, Morii H, Giachelli CM: Phosphate regulation of vascular smooth muscle cell calcification. Circ Res 2000;87:E10–E17.

8 Giachelli CM: Vascular calcification: in vitro evidence for the role of inorganic phosphate. J Am Soc Nephrol 2003;14:S300–S304.

9 K-DOQI Clinical Practice Guidelines for CKD: Evaluation, classification and stratification. Kidney Disease Outcome Quality Initiative. Am J Kidney Dis 2002;39(suppl 1):S1–S246.

10 Wang AY, Woo J, Sea MM, Law MC, Lui SF, Li PK: Hyperphosphatemia in Chinese peritoneal dialysis patients with and without residual renal function: What are the implications? Am J Kidney Dis 2004;43:712–720.

11 Twardowski ZJ Prowant BF, Nolph KD, Khanna R, Schmidt LM, Satalowich RJ: Chronic nightly tidal peritoneal dialysis. ASAIO Trans 1990;36:M584–M588.

12 United States Renal Data System, USRDS, 2007 Annual Data Report: Atlas of End Stage Renal Disease in the United States. Bethesda, National Institute of Health, National Institute of Diabetes and Digestive and Kidney Diseases, 2007.
13 Mujais S, Nolph K, Gokal R, Blake P, Burkart J, Coles G, Kawaguchi Y, Kawanishi H, Korbet S, Krediet R, Lindholm B, Oreopoulos D, Rippe B, Selgas R: Evaluation and management of ultrafiltration problems in peritoneal dialysis. International Society for Peritoneal Dialysis. Ad Hoc Committee on Ultrafiltration Management in Peritoneal Dialysis. Perit Dial Int 2000;20:S5–S21.
14 Graff J, Fugleberg S, Brahm J, Fogh-Andersen N: The transport of phosphate between the plasma and dialysate compartments in peritoneal dialysis is influenced by an electric potential difference. Clin Physiol 1996;16:291–300.
15 Hruska, KA, Mathew S, Lund R, Qiu P, Pratt R: Hyperphosphatemia of chronic kidney disease, Kidney Int 2008;74:148–157.
16 Davies SJ, Brown EA, Reigel W, Clutterbuck E, Heimbürger O, Diaz NV, Mellote GJ, Perez-Contreras J, Scanziani R, D'Auzac C, Kuypers D, Divino Filho JC, EAPOS Group: What is the link between poor ultrafiltration and increased mortality in anuric patients on automated peritoneal dialysis? Analysis of data from EAPOS. Perit Dial Int 2008;28(suppl 2):S42.
17 Granja CA, Francis J, Simon D, Bushinsky D, Finkelstein FO: Calcium balance with automated peritoneal dialysis. Perit Dial Int 2008;28(suppl 2):S38–S46.
18 Juergensen P, Eras J, McClure B, Kliger AS, Finkelstein FO: The impact of various cycling regimens on phosphorus removal in chronic peritoneal dialysis patients. Int J Artif Organs 2005;28:1219–1223.

Fredric O. Finkelstein
136 Sherman Avenue
New Haven, CT 06511 (USA)
Tel. +1 203 787 0117, Fax +1 203 777 3559, E-Mail fof@comcast.net

Ronco C, Crepaldi C, Cruz DN (eds): Peritoneal Dialysis – From Basic Concepts to Clinical Excellence. Contrib Nephrol. Basel, Karger, 2009, vol 163, pp 206–212

The Physiology of Vitamin D Receptor Activation

Jose M. Valdivielso

Laboratorio de Investigación Hospital Universitario Arnau de Vilanova, IRBLLEIDA, Lleida, Spain

Abstract

Vitamin D is a steroid hormone that has long been known for its important role in regulating body levels of calcium and phosphorus, and in mineralization of bone. In addition to its endocrine effects, vitamin D has important autocrine/paracrine roles. The last step in the activation of vitamin D, the hydroxylation on carbon 1, takes place mainly in the kidney. However, extrarenal sites showing 1alpha-hydroxylase activity have been also found. The hormonally active form of vitamin D (1,25(OH)-D_3 or calcitriol) mediates its biological effects by binding to the vitamin D receptor, which then translocates to the nuclei of the cell and binds to specific DNA sites to modify the expression of target genes. After activation of the receptor, the protein changes its tridimensional conformation, this change being the key process in order to exert its nuclear actions. Several steps take place in order to increase or decrease the transcription rate of a target gene. First, homodimerization of the vitamin D receptor or heterodimerization with the retinoic X receptor allows the complex to go into the nucleus and bind to the DNA. Then several proteins are recruited to the complex that either increase or decrease chromatin condensation acting then as corepresors or coactivators, respectively, and decreasing or increasing the target gene transcription. The coactivators bind several extra proteins that build a bridge to the basal transcription machinery. Therefore, little changes in the receptor's tridimensional change elicted by the activator can lead to differences in protein recruitment and, thus, in gene transactivation. Furthermore, differences in the cellular environment can yield different responses to the same activator. This characteristic of the nuclear receptors makes them a good candidate as a valuable therapeutic target.

Vitamin D Metabolism

Vitamin D_3 is a fat-soluble prehormone which plays an important role in many biological functions throughout the body. Most of the vitamin D_3 content of the human body is synthesized from the precursor molecule 7-dehydrocholesterol in the skin by the action of UV light. Then, vitamin D_3 is transported in blood bound to the vitamin D-binding protein (DBP), which carries vitamin D_3 to the liver and kidney for bioactivation. In the first activation step vitamin D_3 is hydroxylated by the enzyme 25-hydroxylase to 25-hydroxyvitamin D_3 ($25OHD_3$). This reaction occurs mainly in the liver and is poorly regulated, which is why the level of $25OHD_3$ in serum increases in proportion to vitamin D_3 intake. In the second step, the biologically active hormone 1,25-dihydroxyvitamin D_3 ($1,25(OH)_2D_3$) is generated by the enzyme 25-hydroxyvitamin D_3-1-α-hydroxylase (1α-hydroxylase) and it occurs mainly in the kidney. The 24-hydroxylation of $1,25\text{-}(OH)_2D_3$ is the first catabolic step in the elimination of active hormone leading to the formation of 1,24,25-trihydroxyvitamin D_3. Further oxidative reactions lead to the production of water-soluble calcitroic acid, which is excreted in urine.

Physiological Actions of $1,25(OH)_2D_3$

$1,25(OH)_2D_3$ regulates several functions in the body by modulating genomic events via its nuclear receptor. Classically, the main role of $1,25(OH)_2D_3$ is the regulation of calcium and phosphorous concentrations in serum via actions in bone, parathyroid gland, kidney and intestine. $1,25(OH)_2D_3$ is able to elevate serum calcium and phosphate levels. Therefore, the circulating concentration of $1,25(OH)_2D_3$ is under tight regulation by controlling its rates of synthesis and degradation. Major regulators of $1,25(OH)_2D_3$ concentration in serum are parathyroid hormone (PTH), calcium, phosphate and $1,25(OH)_2D_3$ itself. In addition, $1,25(OH)_2D_3$ is able to generate several other biological responses (nonclassical actions of vitamin D) that are not related to the control of mineral homeostasis (table 1).

Control of Mineral Homeostasis

The principal function of $1,25(OH)_2D_3$ and PTH is to control the calcium and phosphate status to ensure the availability of the minerals for biological functions as well as skeletal mineralization. This is achieved by coordinated actions of the parathyroid, kidney, intestine and bone.

Table 1. Clasical and nonclassical effects of vitamin D receptor activation on target cells

Tissue/cell	Action
Hematopoietic tissues	differentiation
Immune system	enhancement of immune function to control viral and bacterial infections and tumor growth
Monocyte/macrophages, lymphocyte	immunosuppression
Skin	antiproliferative, differentiation
Muscle	antiproliferative, differentiation
Smooth muscle cell	proproliferative
Myoblast, heart cardiac muscle cell and atrial myocytes	inhibition of antinatriuretic factor synthesis
Pancreas β cells	enhancement of insulin synthesis and secretion
Mammary gland	growth regulation
Cancer cells	antiproliferative, differentiation
Adrenal gland medullary cells	control of catecolamine metabolism
Prostate	antiproliferative, differentiation
Brain hippocampus/selected neurons	neuronal regeneration, enhancement of nerve growth factor and neurotrophin synthesis, control of sphingomyelin cycle
Cartilage chondrocyte	antiproliferative, differentiation
Female reproductiveovarian, myometrial and endometrial cells	antiproliferative, control of foliculogenesis organs
Liver parenchymal cell	enhancement of liver regeneration, control of glycogen and transferrin synthesis
Lung	enhancement of maturation, phospholipid synthesis and surfactant release
Male reproductive organs sertoli/ semminiferus tubule	enhancement of Sertoli cell function and spermatogenesis
Pituitary production	control of T_3-induced growth hormone, prolactin and tyrotropin
Thyroid	inhibition of calcitonin synthesis
Parathyroid	inhibition of parathormone synthesis

The most critical role of $1,25(OH)_2D_3$ in mineral homeostasis is to enhance the efficiency of the small intestine to absorb dietary calcium. To do so, $1,25(OH)_2D_3$ increases the entry of calcium through the plasma membrane into the enterocytes and enhances the movement of calcium through the cytoplasm and across the basolateral membrane into the circulation. Furthermore,

$1,25(OH)_2D_3$ is capable of reducing the excretion of calcium in urine by increasing the reabsorption of calcium in the kidneys. Moreover $1,25(OH)_2D_3$ is able to increase the mobilization of calcium from bone into the circulation through the enhancement of osteoclastogenesis and osteoclastic activity. $1,25(OH)_2D_3$ also has an essential role in bone development, mineralization and maintaining the dynamic nature of bone by controlling the availability of calcium and phosphate and regulating the level of PTH, PTH-related peptide and insulin-like growth factor. In addition, it is also involved in the synthesis of bone matrix proteins such as type I collagen, alkaline phosphatases, osteocalcin, osteopontin and matrix Gla protein [1].

New Functions of $1,25(OH)_2D_3$

In recent years, it has been shown that vitamin D has many additional target cells where it is involved in a wide array of new functions that are unrelated to its actions on mineral metabolism. For instance, vitamin D has been involved in the regulation of vascular smooth muscle cell proliferation and calcification [2, 3], in regulation of the renin-angiotensin system and, in general, in cardiovascular health [4]. It has been also shown that some vitamin D target cells have their own enzyme machinery for the local regulation of $1,25(OH)_2D_3$ concentration that enables a separate regulation of the endocrine function from the auto and paracrine role of vitamin D [5].

Nuclear Vitamin D Receptor

The genomic actions of $1,25(OH)_2D_3$ are mediated by its nuclear receptor the vitamin D receptor (VDR). The human VDR is a product of the single chromosomal gene which locates on chromosome 12. The gene is comprised of 11 exons and spans approximately 75 kb. In man, the VDR protein consists of 427 amino acids, with a molecular mass of ~48 kDa. VDR can be divided by function into several domains. In the amino-terminus there is an A/B domain 20 amino acids long. The DNA-binding domain (DBD), also termed C domain, locates between amino acids 21 and 92. The flexible linker region locates approximately between amino acids 93 and 123, followed by the E- or ligand-binding domain (LBD) between amino acids 124 and 427. VDR regulates transcription by binding to specific genomic sequences known as vitamin D response elements (VDRE). The DNA-binding domain mediates this vital interaction. The C-terminal LBD is a globular multifunctional domain. It is responsible for hormone binding, strong receptor dimerization and interaction with co-repressors and co-activators, which are critical for the regulation of transcriptional activities.

Molecular Mechanism of the Control of Transcription by VDR

Hormone binding to VDR initiates the series of events that leads to active repression or activation of target gene expression. Tens of proteins and several protein-protein and protein-DNA interactions participate in those complicated chain of events. The initial step in transactivation is the binding of VDR to their VDRE within the promoter region of the responsive gene. Ligand binding to the VDR starts the series of events that arrives at the releasing of nucleosomal repression and the initiation of gene transcription. Conformational change induced by ligands has an important role in this process. It promotes the tight association of receptor to its response element, enhances receptor dimerization and generates new surfaces on the receptor that allows the binding of coactivator molecules, which are essential factors in the gene activation cascade [6]. Coactivators do not directly bind to the DNA but are associated with the promoter region via a gene specific activator molecule like VDR. Firstly, coactivators remodel the chromatin structure of the promoter region in order to facilitate binding of other activators and the component of the RNA polymerase II transcriptional machinery. Secondly, coactivators recruit protein complexes that interact with one or more subunits of the RNA polymerase II and enhance the initiation of transcription by stabilizing the preinitiation complex [7].

These protein complexes act as a direct link between the ligand-activated receptor and RNA polymerase II holoenzyme complex and possibly recruit limiting components into the preinitiation complex. Multiple protein-protein interactions formed this way may enhance the stability of the complex and thereby facilitate the initiation of transcription [8].

In summary, in VDR-mediated gene activation various coactivator complexes with distinct activities enter and exit their target promoter in an ordered manner and the action of one complex sets the stage for the arrival of the next one.

VDR Activators and Their Therapeutic Applications

The therapeutic applications of 1,25$(OH)_2D_3$ are limited due to the hypercalcemic and phosphatemic activity of this compound. The elevated level of calcium and phosphate in serum causes soft tissue calcification especially in the kidney, heart, aorta and intestine that can lead to organ failure and death. In order to avoid these unwanted side effects, a lot of work has been done to synthesize analogs or VDR activators (VDRAs) that exhibit weaker effects on calcium metabolism while retaining growth and immune regulating properties.

Pharmacological and Molecular Basis for Differential Actions of VDRAs

Factors that influence the biological profile of the VDRAs can be divided into pharmacokinetic and pharmacodynamic factors. There are two main pharmacokinetic factors which affect the ligand availability for VDR, stability in blood and catabolism of the target cell. Binding of the VDRAs to DBP or other blood molecules such as albumin and lipoproteins affects half-life values of the VDRAs in the blood and the rate of VDRAs uptake by target cell. Thus, VDRAs which have a strong affinity to DBP possess the longest extracellular half-lives, whereas VDRAs with reduced affinity to DBP are metabolized and excreted most rapidly [9].

Catabolic enzymes of target cells are another important factor controlling the concentration of $1,25(OH)_2D_3$ and other VDRAs inside the target cell. Target cells might inactivate VDRAs in different ways or create new compounds which retain significant biological activity [10]. It has also been reported that various cell types have different ability for catabolic vitamin D compounds, causing cell-specific differences in the action of the VDRAs [9].

The pharmacodynamic influences of the VDRAs are based on their ability to modulate VDR functions differently from the natural hormone. VDRAs could use different contact amino acid residues for binding into the ligand-binding cavity of VDR [11]. This may result in prolonged half-lives of activated receptor and longer-lasting effects on gene activation [12]. Furthermore, VDRAs might induce different structural conformation within the hormone-receptor complex [13], which may modulate the receptor dimerization [14], affect the DNA binding properties and even the promoter selectivity of VDR [15]. Moreover, a VDRA-induced different conformation may influence receptor interactions with tissue-specific cofactors and the stability of RXR/VDR/DNA/coactivator complex [16] and interfere with the proteosome-mediated receptor degradation [17].

Thus, different VDRAs can induce differential gene expression profiles in the same cell. This is what we can call VDRAs selectivity. Furthermore, cellular environment can affect the response of a cell to a VDRA. This particularity can be explained by the different expression profile of coactivators-corepresors or in the enzymes responsible for the catabolism of the VDRA. Thus, the same VDRA can also induce different gene expression profile in different tissues. Thus, together with the VDRA selectivity, VDR can also show tissue selectivity. This plasticity of VDR activation makes it a very interesting target for pharmacological intervention.

References

1 White C, Gardiner E, Eisman J: Tissue specific and vitamin D responsive gene expression in bone. Mol Biol Rep 1998;25:45–61.
2 Cardus A, Parisi E, Gallego C, Aldea M, Fernandez E, Valdivielso JM: 1,25-Dihydroxyvitamin D-3 stimulates vascular smooth muscle cell proliferation through a VEGF-mediated pathway. Kidney Int 2006;69:1377–1384.
3 Cardus A, Panizo S, Parisi E, Fernandez E, Valdivielso JM: Differential effects of vitamin D analogues on vascular calcification. J Bone Miner Res 2007;22:860–866.
4 Valdivielso JM, Coll B, Fernandez E: Vitamin D and the vasculature: can we teach an old drug new tricks? Expert Opin Ther Targets 2009;13:29–38.
5 Hewison M, Zehnder D, Bland R, Stewart PM: 1alpha-Hydroxylase and the action of vitamin D. J Mol Endocrinol 2000;25:141–148.
6 Nolte RT, Wisely GB, Westin S, Cobb JE, Lambert MH, Kurokawa R, Rosenfeld MG, Willson TM, Glass CK, Milburn MV: Ligand binding and co-activator assembly of the peroxisome proliferator-activated receptor-gamma. Nature 1998;395:137–143.
7 Urnov FD, Wolffe AP: Chromatin remodeling and transcriptional activation: the cast (in order of appearance). Oncogene 2001;20:2991–3006.
8 Chiba N, Suldan Z, Freedman LP, Parvin JD: Binding of liganded vitamin D receptor to the vitamin D receptor interacting protein coactivator complex induces interaction with RNA polymerase II holoenzyme. J Biol Chem 2000;275:10719–10722.
9 Brown AJ: Mechanisms for the selective actions of vitamin D analogues. Curr Pharm Des 2000;6:701–716.
10 Brown AJ, Dusso A, Slatopolsky E: Vitamin D. Am J Physiol Renal Physiol 1999;277: F157–F175.
11 Gardezi SA, Nguyen C, Malloy PJ, Posner GH, Feldman D, Peleg S: A rationale for treatment of hereditary vitamin D-resistant rickets with analogs of 1alpha,25-dihydroxyvitamin D(3). J Biol Chem 2001;276:29148–29156.
12 Peleg S, Nguyen C, Woodard BT, Lee JK, Posner GH: Differential use of transcription activation function 2 domain of the vitamin D receptor by 1,25-dihydroxyvitamin D_3 and its A ring-modified analogs. Mol Endocrinol 1998;12:525–535.
13 Liu YY, Collins ED, Norman AW, Peleg S. Differential interaction of 1alpha,25-dihydroxyvitamin D_3 analogues and their 20-epi homologues with the vitamin D receptor. J Biol Chem 1997;272:3336–3345.
14 Liu YY, Nguyen C, Ali Gardezi SA, Schnirer I, Peleg S: Differential regulation of heterodimerization by 1alpha,25-dihydroxyvitamin D(3) and its 20-epi analog. Steroids 2001;66:203–212.
15 Quack M, Carlberg C: Selective recognition of vitamin D receptor conformations mediates promoter selectivity of vitamin D analogs. Mol Pharmacol 1999;55:1077–1087.
16 Issa LL, Leong GM, Sutherland RL, Eisman JA: Vitamin D analogue-specific recruitment of vitamin D receptor coactivators. J Bone Miner Res 2002;17:879–890.
17 Jaaskelainen T, Ryhanen S, Mahonen A, Deluca HF, Maenpaa PH: Mechanism of action of superactive vitamin D analogs through regulated receptor degradation. J Cell Biochem 2000;76:548–558.

Dr. Jose M. Valdivielso
Laboratorio de Investigación HUAV-UDL. Hospital Universitari Arnau de Vilanova
Rovira Roure 80
E–25198 Lleida (Spain)
Tel. +34 973 003 650, Fax +34 973 702 213, E-Mail Valdivielso@medicina.udl.es

Ronco C, Crepaldi C, Cruz DN (eds): Peritoneal Dialysis – From Basic Concepts to Clinical Excellence. Contrib Nephrol. Basel, Karger, 2009, vol 163, pp 213–218

Importance of Vitamin D Receptor Activation in Clinical Practice

Mario Cozzolino, Giuditta Fallabrino, Sabina Pasho, Laura Olivi, Paola Ciceri, Elisa Volpi, Maurizio Gallieni, Diego Brancaccio

Renal Division, S. Paolo Hospital, Department of Medicine, Surgery and Dentistry, University of Milan School of Medicine, Milan, Italy

Abstract

Continuously emerging evidence indicates that deficiencies in 25-hydroxyvitamin D and consequently vitamin D receptor (VDR) activation play crucial roles in adversely affecting cardiovascular (CV) health in the general population and those at high risk of CV disease, as well as in patients with chronic kidney disease (CKD). In CKD patients, a lack of VDR activation is one of the main pathophysiological factors contributing to secondary hyperparathyroidism (SHPT). However, this lack of VDR activation has numerous additional implications on CV and renal function, with SHPT being only one symptom of a much more extensive disorder. VDRs are widely expressed throughout the body with manifold activities that involve feedback loops within the CV, immune, and renal systems. Modulation of VDR activator levels results in correlative regulatory effects on mineral homeostasis, hypertension, vascular disease, and vascular calcification, as well as a number of other endpoints in cardiac and renal pathology. Among compounds available for the treatment of SHPT, paricalcitol is a selective VDR activator. The term 'selective' refers to paricalcitol being more selective in affecting VDR pathways in the PTH gland compared with bone and intestine. As such, paricalcitol's selectivity allows for a wider therapeutic window with effects beyond PTH control and mineral management, and may explain, in part, the increased survival advantage with paricalcitol treatment.

Recently, a lot of attention has been given to the beneficial effects of 'vitamin D'. In many cases, these studies refer to the administration of cholecalciferol or ergocalciferol or levels of 25-hydroxyvitamin D (25D) in the blood. However, the eventual biological effect of these substances is to activate the VDR (vitamin D receptor) after being activated to the endogenous hormone 1,25-dihydroxyvitamin D (1,25D). Therefore, the important aspect is the biological pathway by which these substances eventually exert their beneficial effects, and this pathway is VDR activation.

Chronic kidney disease-mineral and bone disorder (CKD-MBD) begins early in the course of kidney disease and is underdiagnosed. Decreased VDR activation is a major pathophysiologic factor in the development of secondary hyperparathyroidism (SHPT) and contributes to chronic kidney disease (CKD) morbidity and mortality. In contrast, VDR inactivation influences bone and cardiovascular disease (CVD) progression and mortality. Some data suggest that treatment choices influence patient survival. These data underline the need for early assessment and clear management strategies. It is important to identify patients at increased risk for a poor prognosis and to better understand whether early treatment will benefit these patients.

VDRs are widely expressed throughout the body with manifold activities that involve feedback loops within the CV, immune, and renal systems. Modulation of VDR activator levels results in correlative regulatory effects on mineral homeostasis, hypertension, vascular disease, and vascular calcification, as well as a number of other endpoints in cardiac and renal pathology.

Cardiovascular System and VDR Activation

Deficiency in 25D has a demonstrated impact on the cardiovascular system in individuals without renal dysfunction. Recent longitudinal data from 1,739 individuals in the Framingham Offspring Study who had no previous cardiovascular disease, identified an association between low 25D levels and incident cardiovascular disease [1]. Reduced serum 25D levels are significantly depleted in patients in urgent need of care for end stage heart failure (i.e. chronic heart failure) compared with those undergoing elective surgery, and may be independently associated with poor clinical outcome in these patients, including a higher risk for myocardial infarction [2].

Diminished levels of 1,25-D have been significantly associated with an increased risk of vascular calcification (VC) in patients with a moderate or high risk of coronary heart disease (excluding patients with hyperparathyroidism, ESRD, or known malignancies) [3]. This association was independent of PTH, osteocalcin, cholesterol, or age. Age, race, diabetic history, and log 1,25-D have been related to arterial mass of coronary calcium and inversely correlated with calcium phosphate mass. Furthermore, a cross-sectional examination from NHANES (2001–2004) patient data demonstrated a significant graded, inverse correlation between serum 25D levels and the prevalence of peripheral arterial disease and was independent of gender, age, race, and multivariable adjustment [4]. The lowest quartile of serum 25D was set at <17.8 ng/ml and

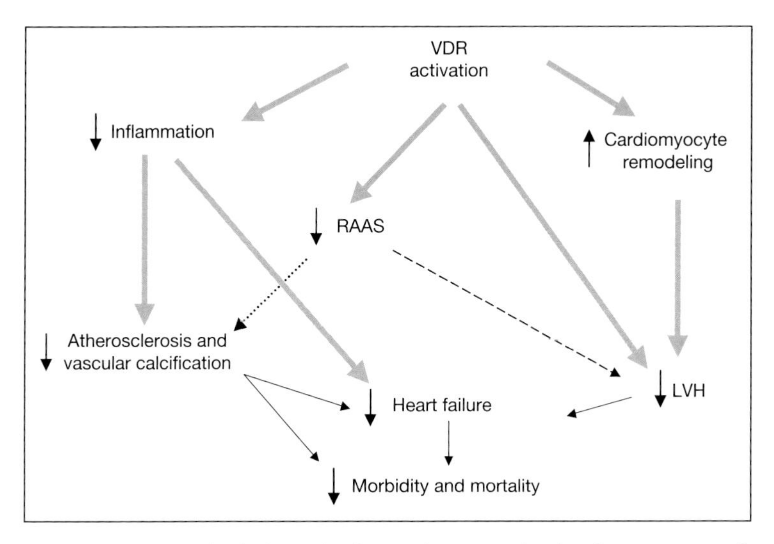

Fig. 1. Hypothetical mechanisms of VDR activation impact on cardiovascular outcome. RAAS = Renin-angiotensin-aldosterone system; LVH = left-ventricular hypertrophy.

demonstrated a prevalence ratio of 1.8 (95% CI, 1.19–2.74) after multivariable adjusted analysis.

In figure 1, we show the hypothetical mechanisms of VDR activation impact on cardiovascular outcome.

Inappropriate activation of the renin-angiotensin-aldosterone system (RAAS), which plays a central role in the regulation of blood pressure, electrolyte, and volume homeostasis, may represent a major risk factor for hypertension, heart attack, and stroke. Evidence from clinical studies has demonstrated an inverse relationship between circulating vitamin D levels and the blood pressure and/or plasma renin activity, but the mechanism is not understood. Renin expression and plasma angiotensin II production increase several fold in VDR-null mice, leading to hypertension, cardiac hypertrophy, and increased water intake. However, the salt- and volume-sensing mechanisms that control renin synthesis are still intact in the mutant mice. In wild-type mice, inhibition of 1,25D synthesis also led to an increase in renin expression, whereas 1,25D injection led to renin suppression. Vitamin D regulation of renin expression was independent of calcium metabolism and 1,25D markedly suppressed renin transcription by a VDR-mediated mechanism in cell cultures. Hence, 1,25D is a novel negative endocrine regulator of the renin-angiotensin system. Its apparent critical role in electrolytes, volume, and blood pressure homeostasis suggests that VDR activators could help prevent or ameliorate hypertension.

Role of VDR Activation on SHPT and Cardiovascular Outcome

For many years, the administration of calcitriol has been the mainstay of treatment for SHPT in CKD patients [5]. However, several novel findings and issues have substantially influenced nephrologists' attitudes towards VDRAs administration in recent years. High serum calcium (Ca) and phosphate (P) levels have been convincingly associated with reduced survival in CKD patients [6]. Moreover, Tonelli et al. [7] investigated the impact upon outcome of even small increases in serum P levels at the time of an acute myocardial infarction. They found an association of higher – but still within the normal range – P levels with the occurrence secondary cardiovascular events in non-dialysis patients. VC may result from high dosages of vitamin D administration, as seen in several animal models with renal insufficiency [8]. Moreover, some evidence exists that previous calcitriol treatment in humans with advanced CKD is one of the VC-promoting factors [9].

There are consistent observational data available that the administration of active vitamin D in dialysis patients and patients with advanced renal failure is associated with improved survival, irrespective of underlying P and Ca levels [10]. This improvement in survival may be attributable to both the traditional bone and mineral actions of vitamin D as well as to the pleiotropic actions. As a consequence of these divergent statements, and since prospective, randomized data are missing, nephrologists might well get 'lost in translation' if they try to transfer all the available experimental and observational data into every-day patient care. Currently, the bedside treatment decision for SHPT in CKD is even more complex: In the beginning of active vitamin D treatment in ESRD there used to be only the simple question: to give calcitriol or not to give calcitriol. In contrast, nowadays, there are several so-called vitamin D receptor activators (VDRAs) available: 1,25-dihydroxy-22-oxavitamin D_3 (22-oxacalcitriol, OCT), 1,25-dihydroxy-19-norvitamin D_2 (19-norD_2, paricalcitol), 1α-hydroxyvitamin D_2 (1αOHD_2). These novel alternative agents all claim to imitate the typical calcitriol action, i.e. reduction of SHPT. On the other hand, they deny being comparable to vitamin D regarding some other less desirable actions such as induction of hypercalcaemia or hyperphosphataemia [5].

Numerous observational studies show a clinical advantage for VDR activator therapy. To date, no studies have demonstrated increased mortality in patients receiving VDR activator therapy. Intravenous VDR activator therapy confers a survival advantage in patients on dialysis [10]. Kalantar-Zadeh et al.

[11] have shown an association between any time-varying administered dose of paricalcitol and relative risk of death in over 58,000 maintenance HD patients over 2 years. Moreover, Wolf et al. [12] have shown in the ArMORR prospective, cross-sectional studies that 1,25-D and 25D deficiencies in incident HD patients are associated with an increased mortality risk that is reduced following the introduction of VDR activator therapy given in conjunction with dialysis. The association of improved survival with VDR activation therapy can already be observed in patients with moderate renal impairment. VDR activators directly affect cardiovascular outcomes, apparently by mechanisms independent of SHPT. These include effects on the immune system, RAS, and development of atherosclerosis, cardiac remodeling, and LVH.

Role of Paricalcitol in Peritoneal Dialysis Patients

Recently, Ross et al. [13] investigated the role of paricalcitol capsule treatment in both HD and peritoneal dialysis (PD) patients. In this study, serum Ca levels remained in the normal range throughout the treatment phase and were statistically significantly different between treatment groups at each time point, except for weeks 1, 2 and 5. No difference between treatment groups was detected in serum P levels values throughout the study. Furthermore, paricalcitol capsules provide a rapid and sustained reduction of PTH in both HD and PD patients with minimal effect on serum Ca and P and no significant difference in adverse events compared with placebo. Initial dosing (based on iPTH/60) and subsequent dose titration of oral paricalcitol based on severity of SHPT was safe and effective in the management of SHPT. In addition, the monthly average dose data indicated that in clinical practices in which monitoring of PTH, Ca, or P could occur less frequently than weekly, a more modest initial dose and subsequent titration ratio may be warranted.

Paricalcitol provided significant decreases in bone biochemical markers compared with placebo, including bone alkaline phosphatase, which has been associated with increased mortality [11]. No adynamic bone disease developed in paricalcitol-treated patients.

Conclusions

SHPT is a consequence of VDR inactivation. VDR activation is involved in the regulation of many biologic processes and very important for cardiorenal outcomes in CKD. The VDR can be activated by 1,25D (calcitriol, alfacalcidol, doxercalciferol) and by selective VDR activators (paricalcitol or maxacalcitol).

Based on gene expression profiles, there are tissue-specific differences between vitamin D (1,25D) and selective VDR activator compounds. Selective VDR activation has multiple beneficial effects for cardiovascular and renal outcomes in preclinical studies, which could help explain earlier findings on patient survival. First clinical results confirm the preclinical findings and randomized clinical trials are being conducted to further investigate the benefits of selective VDR activation with paricalcitol.

References

1 Wang TJ, Pencina MJ, Booth SL, et al: Vitamin D deficiency and risk of cardiovascular disease. Circulation 2008;117:503–511.
2 Giovannucci E, Liu Y, Hollis BW, Rimm EB: 25-hydroxyvitamin D and risk of myocardial infarction in men: a prospective study. Arch Intern Med 2008;168:1174–1180.
3 Watson KE, Abrolat ML, Malone LL, et al: Active serum vitamin D levels are inversely correlated with coronary calcification. Circulation 1997;96:1755–1760.
4 Melamed ML, Muntner P, Michos ED, et al: Serum 25-hydroxyvitamin D levels and the prevalence of peripheral arterial disease: results from NHANES 2001 to 2004. Arterioscler Thromb Vasc Biol 2008;28:1179–1185.
5 Cozzolino M, Galassi A, Gallieni M, Brancaccio D: Pathogenesis and treatment of secondary hyperparathyroidism in dialysis patients: the role of paricalcitol. Curr Vasc Pharmacol 2008;6:148–153.
6 Ganesh SK, Stack AG, Levin NW, et al: Association of elevated serum PO(4), Ca x PO(4) product, and parathyroid hormone with cardiac mortality risk in chronic hemodialysis patients. J Am Soc Nephrol 2001;12:2131–2138.
7 Tonelli M, Sacks F, Pfeffer M, et al: Relation between serum phosphate level and cardiovascular event rate in people with coronary disease. Circulation 2005;112:2627–2633.
8 Henley C, Colloton M, Cattley RC, et al: 1,25-Dihydroxyvitamin D_3 but not cinacalcet HCl (Sensipar/Mimpara) treatment mediates aortic calcification in a rat model of secondary hyperparathyroidism. Nephrol Dial Transplant 2005;20:1370–1377.
9 Civilibal M, Caliskan S, Adaletli I, et al: Coronary artery calcifications in children with end-stage renal disease. Pediatr Nephrol 2006;21:1426–1433.
10 Tentori F, Hunt WC, Stidley CA, et al: Mortality risk among hemodialysis patients receiving different vitamin D analogs. Kidney Int 2006;70:1858–1865.
11 Kalantar-Zadeh K, Kuwae N, Regidor DL, et al: Survival predictability of time-varying indicators of bone disease in maintenance hemodialysis patients. Kidney Int 2006;70:771–780.
12 Wolf M, Betancourt J, Chang Y, et al: Impact of activated vitamin D and race on survival among hemodialysis patients. J Am Soc Nephrol 2008;19:1379–1388.
13 Ross EA, Tian J, Abboud H, et al: Oral paricalcitol for the treatment of secondary hyperparathyroidism in patients on hemodialysis or peritoneal dialysis. Am J Nephrol 2008;28:97–106.

Mario Cozzolino, MD, PhD
Renal Division, Azienda Ospedale San Paolo
Via A. di Rudinì, 8
IT–20142 Milano (Italy)
Tel. +39 02 8184 4381, Fax +39 02 8912 9989, E-Mail mariocozzolino@hotmail.com

Ronco C, Crepaldi C, Cruz DN (eds): Peritoneal Dialysis – From Basic Concepts to Clinical Excellence. Contrib Nephrol. Basel, Karger, 2009, vol 163, pp 219–226

New Acquisitions in Therapy of Secondary Hyperparathyroidism in Chronic Kidney Disease and Peritoneal Dialysis Patients: Role of Vitamin D Receptor Activators

Diego Brancaccio[a,b], *Mario Cozzolino*[a], *Sabina Pasho*[a], *Giuditta Fallabrino*[a], *Laura Olivi*[a], *Maurizio Gallieni*[a]

[a]Chair of Nephrology, University of Milan, c/o Renal Division, Ospedale San Paolo, and [b]Scientific Department, Renal Unit Simone Martini, Milan, Italy

Abstract

Secondary hyperparathyroidism is a serious complication of chronic renal disease when function decline and is characterized by abnormalities in serum calcium and phosphate profile, along with a decline in calcitriol synthesis. A reduced density of specific receptors for vitamin D and calcium in several tissues and organs are also present, thus contributing to parathyroid hyperplasia and abnormal parathyroid hormone synthesis and secretion. This metabolic derangement is observable early in the course of chronic renal failure (stages 3 and 4) and on this basis it should also be treated early in order to avoid important clinical consequences. To afford secondary hyperparathyroidism, several strategies should be considered: phosphate oral intake control (diet and phosphate binders), adequate calcium oral intake, vitamin D receptor activation. More specifically, the concept of selective vitamin D receptor activation will be considered as well as its biological effects, the use of paricalcitol (a selective vitamin D receptor activator) given orally to patients on peritoneal dialysis, and stages 3 and 4 of chronic renal failure. Finally, we will consider a series of nonclassical interesting potential mechanisms of selective vitamin D receptor activation leading to reduced cardiovascular and all-cause mortality.

Secondary hyperparathyroidism represents a common and serious complication of chronic kidney disease (CKD) mainly when renal failure occurs progressively. This condition is characterized by abnormalities in calcium and phosphorus serum profile, decline in calcitriol synthesis and also reduced vitamin D receptor and CaS receptor density in parathyroid glands and well as in several other tissues and organs.

These conditions lead to parathyroid gland hyperplasia and elevated serum parathyroid hormone (PTH) whose clinical consequences are a systemic disorder of mineral and bone metabolism, characterized by bone abnormalities related to turnover, mineralization, volume, linear growth and strength, often accompanied by arterial and soft tissue calcifications [1, 2] thus increasing cardiovascular morbidity and mortality.

Several contributions indicate that this metabolic derangement is present early in the course of chronic renal failure at stages 3 and 4; Levin et al. [3] showed that when glomerular filtration rate (GFR) declines below 60 ml/min, serum calcitriol levels also decline while no changes in serum calcium and phosphorus are observable until terminal uremia occurs. However, when GFR further declines (<40 ml/min), serum PTH profile tends to rise and this has been considered as an adaptive process for phosphorus homeostasis. On this basis, it could be assumed that CKD represents a condition of metabolic derangement to be treated early in order to avoid important clinical consequences.

Phosphate Control

One of the early triggers of a series of negative metabolic events is represented by the expanded pool of inorganic phosphate due to the reduced renal excretion capacity to adapt to phosphate oral intake.

In an early phase of CKD, both serum PTH and FGF23 are elevated in order to maximize phosphaturia in the residual nephrons [4]. However, when GFR declines below 30 ml/min, there are actual possibilities to observe increased serum phosphate levels; in the majority of these patients, a low phosphate diet should be successfully prescribed, occasionally along with oral phosphate binders.

The most traditional combined approach to this condition is the control of serum phosphorus levels, by reducing oral phosphate intake combined with oral phosphate binders, associated with the administration of calcitriol in order to prevent or to treat parathyroid hyperplasia and decrease PTH serum levels.

After the 'aluminum gel era' (aluminum salts given on a long-term basis were abandoned in most countries as phosphate binders), calcium carbonate and calcium acetate proved to be efficacious binders and widely adopted, limited however because possible responsible of increased body calcium pool. Among the 3rd generation of phosphate binders, sevelamer appears to be a weak binder and the fairly high number of pills often needed to control serum phosphate poses problems of compliance. During the last decade, lanthanum carbonate appears to be a much more promising new opportunity for countering hyperphosphatemia in dialysis patients because this drug has an efficacious profile similar to aluminum salts without any apparent evidence of bone or organ toxicity. A survey on long-term-treated patients for 6 years has been recently published by Hutchison et al. [5] who showed no evidence of toxicity at any level and specifically at the liver where lanthanum is handled before its removal through biliary excretion.

Vitamin D Receptor Activators

Calcitriol has been widely used in the past decade for controlling PTH secretion and preventing parathyroid hyperplasia. However, its use – as well as the use of other nonselective agents – might be limited by its potential effects on serum calcium levels, hypercalciuria and hyperphosphatemia. This condition, in patients undergoing regular dialysis treatment but also in patients at stages 3, 4 and 5 predialysis, can be related to the development and progression of vascular calcification, thus any therapy that could expand calcium load can potentially exacerbate this process [6, 7].

Therefore targeting parathyroid glands with selective vitamin D receptor activators that also has minimal effects on intestinal calcium absorption should be considered with priority in uremic patients.

Concept of Selective Vitamin D Receptor Activation

Early and new generation vitamin D receptor activators are believed to be related to the differential interaction with the vitamin D receptors, coactivator recruitment and cell/tissue selectivity.

These processes at different levels imply a more efficient PTH suppression and minimal effects on intestinal calcium absorption while along minimizing the negative effects on bone [8, 9].

As an example of the concept of selectivity, studies in uremic rats showed that while treatment with calcitriol was followed by an increased expression of intestinal vitamin D receptors, treatment with paricalcitol was not. Additionally, a 10-fold dose of paricalcitol, compared with calcitriol, was needed to produce the same increases in serum calcium levels. The same studies showed no increase in serum phosphate at any paricalcitol dose, whereas calcitriol was associated with significant increases [10, 11].

In summary, paricalcitol biological effects from preclinical studies can be summarized as follows (table 1):

Table 1. Paricalcitol biological effects

Biological effects	Potency versus calcitriol
Suppression of PTH	1/3
Capacity to raise serum calcium	1/10
Capacity to raise serum phosphate	1/10

These data were more recently confirmed by Lund et al. [12], who showed a 17% overall decrease in calcium absorption relative to calcitriol in a group of hemodialysis patients in a double-blind crossover study. This study also showed that the appropriate paricalcitol/calcitriol ratio to suppress PTH equally in both groups was 3:1.

Several other investigators also showed similar findings, clarifying the different mechanisms of the reduced intestinal calcium absorption likely due to the different effect of calcitriol and paricalcitol in increasing calbindin expression, which plays a role in modulating calcium absorption [13].

Selective vitamin D receptor activator therapy also showed beneficial effects on bone metabolism. Paricalcitol showed less stimulation of bone resorption than calcitriol and doxercalciferol, as measured by less increases in serum phosphorus levels [14, 15].

Oral Paricalcitol in Patients on Peritoneal Dialysis

The oral preparation became available in 2006 in the US and in the following years in several European countries. Its use is basically the prevention and the treatment of secondary hyperparathyroidism in patients with chronic renal failure stages 3, 4 and 5 predialysis as well as in patients on peritoneal dialysis (PD).

A recent prospective randomized placebo-controlled, double-blind multicenter study was designed to evaluate the safety and efficacy of paricalcitol capsules in both peritoneal and hemodialysis patients and the results were published last year by Ross et al. [16].

The primary endpoints were the achievement of 2 consecutive >30% decreases from baseline iPTH levels and the achievement of 2 consecutive serum calcium values >11.0 mg/dl. Patients were randomized to receive placebo (n = 27) or paricalcitol (n = 61) for a 12-week period.

The patients were treated initially with a doses determined by baseline iPTH/60 (more elevated than that suggested for patients treated with intrave-

Table 2. Calcium and phosphate elevations by dialysis type [adapted from ref. 16]

	All treated subjects		HD regimen		PD regimen		
	paricalcitol (n = 61)	placebo (n = 27)	paricalcitol (n = 42)	placebo (n = 20)	paricalcitol (n = 19)	placebo (n = 6)	p value
Calcium >11.0 mg/dl	1 (1.6)	0 (0)	0 (0)	0 (0)	1 (5.3)	0 (0)	>0.999 NS >0.999
Phosphorus >5.5 mg/dl	36 (59)	11 (40.7)	28 (66.7)	9 (45.0)	8 (42.1)	2 (33.3)	NS 0.165 >0.999

p value for test of odds ratio homogeneity from the Breslow-Day test. Values in parentheses are percentages.

nous paricalcitol) up to a maximum dose of 32 μg. Doses were titrated thereafter based on iPTH, calcium and phosphate values. It should be underlined that both PD or hemodialysis (HD) patients, whose mean basal iPTH was 730 pg/ml, showed a rapid control of iPTH levels during the first 2 weeks of therapy (>40%) so that paricalcitol doses were drastically reduced from a mean of 33.7 pg weekly doses to 16.9 pg after 4 weeks of therapy.

Additionally, a statistically greater proportion of both HD and PD paricalcitol patients [83% (33/40) and 100% (18/18), respectively] achieved 2 consecutive >30% decreases in iPTH. The treatment groups were not statistically different in terms of hypercalcemia safety end point. Phosphate binder use and mean serum phosphorus levels were not different between the treatment groups. Finally, the marker of bone activity improved in the treated subjects and worsened in those on placebo.

This study showed that paricalcitol provides a rapid and important reduction of PTH in both HD and PD patients, with minimal effects on serum calcium and phosphate. No adverse effects were observed as compared with placebo (table 2).

Vitamin D Receptor Activators in Chronic Renal Failure: Rationale for Their Use

Several data from the recent literature indicate that vitamin D receptor activator therapy, and in particular selective vitamin D receptor activators, is

Table 3. Nonclassical vitamin D receptor activator potential mechanisms leading to reduced cardiovascular and all-cause mortality

Heart – reduced myocyte proliferation/calcification/arterial thickening/arterial stiffening (= regression of left ventricular hypertrophy)
Kidney – reduced glomerular cell growth/differentiation/fibrosis/proteinuria (= control of damage progression)
Renin-angiotensin system suppression
Anti-inflammation effects
Anti-atherogenesis effects
Anti-thrombosis effects

associated with improved cardiovascular and all-cause mortality outcomes in patients with stage 5 CKD [17–19].

Several researches are also elucidating a variety of nonclassical vitamin D receptor activator effects in cardiovascular tissues in terms of control of vascular cell proliferation, inflammation, and vascular calcification (table 3). More specifically, calcitriol and mainly paricalcitol proved to have inhibitory effects on cardiac myocyte hypertrophy, suppression of the renin-angiotensin system, anti-inflammatory, antiatherogenic and antithrombotic effects [20]. In addition, direct beneficial effects of selective vitamin D receptor activators on glomerular remodeling and on proteinuria have also been documented, even in patients already treated with ACE inhibitors, likely relating to modulation of structural proteins, growth factors and cytokines [21]. These data were more recently confirmed by Alborzi et al. [22] who conducted a pilot study in which 24 patients, affected by chronic renal failure at stages 2 and 3, were allocated to daily receive 0, 1 or 2 μg of oral paricalcitol for 1 month. These authors showed that even a small daily dose of paricalcitol (1 μg) was able to significantly reduce albumin excretion rate along with a significant reduction of the levels of high sensitivity C-reactive protein. All patients showed no change in iothalamate clearance, 24-hour blood pressure or parathyroid hormone with treatment or on washout.

All these pleiotropic effects related to the modulation of gene expression have the potential to ameliorate the development and progression of cardiovascular disease and reduce mortality in patients with any level of chronic renal failure.

An evaluation of both the classical and nonclassical effects of vitamin D receptor activators in patients with CKD indicates that a selective vitamin D receptor activation should be the treatment of choice with regards to maximize the control of serum PTH while minimizing the possible deleterious calcemic and phosphatemic effects, as well as providing benefits on cardiovascular comorbidity and renal disease progression.

References

1 Slatopolsky E, Brown A, Dusso A: Pathogenesis of secondary hyperparathyroidism. Kidney Int 1999;56(suppl 73):S14–S19.
2 Block GA, Port FK: Re-evaluation of risks associated with hyperphosphatemia and hyperparathyroidism in dialysis patients: recommendations for a change in management. Am J Kidney Dis 2000;35:1226–1237.
3 Levin A, Bakris GL, Molitch M, et al: Prevalence of abnormal Vitamin D, PTH, calcium and phosphorus in patients with chronic kidney disease: results of the study to evaluate early kidney disease. Kidney Int 2007;71:31–38.
4 Quarles LD: Endocrine functions of bone in mineral metabolism regulation. J Clin Invest 2009;119:421–428.
5 Hutchison AJ, Barnett ME, Krause R, et al: Long term and safety of lanthanum carbonate: results for up to 6 years of treatment. Nephron Clin Pract. 2008;110:15–23.
6 Chertow GM, Burke SK, Raggi P: Sevelamer attenuates the progression of coronary and aortic calcification in hemodialysis patients. Kidney Int 2002;62:245–252.
7 Goodman WG, Goldin J, Kuizon BD, et al: Coronary-artery calcification in young adults with end-stage renal disease who are undergoing dialysis. N Engl J Med 2001;342:1478–1483.
8 Issa LL, Leong GM, Sutherland RL, et al: Vitamin D analogue-specific recruitment of vitamin D receptor coactivators. J Bone Miner Res 2002;17:879–890.
9 Takeyama K-I, Masuhiro Y, Fuse H, et al: Selective interaction of Vitamin D receptor with transcriptional coactivators by a vitamin D analog. Mol Cell Biol 1999;19:1049–1055.
10 Slatopolsky E, Finch J, Ritter C, et al: Effects of 19-nor-1,25 $(OH)_2D_2$, a new analogue of calcitriol, on secondary hyperparathyroidism in uremic rats. Am J Kidney Disease 1998;32(suppl 2):S40–S47.
11 Slatopolsky E, Cozzolino M, Finch JL: Differential effects of 19-nor-1,25-$(OH)_2D_3$ and 1a-hydroxyvitamin D_2 on calcium and phosphorus in normal and uremic rats. Kidney Int 2002;62:1277–1284.
12 Lund R, Tian J, Melnick J, et al: Differential effects of paricalcitol and calcitriol on intestinal calcium absorption in hemodialysis patients (abstract SP-607). XLIII ERA-EDTA Congress Glasgow, 2006.
13 Ma JN, Osinski M, Rose M, et al: Effects of VDR activators on intestinal calcium (abstract SA-PO613). J Am Soc Nephrol 2004;115:437A.
14 Coyne DW, Grieff M, Ahya SN, et al: Differential effects of acute administration of 19-nor-1,25-dihydroxy-vitamin D_2 and 1,25-dihydroxy-vitamin D_3 on serum calcium and phosphorus in hemodialysis patients. Am J Kidney Dis 2002;40:1283–1288.
15 Joist HE, Ahya SN, Giles K, et al: Differential effects of very high doses of deoxercalciferol and paricalcitol on serum phosphorus in hemodialysis patients. Clin Nephrol 2006;65:335–341.
16 Ross EA, Tian J, Abboud H, et al: Oral paricalcitol for the treatment of secondary hyperparathyroidism in patients on hemodialysis or peritoneal dialysis. Am J Nephrol 2008;28:97–106.
17 Teng M, Wolf M, Lowrie, et al: Survival of pateints undergoing hemodialysis with paricalcitol or calcitriol therapy. N Engl J Med 2003;349:446–456.
18 Teng M, Wolf M, Ofsthun MN, et al: Activated injectable vitamin D and hemodialysis survival: a historical cohort study. J Am Soc Nephrol 2005;16:1115–1125.
19 Kalantar-Zadeh K, Kuwae N, Regidor DL, et al: Survival predictability of time-varying indicators of bone disease in maintenance hemodialysis patients. Kidney Int 2006;70:771–780.
20 Andress D: Nonclassical aspects of differential vitamin D receptor activation. Implications for survival in patients with chronic kidney disease. Drugs 2007;67:1999–2012.

21 Agarwal R, Acharya M, Tian J, et al: Antiproteinuric effect of oral paricalcitol in chronic kidney disease. Kidney Int 2005;68:2823–2828.
22 Alborzi P, Patel NA, Peterson C, et al: Paricalcitol reduces albuminuria and inflammation in chronic kidney disease: a randomized double-blind pilot trial. Hypertension 2008;52:249–255.

Prof. Diego Brancaccio
Renal Division, Ospedale San Paolo
8 Via di Rudinì
IT–20142 Milan (Italy)
Tel. +39 02 8184 4215, Fax +39 02 8912 9989, E-Mail diego.brancaccio@unimi.it, diego.brancaccio@tiscalinet.it

Ronco C, Crepaldi C, Cruz DN (eds): Peritoneal Dialysis – From Basic Concepts to Clinical Excellence. Contrib Nephrol. Basel, Karger, 2009, vol 163, pp 227–236

What Can We Learn from Registry Data on Peritoneal Dialysis Outcomes?

Norbert Lameire, Wim Van Biesen

Renal Division, Department of Medicine, University Hospital, Ghent, Belgium

Abstract

Registries, collecting observational data, can be of great value in our understanding of clinical questions, since they reflect 'real practice in a real world'. This paper provides an overview of what registry data from different regions of the world have taught us on survival outcomes of peritoneal dialysis patients compared to hemodialysis patients. The majority of the comparative studies show that patient survival on peritoneal dialysis is either similar or even better than on hemodialysis, at least for the first few years after starting dialysis. These results support a policy of peritoneal dialysis as the first dialysis modality in a program of integral care.

Although randomized controlled clinical trials (RCCT) are considered the mainstay of 'evidence-based medicine', also registry data (observational data) are of great value in advancing our understanding of certain clinical questions. Whereas RCCTs are outstanding to answer very-well-defined interventions (e.g. compare effects of two different medications), they become cumbersome when 'treatment policies' come into play, either because the intervention is difficult to be strictly defined, or because the treatments are so different that randomization becomes difficult. In addition, whereas RCCTs mostly are confined to well-defined interventions, registry data reflect 'real practice in a real world', and as such do not evaluate 'efficacy' (does it work?) but 'efficiency' (does it work in the real world?).

The main advantage of an RCT is that the randomization procedure helps to prevent selection bias by the clinician by breaking the link between the clinician's therapy prescription and the patient's prognosis [1]. Within observational studies, however, selection by the clinician may occur, and, even after adjustment for potential confounders in the statistical analysis, it may not be possible to make a fair comparison between the groups. Usually, results from observational studies are needed to come to a hypothesis that can subsequently be tested within an RCT. Moreover, observational data are most often more useful than RCTs for nontherapeutic studies.

The number of RCTs published in nephrology is fewer than all other specialties of internal medicine with the proportion of all citations which are RCT being the third lowest (1.15%). Despite that there has been an increase in both indices over the years, this occurred not at a greater rate than in other specialties [2].

A RCT of sufficient power addressing the question whether the outcome of PD is better or worse than the outcome of HD has not been feasible to complete due mainly to unwillingness of patients to be randomized [3]. The single RCT addressing this question has been performed in the Netherlands [4] and due to the low inclusion rate, the trial was prematurely stopped after which only 38 patients had been randomized: 18 patients to HD and 20 to PD. After 5 years of follow-up, and after adjustment for age, comorbidity, and primary kidney disease, the hasard ratio (HR) for mortality was 3.6 (0.8–15.4) in disfavour of HD.

It is thus not surprising that nephrology in general and peritoneal dialysis in particular continue to rely on retrospective data, often from renal registries or small clinical trials, to guide changes in practice [5].

However, the value of registries strongly depends on the quality of the data contained in the registry [6]. To optimize the quality of medical registry data, participating centers should follow certain procedures designed to minimize inaccurate and incomplete data. A recent paper has tried to identify causes of insufficient data quality and has made a list of procedures for data quality assurance in medical registries and put them in a framework [6].

This problem of data quality is buffered to some extent by the power of the huge number of records usually present in registry databases.

Other sources of bias may, however, threaten the reliability of observational studies on outcomes of HD, compared with PD and these may explain the conflicting results in different studies.

First, differences in clinically important characteristics may exist between patients who start with HD and patients who start with PD (selection bias, bias by indication). Although these differences are mostly accounted for by using appropriate statistical methodology, at the end apples may still be compared with pears and differences in clinical reality of patients can simply not be captured in a 'correction' formula. This is of particular importance in the comparison HD versus PD, as these are by definition strongly different clinical settings with most likely different types of patients. Furthermore, large center differences exist in the utilization of PD as starting modality and thus with large differences in PD experience. These center differences are difficult to correct for but are real. This can be illustrated by the fact that in the USRDS database there is a clear trend for improved PD outcome, despite a decline in the number of centers offering PD. This observation suggests that center experience markedly influences outcome.

Second, various studies have indicated that the relative risk of death (RR) among HD patients, compared with PD patients, decreases with time after the initiation of dialysis (i.e. in favour of HD). This finding can be the consequence of several different mechanisms, which can best be described as 'competing risks'. In the first months of RRT, the RR of mortality is higher in HD as compared to PD. This results in a 'survival of the fittest' in the HD group, a problem that is especially present in the USRDS database, where patients are only included after 90 days of treatment. In a later stage, there is competing risk between drop out because of death or transplantation. As PD is more often started in younger patients, there is a higher drop out due to transplantation, resulting in older patients staying on PD.

Third, various studies have demonstrated significant differences in the RR for HD patients, compared with PD patients, among different subgroups of patients defined on the basis of age and diabetes mellitus status [7]. Another source of confusion is the method of handling data for patients who switch from one modality to the other. It has been demonstrated that different strategies for censoring the survival times of patients who switch therapeutic modalities may influence the estimated RR for HD patients, compared with PD patients [7, 8].

All these aspects make the interpretation of registry data not straightforward and as such they cannot be considered 'hard evidence'. We believe, however, that with correct interpretation registry data are based on 'basic truths' and reflect a tremendous source of experience. As such, these data reflect daily life and do take into account patient preferences.

This paper provides an overview of what registry data have taught us on survival outcomes of PD patients compared to HD patients.

Outcome Results in the Registries

Table 1 modified and updated from Davies [9] is an attempt to summarize data from those registries that have compared survival outcome in both HD and PD patients. The data have been grouped according to the region of the registry and include data from the USRDS, the Canadian registry (CORR), the Danish registry, the Dutch registries NECOSAD and RENINE and the Australian-New Zealand registry.

Table 1. Registry data comparing PD with HD patient outcome

Study population	Size, n (% on PD)	Group	RR of death (PD vs. HD)	Comment
North America				
Bloembergen 1995, USRDS [22]	170,700 (15)	all	1.19	prevalent, not incident
		diabetics	1.38	
		nondiabetics	1.11	
Fenton, 1997, CORR [3]	19,663 (38)	all	0.73	incident cohort; relative risk changes with time, advantage for PD in first 2 years
Collins, 1999, USRDS reworked [23]	117,158 (15.5)	age <55, nondiabetic	0.61–0.72	relative risk changes with time; only group worse on PD was F >55 years with DM
		age >55, nondiabetic	0.87	
		age <55, diabetic	0.86–0.88	
		M age >55, diabetic	1.03	
		F age >55, diabetic	1.21	
Stack, 2003/Ganesh, 2003 USRDS [24;25]	107,922 (13)	nondiabetics	0.97	not corrected for age/ comorbidity/modality interactions
		diabetics	1.11	
		heart failure	1.3	
		ischemic heart disease	1.2–1.23	
Vonesh, 2004, USRDS [26]	398,940 (11.6)	nondiabetics,no comorbidity	0.76–0.89	corrected for age/ comorbidity/modality interactions
		non-diabetics, comorbidity	0.81–1.04	
		diabetics, no comorbidity	0.88–1.08	
		diabetics, comorbidity	0.9–1.2	
Jaar, 2005, CHOICE study [12]	1041 (26%)	all at year 1	1.06	found worse outcomes for PD in younger DM patients; laboratory data collected after start of dialysis
		all at year 2	0.84	
		including lab results, year 1	1.39	
		including laboratory results, year 2	2.34	
Europe				
Heaf, 2001, Danish registry, (1990-99) [17]	4921 (50)	all	0.86	early PD advantage NS for diabetes; corrected for transplant candidacy
Termorshuizen, NECOSAD 2003 [8]	1,222 (39)	3–12 months	0.56	adjusted for RRF at start and throughout treatment
		12–24 months	0.96	
		24–36 months	1.47	
		36–48 months	1.71	

Table 1. (Continued)

Study population	Size, n (% on PD)	Group	RR of death (PD vs. HD)	Comment
Liem, 2007, RENINE Included between 1987 and end 2002 [18]	16,643 (34.9)	>3–6 months		all incident patients; study start at day 91; censored for transplantation; adjusted for age, gender, primary renal disease, center of dialysis, year of start dialysis
		age 40 diabetes yes	0.40	
		diabetes no	0.26	
		age 50 diabetes yes	0.53	
		diabetes no	0.35	
		age 60 diabetes yes	0.71	
		diabetes no	0.46	
		age 70 diabetes yes	0.95	
		diabetes no	0.62	
		> 6–15 months		
		age 40 diabetes yes	0.59	
		diabetes no	0.51	
		age 50 diabetes yes	0.72	
		diabetes no	0.62	
		age 60 diabetes yes	0.87	
		diabetes no	0.75	
		age 70 diabetes yes	1.07	
		diabetes no	0.92	
		>15 months		
		age 40 diabetes yes	1.06	
		diabetes no	0.86	
		age 50 diabetes yes	1.17	
		diabetes no	0.95	
		age 60 diabetes yes	1.29	
		diabetes no	1.05	
		age 70 diabetes yes	1.42	
		diabetes no	1.16	
Australia/New Zealand				
McDonald, 2009, ANZDATA, included 1 Oct 1991 and end 2005 [19]	25,277 (41.7)	all; mulivariate COX		all incident patients; several statistical methods applied
		90–365 days	0.80	
		≥366 days;	1.32	
		multivariate COX (modality as time-varying factor		
		90–365	0.82	
		≥366 days	1.29	

Table 1. (Continued)

Study population	Size, n (% on PD)	Group	RR of death (PD vs. HD)	Comment
Asia				
Taiwan Renal Registry, 2008 Huang [20]	48,089 (5.8)	all		intention to treat; no correction for co-morbidities or laboratory data; overall no significant survival differences
		1-year survival PD	89.8%	
		1-year survival HD	87.5%	
		5-year survival PD	55.5%	
		5-year survival HD	54.3%	
		10-year survival PD	35%	
		10-year survival HD	33.8%	

North-American Registries

Virtually all registry studies based on data of the USRDS to date have produced similar results: patients who are treated with PD have a lower risk for death compared with HD patients in the first few years of dialysis therapy [7, 10]. This early survival advantage diminishes over time; in some subgroups, such as elderly individuals with diabetes, where PD patients develop a higher risk for death.

Although not based on a national registry the most recent comparative data on outcome of PD versus HD in Canada have been reported by Murphy et al. [11]. They concluded that the previously observed survival advantage of PD in Canada [3] which were based on Canadian Organ Replacement Register (CORR) data, was due to lower comorbidity and a lower burden of acute onset end-stage renal disease at the inception of dialysis therapy.

The important results of the CHOICE study [12] should be mentioned briefly. This prospective, cohort study, although not using registry data, showed that, after adjustment for demographics, case mix, and laboratory parameters, the relative risk for death for HD and PD patients was similar during the first year but was significantly higher among PD patients during the second year of treatment. These results are contrary to many previously published results from different national registries, including the studies discussed above that showed an earlier survival advantage for PD patients, particularly among those without diabetes and young patients with diabetes. The study design of CHOICE has received some criticism [13, 14] because PD patients were enrolled from substantially fewer clinics than HD patients. In addition, the statistical models may have been overadjusted by inclusion of variables that may have been a consequence of the modality. The study controlled for residual renal function and employment status – both benefits that are attributable to selection of PD – and therefore was probably inappropriate.

European Registries

The registries from Europe (Danish Registry, NECOSAD, RENINE) and Canada suggest that outcomes in PD and HD are overall equal, with a slight advantage for PD in the first 1–2 years. Recently, Burkart et al. [15] have discussed several studies performed on mortality comparisons between HD and PD patients in individual European countries. Most studies reported no differences in mortality between the two treatment modalities, but two favoured PD when the whole duration of follow-up was taken into account [16, 17]. No study found superiority of HD.

The 2 most comprehensive European studies have been performed in the Netherlands, based on the NECOSAD (Netherlands Cooperative Study on the Adequacy of Dialysis) and RENINE (Dutch End-Stage Renal Disease Registry).

In the NECOSAD study [8], no statistically significant differences in adjusted mortality rates between HD and PD patients were observed during the first 2 years of dialysis. In the years thereafter, increases in mortality rates for PD patients and resulting decreases in RR in favour of HD were observed.

The RENINE data [18] also allow to conclude that the survival advantage for PD compared with HD patients decreases over time, with age and in the presence of diabetes as primary disease. Like many studies before, also this study showed a substantial benefit to PD in early treatment among younger people without diabetes, with lesser benefit among other groups, with a subsequent change to increased longer term mortality associated with PD.

Australia and New Zealand Dialysis and Transplantation Registry (ANZDATA)

In the recent study of the ANZDATA [19] using data from 27,015 patients, overall mortality rates were significantly lower during the 90- to 365-day period among those being treated with PD at day 90 (adjusted HR 0.89; 95% CI 0.81–0.99; $p < 0.001$). This effect, however, varied in direction and size with the presence of co-morbidities: younger patients without co-morbidities had a mortality advantage with PD treatment, but other groups did not. After 12 months, the use of PD at day 90 was associated with significantly increased mortality (adjusted

HR 1.33; 95% CI 1.24 to 1.42; $p < 0.001$). In a supplementary as-treated analysis, PD treatment was associated with lower mortality during the first 90 days (adjusted HR 0.67; 95% CI 0.56 to 0.81; $p < 0.001$), supporting the 'survival of the fittest' bias, also present in the USRDS database.

Asian Registries

As far as we know, only one Asian registry (the Taiwan Renal Registry) has recently analyzed the comparative survival between HD and PD patients in an intent-to-treat analysis [20]. After adjusting for both demographic and clinical case-mix differences, PD and HD patients had a similar long-term survival. However, subgroup analysis revealed that, among diabetic patients and patients older than 55 years, those on HD experienced better survival than those on PD.

Le Régistre de Dialyse Péritonéale de Langue Française (RDPLF) (French Language Peritoneal Dialysis Registry)

Although this register does not include comparative groups of HD patients it is a 'specific' PD-related registry and therefore interesting to discuss it briefly [21].

Of 11,744 incident patients with a median age of 71 years, 21.5% were over 80 years of age and 56% were not able to perform PD treatment at home without assistance. Eighty-six percent of the latter group received external assistance from a private nurse and 14% were aided by their family. Of all the continuous ambulatory peritoneal dialysis (CAPD) patients, 61.7%, and among automated peritoneal dialysis (APD) patients 23%, are assisted at home for their bag exchanges and connections. These results demonstrate that PD may be successfully prescribed for older patients who receive assistance either from their family or from a nurse. The system of nurse-assisted home PD is unique for France.

Conclusions

In summary, taking all the caveats of observational studies in mind which have been discussed in the first part of this review, the majority of comparative studies show that patient survival on PD is either similar or even better than on HD, at least for the first few years. The survival results during longer follow-up periods are equivocal, probably because of inevitable differences between the populations opting for either modality. These results strongly support a policy of starting with PD as first dialysis therapy in a program of integrated care. The conceptual basis for such a program should be discussed with patients during their predialysis phase, and the patients should be offered free choice between modalities as there is no evidence that either of the treatments would jeopardize their outcome.

As such, registries teach us that the million dollar question 'Is HD better than PD?' is rather irrelevant as the answer depends upon the individual patient.

References

1 Stel VS, Jager KJ, Zoccali C, Wanner C, Dekker FW: The randomized clinical trial: an unbeatable standard in clinical research? Kidney Int 2007;72:539–542.
2 Strippoli GF, Craig JC, Schena FP: The number, quality, and coverage of randomized controlled trials in nephrology. J Am Soc Nephrol 2004;15:411–419.
3 Fenton SS, Schaubel DE, Desmeules M, Morrison HI, Mao Y, Copleston P, Jeffery JR, Kjellstrand CM: Hemodialysis versus peritoneal dialysis: a comparison of adjusted mortality rates. Am J Kidney Dis 1997;30:334–342.
4 Korevaar JC, Feith GW, Dekker FW, van Manen JG, Boeschoten EW, Bossuyt PM, Krediet RT: Effect of starting with hemodialysis compared with peritoneal dialysis in patients new on dialysis treatment: a randomized controlled trial. Kidney Int 2003;64:2222–2228.
5 Moist LM: Building the evidence in peritoneal dialysis: use of randomized controlled trials, and observational and registry data. Perit Dial Int 2001;21(suppl 3):S263–S268.
6 Arts DG, De Keizer NF, Scheffer GJ: Defining and improving data quality in medical registries: a literature review, case study, and generic framework. J Am Med Inform Assoc 2002;9:600–611.
7 Vonesh EF, Snyder JJ, Foley RN, Collins AJ: Mortality studies comparing peritoneal dialysis and hemodialysis: what do they tell us? Kidney Int Suppl 2006;S3–S11.
8 Termorshuizen F, Korevaar JC, Dekker FW, van Manen JG, Boeschoten EW, Krediet RT: Hemodialysis and peritoneal dialysis: comparison of adjusted mortality rates according to the duration of dialysis: analysis of The Netherlands Cooperative Study on the Adequacy of Dialysis 2. J Am Soc Nephrol 2003;14:2851–2860.
9 Davies SJ: Comparing outcomes on peritoneal dialysis and hemodialysis: a case study in the interpretation of observational studies. Saudi J Kidney Dis Transplant 2007;18:24–30.
10 Khawar O, Kalantar-Zadeh K, Lo WK, Johnson D, Mehrotra R: Is the declining use of long-term peritoneal dialysis justified by outcome data? Clin J Am Soc Nephrol 2007;2:1317–1328.
11 Murphy SW, Foley RN, Barrett BJ, Kent GM, Morgan J, Barre P, Campbell P, Fine A, Goldstein MB, Handa SP, Jindal KK, Levin A, Mandin H, Muirhead N, Richardson RM, Parfrey PS: Comparative mortality of hemodialysis and peritoneal dialysis in Canada. Kidney Int 2000;57:1720–1726.
12 Jaar BG, Coresh J, Plantinga LC, Fink NE, Klag MJ, Levey AS, Levin NW, Sadler JH, Kliger A, Powe NR: Comparing the risk for death with peritoneal dialysis and hemodialysis in a national cohort of patients with chronic kidney disease. Ann Intern Med 2005;143:174–183.
13 Piraino B: The choice study. Perit Dial Int 2006;26:423–425.
14 Piraino B, Bargman J: Does the risk of death differ between peritoneal dialysis and hemodialysis patients? Nat Clin Pract Nephrol 2006;2:128–129.
15 Burkart J, Piraino B, Kaldas H, Lee JY, Bender FH, Krediet RT, Boeschoten EW, Dekker FW, Vonesh E, Mujais S: Why is the evidence favoring hemodialysis over peritoneal dialysis misleading? Semin Dial 2007;20:200–202.
16 Davies SJ, Van Biesen W, Nicholas J, Lameire N: Integrated care. Perit Dial Int 2001;21(suppl 3):S269–S274.

17 Heaf JG, Lokkegaard H, Madsen M: Initial survival advantage of peritoneal dialysis relative to haemodialysis. Nephrol Dial Transplant 2002;17:112–117.
18 Liem YS, Wong JB, Hunink MG, de Charro FT, Winkelmayer WC: Comparison of hemodialysis and peritoneal dialysis survival in The Netherlands. Kidney Int 2007;71:153–158.
19 McDonald SP, Marshall MR, Johnson DW, Polkinghorne KR: Relationship between dialysis modality and mortality. J Am Soc Nephrol 2009;20:155–163.
20 Huang C-C, Cheng K-F, Wu H-D: Survival analysis: comparing peritoneal dialysis and hemodialysis in Taiwan. Perit Dial Int 2008;28(suppl 3):S15–S20.
21 Verger C, Ryckelynck JP, Duman M, Veniez G, Lobbedez T, Boulanger E, Moranne O: French peritoneal dialysis registry (RDPLF): outline and main results. Kidney Int Suppl 2006;S12–S20.
22 Bloembergen WE, Port FK, Mauger EA, Wolfe RA: A comparison of mortality between patients treated with hemodialysis and peritoneal dialysis. J Am Soc Nephrol 1995;6:177–183.
23 Collins AJ, Hao W, Xia H, Ebben JP, Everson SE, Constantini EG, Ma JZ: Mortality risks of peritoneal dialysis and hemodialysis. Am J Kidney Dis 1999;34:1065–1074.
24 Ganesh SK, Hulbert-Shearon T, Port FK, Eagle K, Stack AG: Mortality differences by dialysis modality among incident ESRD patients with and without coronary artery disease. J Am Soc Nephrol 2003;14:415–424.
25 Stack AG, Molony DA, Rahman NS, Dosekun A, Murthy B: Impact of dialysis modality on survival of new ESRD patients with congestive heart failure in the United States. Kidney Int 2003;64:1071–1079.
26 Vonesh EF, Snyder JJ, Foley RN, Collins AJ: The differential impact of risk factors on mortality in hemodialysis and peritoneal dialysis. Kidney Int 2004;66:2389–2401.

Norbert Lameire
Renal Division, Department of Medicine, University Hospital
185 De Pintelaan
BE–9000 Ghent (Belgium)
Tel. +32 9332 4402, Fax +32 9332 4403, E-Mail norbert.lameire@ugent.be

Ronco C, Crepaldi C, Cruz DN (eds): Peritoneal Dialysis – From Basic Concepts to Clinical Excellence. Contrib Nephrol. Basel, Karger, 2009, vol 163, pp 237–242

How to Persuade Peritoneal Dialysis – Skeptical Hemodialysis Fans

Anabela Rodrigues

Division of Nephrology, Hospital de Santo António, Porto, Portugal

Abstract

Already from its early decades, peritoneal dialysis (PD) has proved to be efficient and able to confer similar or better chronic patient survival in comparison with hemodialysis (HD). More recent years allowed many PD therapy advances with further outcomes improvement: mortality, hospitalizations and clinical complications all have been reduced across patient's vintages. Adequacy parameters of PD also compare advantageously with the erroneously named 'high-efficiency' HD which is now facing the limitations of intermittent procedures, frailty of KT/V as measure of adequacy, importance of sustained fluid removal and time of dialysis. Adequacy should also include life satisfaction and PD also compares favorably as a home therapy. The best approach, also the most intelligent and cost-effective, would be not to underestimate a different therapy, but discover how complementary it can be for success of long term patient treatment.

The prevalent dialysis population keeps increasing steadily. According to the United State Renal Data System (USRDS) 2008 Annual Report, hemodialysis accounted for nearly 92% of incident patients and the number of new patients placed on peritoneal dialysis (PD) has fallen by nearly 3% in 2006, while the number of prevalent PD patients grew by only 3.65% from 2000 to 2006. This is occurring in spite that North-American nephrologists considered overall medical and psychological eligibility for PD to be present in 78% of uremic incident patients [1]. The underutilization of PD also occurs in some countries of Europe, but not in Asia where massive adoption of the PD-first concept proved to be a feasible and successful strategy [2]. HD-maximizing private health care systems, PD discouraging financial reimbursements, and insufficient level of continuous medical formation and fellow training investment are main reasons for such discrepancies in PD penetration [3]. Clinicians' unfounded skepticism must be effaced as well because incomplete presentation of treatment options is also an important reason for underutilization of home dialysis therapies [4].

The main areas of critics and misleading data concerning PD will be briefly discussed.

Patient Survival

Patient survival is the ultimate marker of treatment efficacy, while quality of life is a more difficult to measure therapy target. According to the literature, we can conclude that patient survival is similar with both hemodialysis and PD [5]. Even without statistical expertise, it is recognized how difficult and fallacious it often is to compare such incomparable populations. If, however, a '$p < 0.001$' is requested to guide a dialysis prescription, its exact meaning for the individual patient must be questioned [6]. Analysis with huge numbers from national patient registries often suffer from relevant methodological limitations and reveal statistically significant patient survival differences that in the end mean only a few extra months of survival. Additionally, some common errors in comparing HD and PD survival have been mentioned consistently: patient recruitment bias, eliminating the initial period of dialysis, using prevalent rather than incident patients, including patients started on HD and transferred to PD as 'PD patients' only; misspecification of important risk factors and confounding factors describing case-mix differences, besides not taking into account regional and national differences. This late aspect is indeed notorious: the Canadian Registry was the first to report a survival advantage for PD patients and European registries such as those from the Netherlands, UK, Spain or Italy, generally reproduce similar patient survival, if not better, at least in the first years of treatment. On the contrary, results from the USA are more often negative and it is discussed that worse results account for more comorbid patient characteristics. However, even after adjustment for such variables there are important differences among countries and continents, most presumably dependent on different therapy policies [5, 7].

In the last decade, advances in CKD therapy were reflected in the overall decreased adjusted mortality rates across all dialysis populations.

From the USRDS 2008 report, there is evidence that even with detailed adjustments for severity of disease, the first-year mortality for hemodialysis patients shows little change over the last decade, while rates for PD have declined over the same period (6% in hemodialysis patients for 5 or more years, to more than 26% for the newest PD patients).

First-year mortality in hemodialysis decreased by 4.85% since 2000 reaching 254 deaths per 1,000 patient years at risk. In PD, in contrast, the rate has decreased 64.8% since 1980 and 25.5% since 2000 standing now at 162 deaths per 1,000 patient's years at risk. Adjusted 5-year survival probabilities increased also slightly across all modalities. In the first years, patients on PD have a moderate survival advantage, but even later, at 5 years, their adjusted probability of survival is similar to HD patients in the USA. Some previous negative reports reflect old treatment schedules and less PD experience, besides methodological pitfalls of observational studies [8, 9]. Indeed PD mortality rates have fallen consistently across patients' vintages, also in risk patients such as older, diabetic, fast transporters and anurics. Better cumulative center experience, availability of new technology advances and icodextrine showed improved results that might overcome the time-dependent limitations of PD treatment in at-risk patients [9, 10]. One should also look at the total patient survival and not be limited to an individual therapy survival because patient survival varies in time differently between the therapies. So a patient benefits more from a therapy that preserves residual renal function, preserves vascular capital, offers good quality of life and still leaves renal transplantation and hemodialysis as efficient therapies for the hopefully long renal replacement track. PD as first modality protects global patient survival even after PD technique failure.

Access-Related Morbidity and Technique Survival

In the hemodialysis population, hospitalization due to infection and access problems remain a major clinical and financial burden. Admissions per patient year have decreased lately in PD by 11% (only 2.5% for HD). But marked regional differences exist and investment on ambulatory management of peritoneal catheter implantation, peritonitis treatment and training are important strategies to further reduce PD hospitalization.

It is true that PD technique survival is shorter than in hemodialysis mainly to due access-related infection. But access infections are not less serious in HD, with a higher rate of bacteremia and associated mortality. From a recent USRDS report, 62.3% of new patients starting with hemodialysis in 2006 did so with a catheter as their primary access. Catheter use continues to rise in the prevalent population, and surprisingly not only in aged patients. Access-related complications have been better managed in late years but sepsis events are generally on the rise reaching 2.13 events per patient year in those with catheters (0.44 and 0.77, for fistulas and grafts, respectively).

On the contrary, rates of PD catheter-related complications have fallen since 1998, with a low rate of sepsis, similar to the rate with native fistulas, and decreasing peritonitis events. While in the USA the rate of peritonitis is 2.3 events per patient year, this varies widely according to the country or the center and a lower rate of 0.5 is generally achieved.

Change of therapy certainly impacts on patient quality of life and health costs. However, change is part of an individualization strategy to optimize therapy. In hemodialysis serial fistulas and re-implanted catheters are usually not accounted as failure and technique survival is still allowed in spite of comorbidity and cost. Instead, a patient who manipulates his unique peritoneum is at risk to suffer from a more serious peritonitis or loss of membrane function, but will not have an extra peritoneum. However, PD technique survival has also increased lately due to better treatment protocols, the use of APD and icodextrin. Additionally, the most feared complication, sclerosing peritonitis, while rare, has also been managed better with preventive strategies (adequate membrane function monotorization, promising low GDPs solutions, peritoneal rest, tamoxifen) and curative treatments (imunossupression and surgical skills), with reduced mortality. We must level sepsis risk in hemodialysis versus the rare menace of peritoneal encapsulating peritonitis.

Broad Concept Adequacy

The erroneous criticism that PD is unable to offer adequate dialysis shows to be unfounded since no evidence exists for better performance of standard hemodialysis. Dose of dialysis has been limited to the measurement of water-soluble small solutes, normalized for the volume of distribution, more often as KT/V urea. But a number of questions still remain in the field of hemodialysis: Should dialysis be normalized by urea distribution volume? Should KT/V be measured as equilibrated or single-pool KT/V? Should solute removal be preferably normalized to metabolic rate? [11] This points to the frailty of such a measure. Additionally, the HEMO study showed that increasing the dose of dialysis from eKT/V 1.05 (as the standard) to eKT/V 1.45 (high-dose group) did not impact, or using high-flux membrane did not decrease all-cause mortality in the entire cohort. The global attained survival of hemodialysis patients was around 70% at 24 months and remains not better in many other reports. This is an unreasonable fate, facing the expectations with the high efficiency of a hemodialysis apparatus. More frequent sessions do not solve hemodialysis limitations in terms of medium molecules and volume removal [12]. Phosphate control is hardly achieved with our current praxis and even increasing dialysis sessions do not increase its removal significantly [13]. Treatment time, chronic inflammation and hemodynamic stability are certainly overlooked parameters in hemodialysis quantification justifying the poor patient survival offered by such a high efficiency small solute removal therapy [14]. To such problems, we can add the cardiovascular effects of arteriovenous fistulas, the morbidity and the mortality associated with central venous catheters infections [15] and the high risk of malnutrition, not to mention the burden of regular and rigid dislocations to the clinics and the stress of access puncture. Adequacy means all this, and PD does not shape as less efficient, in many aspects looking advantageous.

In PD, a so-called 'low-efficiency' dialysis as measured by small solute removal per minute, some limitations exist but we generally find similar or better outcomes and opportunities to offer improved dialysis beyond KT/V: continuous dialysis allows peritoneal removal of nonmeasured uremic toxins in parallel with preservation of residual renal function. This probably counteracts the potential menace of the systemic effects of cumulative glucose absorption. In many cases, this type of glucose load may instead decrease the risk of malnutrition so often seen in dialysis patients. Increase of fat mass may occur and we still debate if it is good or bad, although extreme obesity is surprisingly rare in spite of caloric load. Phosphate control is generally similar or better achieved under PD [16] and no evidence exists that cardiovascular events occur more under PD, being much more dependent on predialysis morbidity than on dialysis modality. Besides, acid-base control is usually more easily maintained with bicarbonate PD solutions and anemia treatment usually needs a lower dose of erythropoietin, all these favorable aspects to PD being part of adequacy as a broad concept, beyond KT/V.

Conclusion

Domiciliary dialysis indeed confers the best patient outcomes. PD is a simple, safe, and easy to handle home therapy that can confer adequate support to a great number of uremic patients. The best approach, also the more intelligent and cost-effective, would be not to underestimate a different therapy, but discover how complementary it can be for the success of a long-term patient treatment.

References

1 Mendelssohn DC, Mujais SK, Soroka SD, Brouillette J, Takano T, Barre PE, Mittal BV, Singh A, Firanek C, Story K, Finkelstein FO: A prospective evaluation of renal replacement therapy modality eligibility. Nephrol Dial Transplant 2009;24:555–561.

2 Li PK, Szeto CC: Success of the peritoneal dialysis programme in Hong Kong. Nephrol Dial Transplant 2008;23:1475–1478.
3 Khawar O, Kalantar-Zadeh K, Lo WK, Johnson D, Mehrotra R: Is the declining use of long-term peritoneal dialysis justified by outcome data? Clin J Am Soc Nephrol 2007;2:1317–1328.
4 Mehrotra R, Marsh D, Vonesh E, Peters V, Nissenson A: Patient education and access of ESRD patients to renal replacement therapies beyond in-center hemodialysis. Kidney Int 2005;68:378–390.
5 Burkart J, Piraino B, Kaldas H, Lee JY, Bender FH, Krediet RT, Boeschoten EW, Dekker FW, Vonesh E, Mujais S: Why is the evidence favoring hemodialysis over peritoneal dialysis misleading? Semin Dial 2007;20:200–202.
6 Bargman JM: Is there more to living than not dying? A reflection on survival studies in dialysis. Semin Dial 2007;20:50–52.
7 Goodkin DA, Young EW, Kurokawa K, Prutz KG, Levin NW: Mortality among hemodialysis patients in Europe, Japan, and the United States: case-mix effects. Am J Kidney Dis 2004;44:16–21.
8 Davies SJ: Comparing outcomes on peritoneal and hemodialysis: a case study in the interpretation of observational studies. Saudi J Kidney Dis Transpl 2007;18:24–30.
9 Davies SJ: Mitigating peritoneal membrane characteristics in modern peritoneal dialysis therapy. Kidney Int Suppl 2006;S76–S83.
10 Vonesh EF, Snyder JJ, Foley RN, Collins AJ: The differential impact of risk factors on mortality in hemodialysis and peritoneal dialysis. Kidney Int 2004;66:2389–2401.
11 Leypoldt JK, Cheung AK: Revisiting the hemodialysis dose. Semin Dial 2006;19:96–101.
12 Eloot S, van Biesen W, Dhondt A, de Smet R, Marescau B, De Deyn PP, Verdonck P, Vanholder R: Impact of increasing haemodialysis frequency versus haemodialysis duration on removal of urea and guanidino compounds: a kinetic analysis. Nephrol Dial Transplant 2009, Epub ahead of print.
13 Kooienga L: Phosphorus balance with daily dialysis. Semin Dial 2007;20:342–345.
14 Zsom L, Zsom M, Fulop T, Flessner MF: Treatment time, chronic inflammation, and hemodynamic stability: the overlooked parameters in hemodialysis quantification. Semin Dial 2008;21:395–400.
15 Wasse H: Catheter-related mortality among ESRD patients. Semin Dial 2008;21:547–549.
16 Yavuz A, Ersoy FF, Passadakis PS, et al: Phosphorus control in peritoneal dialysis patients. Kidney Int Suppl 2008;S152–S158.

Anabela S. Rodrigues, MD, PhD
Division of Nephrology, Centro Hospitalar do Porto, Hospital de Santo António
Largo Professor Abel Salazar
PT–4100 Porto (Portugal)
Tel. +351 222 077521, Fax +351 222 077520, E-Mail ar.cbs@mail.telepac.pt

Ronco C, Crepaldi C, Cruz DN (eds): Peritoneal Dialysis – From Basic Concepts to Clinical Excellence. Contrib Nephrol. Basel, Karger, 2009, vol 163, pp 243–249

How to Make Peritoneal Dialysis Affordable in Developing Countries

Georgi Abraham, Pallavi Khanna, Milly Mathew, Poorna Pushpkala, Anurag Mehrotra, Aswin Sairam, Asik Ali Mohamed Ali

Division of Dialysis and Renal Transplantation, Department of Nephrology, Madras Medical Mission Hospital, Chennai, India

Abstract

Peritoneal dialysis (PD) is an underutilized renal replacement therapy in the developing world. It offers advantages of simplicity, reduced need of training, lack of dependence on infrastructure and location. The population is extremely underserved by healthcare and means to achieve it. PD is unavailable in many African nations. We explore the logistics of PD, domestic manufacture of PD fluid and accessories and ways to sustain it. Realization of local factors, ways to reduce peritonitis, reduced dosage in patients with residual renal function and use of generics to treat anemia that help improve the logistics. The role of national government especially in countries where dialysis is rationed and its lack of involvement leaving the billions to fetch for themselves is discussed. Innovative schemes by private insurers have improved PD outcome locally. These include subsidized once-in-a-lifetime PD treatment payment and industry sponsored nurse and technician visits to patients. Finally, the factors preventing nephrologists in delivering PD such as lack of training, reimbursement, infrastructure and affordability are discussed.

The global burden of chronic kidney disease (CKD) is concealed behind statistics, reflecting merely the number of people treated, which is far below those dying of kidney failure and associated cardiovascular complications. An assessment of health care sector in developing countries reveals a poor performance on key dimensions of dialysis coverage, purchasing and delivery. Healthcare coverage is available for less than 15% of the population [1]. Incident end-stage renal disease (ESRD) patients started on dialysis and patients maintained on peritoneal dialysis (PD) vary widely in the developing world. Besides economic factors influencing PD utilization, others include, problems with PD training of physicians, technicians and nurses, the availability of CKD education programmes and issues concerning long-term viability of the therapy.

In developing countries in Asia, PD offers certain clear advantages over hemodialysis (HD) such as simplicity of the therapy, reduced need for trained technicians and nurses, minimal requirement for technical support, lack of dependence on electricity, online water purification, home-based therapy with institutional independence which has potential cost savings. There is an urgent call for dialysis services to be affordable, cost effective, value-added and suited to local circumstances.

Peritoneal Dialysis and Economics

It has been observed that in renal replacement therapy (RRT) including PD, the availability and growth of the treatment modality is directly related to the financial prowess of the nation. However, in parts of Latin America, South Asia and eastern European nations, the growth of PD population is impressive.

In India, 5.2% of the GDP is spent on healthcare with a meager 0.9% by government and the remaining by private companies. The average GDP per capita income of an Indian is USD 705. The treatment rate for dialysis in Asian countries shows a definite relationship with the per capita income: the wealthier the nation, the higher the cost of treatment [2–4].

PD utilization in Africa is strikingly low: of the 53 African countries, PD is available only in 12 and it is being delivered only in the public sector [5]. The current annual cost of PD therapy is USD 11,680 in Sudan which is quite low as compared to the developed Asian, European and North American economies [5]. In many countries, PD therapy does not even exist.

Peritoneal Dialysis and Logistics

There is an unmet need to explore ways of providing high-quality, low-cost RRT services as the expenses of imported consumables contributes to the escalating cost of dialysis in developing economies.

To a greater extent, this has been overcome by domestic manufacture of peritoneal dialysis fluid and accessories in India, China and Latin American countries. Given the scale of local demand, domestic manufacturing can be made feasible by meeting the local need as well as exporting to neighboring countries – an example being India and Southeast Asia. The availability of generic erythropoietin and parenteral iron sucrose at affordable cost in India enables longer survival and optimum management of anemia.

The cost of PD consumables is determined by cost per unit of PD fluid and volume of fluid used. In Asia and Africa, if the dialysate manufacturing companies bring down the cost of their products, a positive effect will be seen on PD utilization. The appropriate use of smaller exchange volumes (6 liters/day) in patients with smaller body size or those with residual renal function can positively impact the daily cost of PD without comprising adequate dialysis [6].

A reduction in complications of PD by implementing better 'connectology' techniques that aim at decreasing peritonitis rates or using newer PD solutions such as icodextrin may reduce the associated hospitalization and treatment costs [6, 7]. If newer solutions such as icodextrin, amino acid-containing solutions and those with higher pH are to be widely used in developing countries, the cost differential between the current and the newer products should be negligible.

Malnutrition, highly prevalent in some sections in PD patients, in the developing countries is associated with higher rate of peritonitis especially using subjective global assessment [8].

Change within individual countries in the developing world depends on a facultative international environment whereby professional organizations, government, nongovernmental bodies and industry provide support for drugs and dialysis consumables [9].

Peritoneal Dialysis and Government

A framework for improving the global equity of access to PD therapy and the development of appropriate government policy has a genuine potential to bring PD therapy to many more in need.

National planning for RRT delivery is essential to contain cost and promote equity in resource allocation. As rationing dialysis is the norm for low and middle income countries (LMIC), clear policies regarding eligibility for PD need to be formulated. In publicly funded countries of Africa, Asia and Latin America, rationing has been practiced since the introduction of PD [5, 10]. Patients most likely to be accepted for RRT therapy were aged 20–40 years, white, employed, married, nondiabetic and those who live in proximity to the dialysis center [10].

About 60–90% of the patients cannot afford RRT including PD yielding to social factors related to poverty [3, 4, 10]. In developing countries, socioeconomic factors influence the decision to accept patients more profoundly than medical ones. A look at the Indian CKD registry data in figure 1 shows that 37.3% of patients have a monthly income of less than USD 100, the majority are males (69.5%), 31.4% are diabetics and age range is between 19 and 60 years.

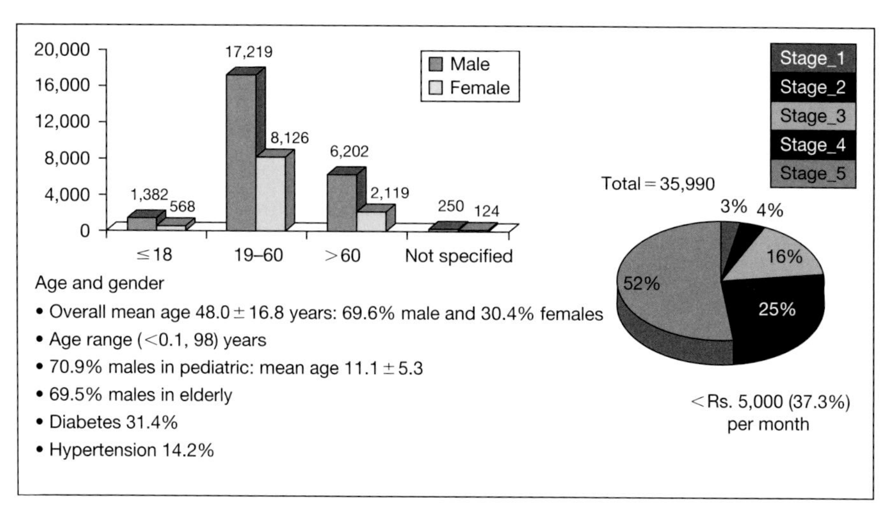

Fig. 1. Indian CKD registry data.

This stark reality with hardly any insurance coverage for Indian PD patients precludes the growth and expansion of PD programmes in developing countries. As the residual renal function declines, a fall in Kt/V below 1.7, in the patients incurring the cost of PD therapy themselves, the number of exchanges drops over a period of time, leading to inadequate dialysis, increased morbidity and mortality.

Peritoneal Dialysis and Insurance

A recent initiative by a medical insurance company involves a payment USD 50 ensuring coverage for peritonitis treatment including hospitalization and antibiotics for PD patients in India is a welcome innovation. Other initiatives such as a once-in-a-lifetime payment by the patient, either upfront or in installments, to the dialysis fluid manufacturing industry assures a lifetime supply of PD fluid and accessories.

Currently, the once-in-a-lifetime cost for three exchanges a day is USD 11,533 and for four exchanges in double bag disposable system is USD 13,037. The current expenses for those out of the umbrella of this lifetime scheme are USD 375 to USD 500 per month. Similar schemes are also available for automated PD.

Once the patient enrolls in a lifetime continuous ambulatory peritoneal dialysis (CAPD) scheme they are provided with a month's supply of fluid and acces-

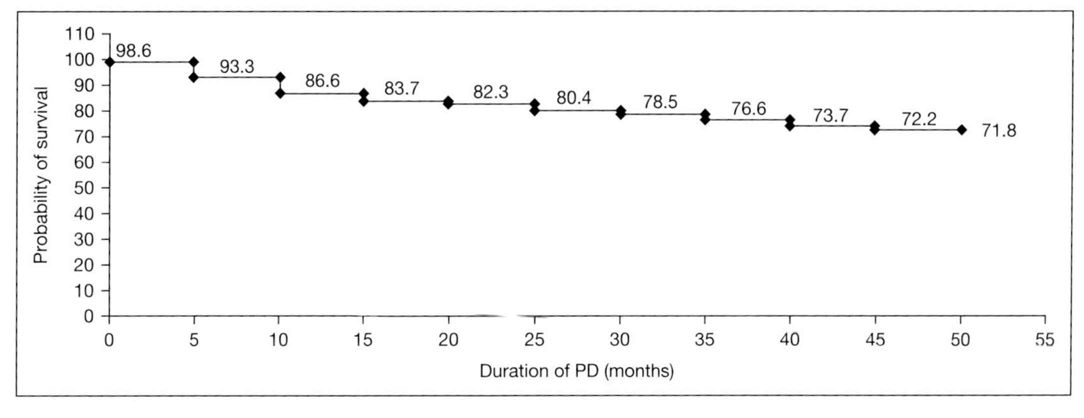

Fig. 2. Overall survival curve of PD patients (n = 209).

sories such as minicaps. Under circumstances such as a switch to HD or renal transplantation or death, the remaining credit amount is refunded to the patient or family after deducting a nonrefundable amount of USD 1,100. A legal agreement is made between the patient and the industry as per existing laws of the country. The supply of stocks is made available anywhere in India and there is no extra transportation cost. All ESRD patients can avail the once-in-a-lifetime scheme. This system of payment may enable the growth of CAPD using the twin bag system as the expenses for lifetime is assured without any cost escalation.

An unpublished data involving a multicenter retrospective study from South India showed that of a majority of patients who survived on PD for 3 years or more, 46% belonged to once-in-a-lifetime payment scheme and 21% were fully reimbursed from their employers as shown in figure 2. A new initiative by the south Indian provincial (state) government to provide more public funding for dialysis patients is a welcome move.

The choice of treatment modality is a gray zone in developing countries as there seems to be an interaction between patient characteristics, reimbursement policies, economic status and mortality risks [11].

Peritoneal Dialysis and the Provider

A recent survey undertaken in 265 south Asian nephrologists from India, Sri Lanka, Bangladesh, Pakistan and Nepal, as shown in figure 3, considers lack of CKD education, reimbursement, time commitment and infrastructure issues as major limitations for expansion and utilization of PD programmes in these countries.

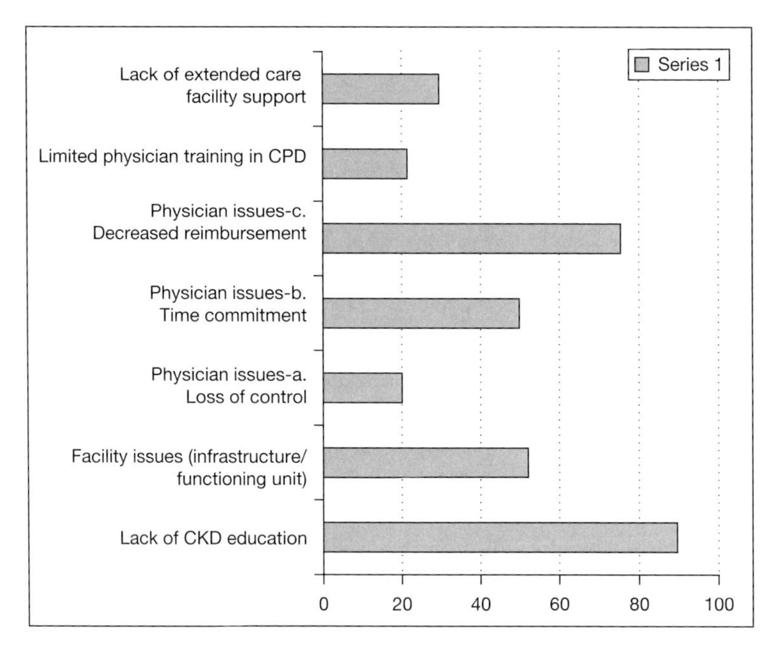

Fig. 3. Limitations in CAPD utilization as perceived by nephrologists (n = 265).

As there is a global shortage of nephrologists, PD can be a better modality in the hands of trained nurses and technicians [12]. Clinical coordinators consisting of trained nurses and technologists employed by the dialysis manufacturing industry who visit PD patients at their place of residence to monitor dialysis practices, educate exit site care and prevention of peritonitis, has reduced the cost of hospitalization and enabled more employment in the peritoneal dialysis population.

References

1 CII and Mckinsey Report: Healthcare in India: The road ahead. 2002:1–2.
2 Li PK, Chow KM: The cost barrier to peritoneal dialysis in the developing world: an Asian perspective. Perit Dial Int 2001;21(suppl 3):S307–S313.
3 Abraham G: The challenges of renal replacement therapy in Asia. Nat Clin Pract Nephrol 2008; 4:643.
4 Abraham G, Pratap B, Sankarasubbaiyan S, Govindan P, Nayak KS, Sheriff R, Naqvi SA: Chronic peritoneal dialysis in South Asia: challenges and future. Perit Dial Int 2008;28:13–19.
5 Elhassan EA, Kaballo B, Fedail H, et al: Peritoneal dialysis in the Sudan. Perit Dial Int 2007; 27:503–510.
6 Thiagarajan T, et al: A correlative study of adequacy parameters using Kt/V and dialysis prescription and survival in continuous ambulatory peritoneal dialysis. Ind J Periton Dial 2007;12:24–27.

7 Abraham G: Are 3 exchanges suitable for Asian patients on peritoneal dialysis. Perit Dial Int 2003; 23:S37–S39.
8 Prasad N, Gupta A, Sharma RK, Sinha A, Kumar R: Impact of nutritional status on peritonitis in CAPD patients. Perit Dial Int 2007;27:42–47.
9 White SL, Chadban SJ, Jan S, Chapman JR, Cass A: How can we achieve global equity in provision of renal replacement therapy? Bull WHO 2008;86:229–237.
10 Moosa MR, Kidd M: The dangers of rationing dialysis treatment: the dilemma facing a developing country. Kidney Int 2006;70:1107–1114.
11 Li PK, Lui SL, Leung CB, et al: Increased utilization of peritoneal dialysis to cope with mounting demand for renal replacement therapy: perspectives from Asian countries. Perit Dial Int 2007;27(suppl 2):S59–S61.
12 Field M: Addressing the global shortage of nephrologists. Nat Clin Pract Nephrol 2008;4:583.

Prof. Georgi Abraham
Department of Nephrology, Madras Medical Mission Hospital
Mogappair
Chennai 600 037 (India)
Tel. +91 984 142 0992, E-Mail abraham_georgi@yahoo.com

Ronco C, Crepaldi C, Cruz DN (eds): Peritoneal Dialysis – From Basic Concepts to Clinical Excellence. Contrib Nephrol. Basel, Karger, 2009, vol 163, pp 250–256

Peritoneal Dialysis and Renal Transplantation

Thierry Lobbedez, Angélique Lecouf, Odette Abbadie, Maxence Ficheux, Bruno Hurault de Ligny, Jean-Philippe Ryckelynck

Néphrologie, Dialyse et Transplantation, CHU Clemenceau, Caen, France

Abstract

Peritoneal dialysis is commonly used in patients awaiting renal transplantation. The occurrence of delayed graft function is lower in CAPD patients than in hemodialysis patients. This could be explained by the fluid expansion observed in CAPD patients before renal transplantation. Acute allograft rejection incidence is similar in peritoneal dialysis patients and hemodialysis patients. There are controversial data regarding the rate of renovascular thrombosis after renal transplantation in peritoneal dialysis patients. The dialysis modality selected prior to transplantation may explain the rate of renovascular thrombosis in peritoneal dialysis patients. There is an increasing number of patients returning to dialysis after transplantation failure. However, peritoneal dialysis is underused in failed transplant patients. There are few data available regarding the impact of dialysis modality on the outcome of failed transplant patients. Immunosuppression and transplant nephrectomy may affect the outcome of these patients on peritoneal dialysis. The aim of this article is to review the use of peritoneal dialysis in patients awaiting renal transplantation and in failed transplant patients.

Survival and quality of life comparisons have shown that renal transplantation is a better overall treatment for end-stage renal disease patients. Peritoneal dialysis (PD) is now widely used for patients awaiting renal transplantation. It is well established that patient survival after renal transplantation does not correlate with pretransplantation dialysis modality [1, 2]. For no clear reason, PD is still underused in patients awaiting renal transplantation [2]. Despite tremendous efforts to increase renal allograft survival, nephrologists have to deal with an increasing number of patients returning to dialysis after allograft failure [3]. This has raised the question of whether or not the choice of the dialysis modality has an impact on patient outcome following renal transplantation failure. In addition, failed transplant patients may have a different outcome on PD than other patients. The aim of this review is to evaluate the use of PD in patients awaiting renal transplantation and after allograft failure.

Transplantation in Peritoneal Dialysis Patients

PD and Delayed Graft Function (DGF)

According to various studies, the occurrence of DGF is lower in CAPD compared to hemodialysis (HD) patients. Perez Fontan et al. [4] compared the results of renal transplantation in a group of patients on CAPD with a matched population of HD patients. The rate of DGF was lower in the group of PD patients. However, patients on HD were dialyzed immediately before surgery more frequently than patients on PD, which could possibly explain a longer cold ischaemia time. Bleyer et al. [5], in a large population, found a significantly lower rate of DGF in the PD group than in the HD group, but patients on PD had a shorter cold ischemia time. However, in the multivariate analysis HD was a risk factor for DGF. In a large United States patient cohort, PD was associated with a lower risk of DGF, the latter being associated with a greater risk of death, a greater risk of graft failure and a greater risk of death censored graft failure [1]. In the North American pediatric renal transplant cooperative study registry, there was no significant difference regarding the rate of DGF between patients on HD and patients on PD [6].

Expanded volume status in CAPD patients may explain in part the lower incidence of DGF in PD patients. In support of this, it has been shown that CAPD patients have a greater weight loss after transplantation than HD patients [7]. Preoperative mean pulmonary artery pressure was high in CAPD patients investigated prior to transplantation [8]. The decline in serum creatinine after transplantation did not differ between HD and PD patients when CAPD patients were submitted to ultrafiltration before transplantation, emphasizing the role of volume status in the occurrence of DGF [9]. This hypothesis was tested by Van Biesen et al. [10]. After adjustment of the volume status, PD was associated with a lower risk of DGF than HD. In a case control study, the rate of DGF was not statistically different in APD compared to HD patients, suggesting that APD may not provide the same advantages as CAPD regarding the rate of DGF [11].

PD and Infections after Transplantation

Only one study found a greater post-transplant infection incidence in PD patients compared to HD patients. Passalacqua et al. [12] studied the incidence of early infectious complications in 156 renal transplant recipients. Patients receiving PD before transplantation had higher rates of infection compared with patients on HD. However, in this study the rate of acute rejection was higher in PD patients. O'Donoghue et al. [13] analyzed the data of 500 renal transplants. Post-transplant peritonitis developed in 10 patients; 6 cases were associated with post-transplant CAPD, 2 with exit-site infection that was present at the time of transplantation. Two case-control studies did not show any significant difference in the rate of infection between PD and HD patients [4, 14]. Bakir et al. [15] retrospectively analyzed the incidence of peritonitis within 90 days after transplantation in 232 patients. There were 30 documented cases of peritonitis, giving a prevalence of 13%. Post-transplant risk factors for peritonitis were technical surgical problems, nonfunctioning graft and urinary leak. Post-transplant peritonitis can be successfully managed with antibiotic therapy [16]. To decrease the risk of infection, some units remove the catheter at the time of transplantation. In the study of Passalacqua et al. [12], PD patients who did not have their catheter removed within 6 days after transplantation were nine times more likely to develop infection within 30 days posttransplantation. However, PD can be undertaken after renal transplantation in the case of DGF if the peritoneal catheter is left in situ [17]. In order to prevent post-transplantation infection, the catheter should be removed in case of exit-site infection or when the DGF risk is low.

PD and Acute Allograft Rejection

In the era of modern immunosuppression, the pretransplant dialysis modality has no effect on the rate of acute allograft rejection. An early study reported a higher number of acute rejections in PD patients compared to HD patients [18]. In the study of Vanholder et al. [19], the number of patients who developed one or more rejection episodes was greater in the CAPD than the HD group. Passalacqua et al. [12] observed that PD patients were more likely to have an episode of rejection during the first 30 days after transplantation than HD patients. On the contrary, in 3 different studies there was no significant difference in the rate of rejection between HD and PD patients [4, 9, 20]. In the studies of both Bleyer et al. [5] and Snyder et al. [1], the risk for rejection, expressed as the percentage of patients, was similar in CAPD and HD patients.

PD and Allograft Renovascular Thrombosis

It should be taken into account that a hypercoagulable state may influence dialysis modality choice, which may bias the results of the studies about PD and renovascular thrombosis after transplantation. In the study of Murphy et al. [21], there were 9 graft thromboses and all occurred in CAPD patients. Multivariate analysis revealed that CAPD was the only factor associated with renal allograft thrombosis in the study of Van der Vliet et al. [22]. In a case-control analysis using the data of the USRDS, PD was a risk factor for renovascular thrombosis [23]. Of the 60 grafts lost because of vascular thrombosis, 60% were in the PD group, 22% in the HD group and 18% in the pre-emptive transplantation group in a pediatric population [6]. In their large study, Snyder et al. [1] were able to obtain data about the cause of graft loss in 30% of patients. Graft thrombosis was more commonly listed as a cause of graft failure among PD patients compared to HD patients.

In contrast, a large retrospective study showed that the frequency of thrombosis was not significantly different in PD patients compared to HD patients [24].

PD in Failed Transplant Patients

The condition of failed transplant patients at dialysis initiation, as well as immunosuppressive therapy and the need for allograft nephrectomy, may have an impact on the outcome of these patients on dialysis [25]. Dialysis initiation is frequently un-planed in failed transplant patients [26]. Failed transplant patient withdrawal of the immunosuppressive therapy may affect peritonitis risk and technique survival.

Two different studies, based on a limited number of patients, showed that patients starting PD after transplantation failure had a similar outcome to those treated by HD after failed transplantation [27, 28]. Although allograft failure is becoming a frequent cause of end-stage renal disease, patients starting PD after failed transplantation are relatively rare. In a study based on the Australian and New Zealand population, allograft failure represented 2.2% of the causes of PD initiation [29]. Patients with a failed transplant accounted for only 2% of the total population in the cohort of incident PD patients extracted from the Baxter Healthcare Corporation On Call database [30]. In addition, only half of patients treated by PD before transplantation restart PD after allograft failure [26].

Three different studies showed that failed transplant did not adversely affect patient outcome on PD [29–31]. Only one single-center retrospective study found that transplanted patients had a worse outcome on PD compared with nontransplanted patients [32]. In two large-scale studies, after adjustment for baseline characteristics, patients starting PD after renal allograft failure had a similar survival on PD to nontransplanted patients [29, 30].

There was no significant difference regarding the technique survival between failed transplanted patients and non-transplanted patients in either the study of Mujais et al. [30] or that of Badve et al. [29]. Two different studies have shown that decline of residual renal function on dialysis is more rapid in the failed transplant group compared with the nontransplanted group [27, 33]. In addition, failed transplant patients are exposed to transplantectomy, which may induce surgical complications and jeopardize the technique survival. Furthermore, patients become anuric following allograft nephrectomy. This may, in turn, provoke a temporary or definitive contraindication to PD.

Only one study has shown that failed transplant patients experienced a poorer survival free of peritonitis than nontransplanted patients [32]. In this study, failed transplant patients were matched with never transplanted patients for age and diabetes. However, there was no adjustment for the comorbid condition or PD modality. Continuation of immunosuppression after transplantation failure is associated with a higher rate of infection in dialysis patients [34]. On the other hand, it has been suggested that continued immunosuppression after allograft failure may prolong patient survival on dialysis [35].

Failed transplant patients are more frequently treated by APD than non-transplanted patients. In the study of Mujais et al. [30], 65% of patients who started PD after allograft failure were treated by APD. This may be due to the patient's choice in view of the mean age of the failed transplant group. Even though this notion is still a matter for debate, it has been hypothesized that failed kidney transplantation is associated with a high peritoneal transport rate, which could partially explain the rate of APD utilization in this group of patients [36].

Conclusion

There is no doubt that PD is a suitable method in patients awaiting renal transplantation. The results of transplantation are not significantly associated with the pretransplant dialysis modality. PD can be used after allograft failure. However, two specific issues must be taken into account in this group of patients: the immunosuppressive therapy and the frequent need for transplant nephrectomy.

References

1 Snyder JJ, Kasiske BL, Gilbertson DT, Collins AJ: A comparison of transplant outcomes in peritoneal and haemodialysis patients. Kidney Int 2002;62:1423–1430.

2 Chalem Y, Ryckelynck JP, Tuppin P, Verger C, Chauve S, Glotz D; French Collaborative Group: Access to, and outcome of, renal transplantation according to treatment modality of end-stage renal disease in France. Kidney Int 2005;67:2448–2453.
3 USRDS 2004 annual data report. Am J Kidney Dis 2005;45:S8–S280.
4 Perez-Fontan M, Rodriguez-Carmona A, Garcia-Falcon T, Moncalian J, Oliver J, Valdes F: Renal transplantation in patients undergoing chronic peritoneal dialysis. Perit Dial Int 1996;16:48–51.
5 Bleyer AJ, Burkart JM, Russell GB, Adams PL: Dialysis modality and delayed graft function after cadaver renal transplantation. J Am Soc Nephrol 1999;10:154–159.
6 Vats AN, Donaldson L, Fine RN, Chavel BM: Pre-transplant dialysis status and outcome of renal transplantation in north american children: a NAPRTCS study. Transplantation 2000;69:1414–1419.
7 Maiorca R, Sandrini S, Cancarini GC, Camerini CC, Scolari F, Cristinelli L, Filippini M: Kidney transplantation in peritoneal dialysis patients. Perit Dial Int 1994;14(suppl 3):S162–S168.
8 Issad B, Mouquet C, Bitker MO, Allouache M, Baumelou A, Rotte J, Jacobs C: Is overhydratation in CAPD patients a contraindication to renal transplantation? Adv Perit Dial 1994;10:68–72.
9 Lambert MC, Bernaert P, Vijt D, de Smet R, Lameire N: CAPD a risk factor in renal transplantation? ARF after renal transplantation. Perit Dial Int 1996;16(suppl 1):S495–S498.
10 Van Biensen W, Vanholder R, Van Loo A, Van der Vennet M, Lameire N: Peritoneal dialysis favorably influences early graft function after renal transplantation compared to haemodialysis. Transplantation 2000;69:508–514.
11 Lobbedez T, Rognant N, Hurault de Ligny B, el Haggan W, Allard C, Ryckelynck JP: Impact of automated peritoneal dialysis on initial graft function after renal transplantation. Adv Perit Dial 2005;21:90–93.
12 Passalacqua JA, Wiland AM, Fink JC, Bartlett ST, Evans DA, Keay S: Increased incidence of postoperative infections associated with peritoneal dialysis in renal transplant recipients. Transplantation 1999;68:535–540.
13 O'Donoghue D, Manos J, Pearson R, Scott P, Bakran A, Johnson R, Dyer P, Martin S, Gokal R: Continuous ambulatory peritoneal dialysis and renal transplantation: a ten year experience in a single centre. Perit Dial Int 1991;12:242–249.
14 Triolo G, Segoloni GP, Salomone M, Piccoli GB, Messina M, Massara C, Boggio-Bertinet D, Vercellone A: Comparison between two dialytic populations undergoing renal transplantation. Adv Perit Dial 1990;6:72–75.
15 Bakir N, Surachno S, Sluiter WJ, Struijk DG: Peritonitis in peritoneal dialysis patients after renal transplantation. Nephrol Dial Transplant 1998;13:3178–3183.
16 Leichter HE, Saluski IB, Ettenger RB, Jordan SC, Hall TL, Marik RN: Experience with renal transplantation in children undergoing peritoneal dialysis (CAPD/CCPD). Am J Kidney Dis 1986;8:181–185.
17 Gokal R, Kost S: Peritoneal dialysis immediately post transplantation. Adv Perit Dial 1999;15:112–115.
18 Guillou PJ, Will EJ, Davidson AM, Giles GR: CAPD a risk factor in renal transplantation. Br J Surg 1984;71:878–880.
19 Vanholder R, Heering P, Van Loo A, Van Biesen W, Lambert MC, Hesse U, Van der Vennet M, Grabensee B, Lameire N: Reduced incidence of acute renal graft failure in patients treated with peritoneal dialysis compared with haemodialysis. Am J Kidney Dis 1999;33:934–940.
20 Joseph JT, Jindal RM: Influence of dialysis on post-transplant events. Clin Transplant 2002;16:18–23.
21 Murphy BG, Hill CM, Middleton D, Doherty CC, Brown JH, Nelson Kernohan RM, Keane PK, Douglas JF, McNamee PT: Increased renal allograft thrombosis in CAPD patients. Nephrol Dial Int 1994;9:1166–1169.
22 Van der Vliet JA, Barendregt WB, Hoitsma AJ, Buskens FGM: Increased incidence of renal allograft thrombosis after continuous ambulatory peritoneal dialysis. Clin Transplant 1996;10:51–54.
23 Ojo AO, Hanson JA, Wolfe RA, Agodoa LY, Leavy SF, Leichtman A, Young EW, Port FK: Dialysis modality and the risk of allograft thrombosis in adult renal transplant recipient. Kidney Int 1999;55:1952–1960.

24 Perez-Fontan M, Rodriguez-Carmona A, Garcia Falcon T, Tresancos C, Bouza P, Valdes F: Peritoneal dialysis is not a risk factor for primary vascular graft thrombosis after renal transplantation. Perit Dial Int 1998;18:311–316.
25 Gill JS, Abichandani R, Khan S, Kausz AT, Pereira BJ: Opportunities to improve the care of patients with kidney transplant failure. Kidney Int 2002;6:2193–200.
26 Lobbedez T, Cousin M, Hurault de Ligny B, Ficheux M, El Haggan W, Ryckelynck JP: Failed transplant patients: dialysis initiation and short-term outcome. Nephrol Ther 2008;Dec 12 [Epub ahead of print].
27 Davies J: Peritoneal dialysis in the patient with a failing renal allograft. Perit Dial Int 2001;21: S280–284.
28 de Jonge H, Bammens B, Lemahieu W, Maes BD, Vanrenterghem Y: Comparison of peritoneal dialysis and haemodialysis after renal transplant failure. Nephrol Dial Transplant 2006;21:1669–1674.
29 Badve SV, Hawley CM, McDonald SP, Mudge DW, Rosman JB, Brown FG, Johnson D: Effect of previously failed kidney transplantation on peritoneal dialysis outcomes in the Australian and New Zealand patients population. Nephrol Dial Transplant 2006;21:776–783.
30 Mujais S, Story K: Patients and technique survival on peritoneal dialysis in patients with failed renal allograft: a case-control study. Kidney Int 2006 70:S133–S137.
31 Duman S, Asci G, Toz H, Ozkahya M, Ertilav M, Sezis M, Ok E: Patients with failed renal transplant may be suitable for peritoneal dialysis. Int Urol Nephrol 2004;36:249–252.
32 Schiffl H, Mucke C, Lang SM: Rapid decline of residual renal function in patients with late renal transplant failure who are re-treated with CAPD. Perit Dial Int 2003;23:398–400.
33 Sasal J, Naimark D, Klassen J, Shea J, Bargman J: Late transplant failure: an adverse prognostic factor at initiation of peritoneal dialysis. Perit Dial Int 2001;21:405–410.
34 Smak Gregoor PJ, Zietse R, van Saase JL, op de Hoek CT, Ijzermans JN, Lavrijssen AT, de Jong GM, Kramer P, Weimar W: Immunosuppression should be stopped in patients with renal allograft failure. Clin Transplant 2001;15:397–401.
35 Jassal SV, Lock CE, Walele A, Bargman J: Continued transplant immunosuppression may prolong survival after return to peritoneal dialysis: results of a decision making analysis. Am J Kidney Dis 2002;40:178–183.
36 Wilmer WA, Pesavento TE, Bay WH, Middendorf DF, Donelan SE, Frabott SM, Mc Elligott RF, Powell SL: Peritoneal dialysis following failed kidney transplantation is associated with high peritoneal transport rates. Perit Dial Int 2001;21:441–443.

T. Lobbedez
Néphrologie, Dialyse et Transplantation, CHU Clemenceau
Avenue G Clemenceau
FR–14033 Caen Cedex (France)
Tel. +33 2 31 27 25 76, E-Mail lobbedez-t@chu-caen.fr

Ronco C, Crepaldi C, Cruz DN (eds): Peritoneal Dialysis – From Basic Concepts to Clinical Excellence. Contrib Nephrol. Basel, Karger, 2009, vol 163, pp 257–260

Decision Making around Dialysis Options

Andrew Mooney

Renal Unit, St James's University Hospital, Leeds, UK

Abstract

Introduction: We have previously shown that information given to patients approaching end stage renal failure to make an informed decision about dialysis modality is frequently incomplete and difficult to comprehend [1]. We have now studied whether there are differences in decisions made about dialysis modality according to the method employed to deliver this information. **Methods:** In an on-line study, 784 participants viewed treatment information about hemodialysis (HD) and continuous cycling peritoneal dialysis (CCPD) and completed a questionnaire. A control group saw only basic information, but otherwise treatment information was varied by format (written or videotaped) and who presented the information (male or female; 'patient' or 'doctor'). The information was carefully controlled to ensure comparable content and comprehensibility. In addition to collection of demographic data, measures included: treatment choice, reasons for treatment choice, decisional conflict, need for affect, need for cognition, decision regret, quality of information, previous knowledge of end-stage renal failure and social comparison. **Results:** There were a number of differences in choices made among subjects who viewed written or video information presented as if by doctors or patients. There was a statistically significant effect that subjects chose the dialysis modality recommended by the patient (whether CCPD or HD). There was no significant effect of the gender of the person presenting information on the modality chosen. However, among participants, females were more satisfied with the information presented, and more likely to choose CCPD (compared to male participants). Subjects' style of information processing (need for cognition/need for affect) had no significant effect on choice of dialysis modality. There was a higher drop-out rate among subjects viewing videotaped information. **Conclusion:** The use of testimonials might bias patients decision making regarding dialysis options and until these effects are understood, they should be used with caution.

All health care professionals treating kidney patients would want to help them make a good decision about the dialysis modality they choose. There is now a substantial scientific basis to inform how we present information to patients and influence their decision-making. However, this is not widely used by members of the renal multi-disciplinary team and so we might inadvertently bias patients and prevent them making a decision that is good for them. This article is designed to help understand how this might happen, and consider strategies to reduce its likelihood of happening.

How Do Patients Make Decisions?

As much as we would like to think we are unique and complex, we all make decisions in one of two ways – heuristically or systematically. A heuristic decision is a quick 'rule of thumb' which works effectively most of the time. When we do this we concentrate on perhaps just one aspect of the decision, and choose the first option that will do without thinking about every aspect – for instance when choosing whether to have red or white wine with our dinner, one does not consider the nutritional aspects of each. The other way to make a decision is to utilise a systematic strategy. When we do this, we consider the risks and benefits of all options, assimilate these with our own beliefs, and choose the most appropriate alternative based on this process.

Although usually a heuristic strategy is satisfactory, when there is an important decision to make, a systematic strategy is more likely to lead to a good decision.

What Is a 'Good Decision'?

A good decision is not easy to define, but it is important to recognise that it is different from a 'good' or 'bad' outcome. For instance, a patient who chooses peritoneal dialysis because it fits best with their lifestyle has not made a bad decision if they develop peritonitis after a few weeks treatment. A good decision is one in which the decision making process is good. In this example, the decision will have been a good one if the peritonitis is not a surprise to the patient and they do not regret their treatment choice (and perhaps seek a change of treatment modality). It is strongly suspected that patients who have made good treatment decisions are more compliant and concordant with their treatment.

How Can Patients Be Helped to Make Good Decisions?

Strategies that help patients utilise systematic decision making processes are more likely to lead to a good decision. There are two important principles: first, a patient needs to be alerted when there is a decision to be made, and, second, giving more information is *not* the same as helping patients make a decision.

Considering the first point – during a consultation many interactions might take place; the patient's blood pressure might be reviewed, an increase in ESA dose might be prescribed, and dialysis modality might be discussed. Clearly the first two need no decisions to be made on the part of the patient but the last one does. By indicating to the patient that a decision needs to be made (and by realising ourselves that we are undertaking a different activity to simply changing a prescription), the patient receives a prompt to engage with the information presented and utilise systematic strategies to process it.

Considering the second point, simply providing more information can 'overload' the patient and make it more likely that they use heuristic strategies. By providing excessive information, a patient will start trying to use memory functions (much like a computer with several programs running simultaneously), and fail to engage fully with the task of decision-making and instead take the first option that will do (i.e. take a heuristic decision). Also, the information we give patients at different times has different purposes. For example the information given to patients to support them using aseptic technique to prevent peritonitis, or to describe what ward to report to when arranging PD tube insertion, is irrelevant to dialysis modality decision-making, but very important when preparing patients for their procedures. Such details, therefore, need not be included in information designed to help decision making. Furthermore, the format in which the information is given when asking patients to make decisions can make this process easier for patients and this is described in the last section.

How Can We Present Information to Facilitate Decision-Making?

There are several established principles of presenting information to patients which help them make decisions between treatment options.

(1) Facts should be presented in neutral terms. Words such as 'good' and 'bad' should be avoided. Language should be simple and readable (short sentences, short words with few syllables).

(2) Risks should be presented in absolute numbers rather than verbal descriptions (e.g. 'eight out of ten', rather than 'high proportion').

(3) Equal amounts of information should be provided about each option (i.e. HD and PD), and equal numbers of 'positive' and 'negative' factors about each option should be given. A table or decision tree might be helpful.

(4) Medical terms with explanations should be used rather than vague terms or euphemisms (e.g. 'peritoneal cavity, which is the space inside the abdomen', rather than 'stomach').

(5) Providing written, audio or video recorded information can be helpful as memory aids. However, the use of patient stories might cause patients to concentrate on the characteristics of the patient (age, gender, race, etc.), rather than the information given (and thus make a heuristic decision).

(6) Care should be taken to avoid 'information leakage'. This is a process by which words used can suggest an opinion by the person giving the information (e.g. 'risk' versus 'chance' of an event occurring; this might cause the listener to conclude the speaker does not favor that choice).

Conclusion

Although the principles (above) are simple, studies show that many renal units and renal charities fail to provide information in this way [1, 2]. However, by providing information to patients in a way which enables active systematic decision making, we increase our chances of providing true patient centred treatment, and probably create more concordant, less anxious patients.

Acknowledgement

The author wishes to acknowledge the contributions of Prof. Mark Conner, Dr. Hilary Bekker and Anna Winterbottom to the preparation of this paper.

References

1 Winterbottom A, Conner M, Mooney A, Bekker HL: Evaluating the quality of patient information provided by Renal Units across the UK. Nephrol Dial Transplant 2007;22:2291–2296.

2 Finkelstein FO, Story K, Firanek C, Barre P, Takano T, Soroka S, Mujais S, Rodd K, Mendelssohn D: Perceived knowledge among patients cared for by nephrologists about chronic kidney disease and end-stage renal disease therapies. Kidney Int 2008;74:1178–1184.

Dr. Andrew Mooney, Consultant Renal Physician
St James's University Hospital
Leeds LS9 7TF (UK)
Tel. +44 113 2066869, Fax +44 113 2066216, E-Mail andrew.mooney@leedsth.nhs.uk

Ronco C, Crepaldi C, Cruz DN (eds): Peritoneal Dialysis – From Basic Concepts to Clinical Excellence. Contrib Nephrol. Basel, Karger, 2009, vol 163, pp 261–263

Unplanned Start on Assisted Peritoneal Dialysis

Johan V. Povlsen

Department of Renal Medicine C, Aarhus University Hospital, Skejby, Aarhus N, Denmark

Abstract

The present paper describes a program for an unplanned start on assisted automated peritoneal dialysis for late referred patients with chronic kidney disease stage V and urgent need for initiation of dialysis. Using a standard prescription for 12 h overnight APD right after PD catheter placement, analysis of our data showed that unplanned start on APD has no detrimental effects on patients, combined patient and technique, peritonitis-free survivals or the risk of infectious complications, while the risk of mechanical complications and the need of replacement of displaced or malfunctioning PD catheters may be increased. Unplanned start on APD right after PD catheter insertion is a feasible, safe and efficient procedure.

A major challenge facing the dialysis community is the high number of late referred patients, with associated detrimental effects on mortality, morbidity and need for hospitalization. In our center, approximately 50% of all incident patients with chronic kidney disease (CKD) stage V are late referred with urgent need for acute initiation of dialysis [1]. In most dialysis units, acute initiation of hemodialysis (HD) via a temporal central venous access is the preferred method in this scenario, despite its associated risks of septicemias and venous thrombosis and stenosis.

Methods

Accordingly, we have developed a program for unplanned start on assisted peritoneal dialysis (APD) for late referred patients with urgent need for initiation of dialysis [1]: coiled, double-cuff Tenckhoff peritoneal dialysis (PD) catheters were inserted by open surgery under

Table 1. Standard prescription for unplanned start on APD

	Bodyweight	
	<60 kg	>60 kg
Time overnight, h	12	12
Total volume, liters	10	14
Max dwell volume, liters	1.2	1.5
Tidal volume	50–75	50–75

local anesthesia. One gram of intravenous vancomycin was used as prophylaxis. Patients with urgent need for dialysis were started immediately on APD (HomeChoice cyclers, Baxter Healthcare) hours to few days after PD catheter insertion. Their regimens consisted of a standard prescription for 12 h overnight APD (table 1). The cyclers were pre-programmed using a chip (ProCard, Baxter Healthcare). After 1–2 weeks, the patients were converted to a standard 8-hour APD program, with or without wet day, with Extraneal (Baxter Healthcare) and discharged from hospital.

Severe hypertension (diastolic blood pressure >120 mm Hg), severe over hydration or pulmonary edema, severe hyperkalemia (serum K^+ >6.5 mmol/l) and signs of uremic pericarditis or colitis were regarded as contraindication for unplanned start on APD.

Results

Based on retrospective analysis of our data, and after adjustment for differences in baseline characteristics using a multiple-variable Cox proportional hazards model, we have shown that [1, 2]:

- Unplanned start on APD has no detrimental effects on patient, combined patient and technique, and peritonitis-free survivals.
- Unplanned start on APD does not increase the risk of infectious complications.
- Unplanned start on APD may increase the risk of mechanical complication and the need for replacement of displaced and malfunctioning PD catheters.

Conclusions

Unplanned start on APD right after PD catheter insertion is a feasible, safe and efficient procedure with no detrimental effect on patient survival, PD technique survival or risk of peritonitis.

References

1 Povlsen JV, Ivarsen P: How to start the late referred ESRD patient urgently on chronic APD. Nephrol Dial Transplant 2006;21(suppl 2):ii56–ii59.
2 Povlsen JV, Ivarsen P: Assisted peritoneal dialysis: also for the late referred elderly patient. Perit Dial Int 2008;28:461–467.

Johan V. Povlsen
Department of Renal Medicine C, Aarhus University Hospital
Skejby
DK–8200 Aarhus N (Denmark)
Tel. +45 8949 5703, Fax +45 8949 6003, E-Mail jvpovlsen@stofanet.dk

Ronco C, Crepaldi C, Cruz DN (eds): Peritoneal Dialysis – From Basic Concepts to Clinical Excellence. Contrib Nephrol. Basel, Karger, 2009, vol 163, pp 264–269

Peritoneal Dialysis in the Elderly

Edwina A. Brown

Imperial College Kidney and Transplant Institute, Hammersmith Hospital, London, UK

Abstract

The elderly on dialysis have unique needs and characteristics and their outcomes vary from those of their younger counterparts. Comparatively fewer will start or be maintained on peritoneal dialysis (PD) compared to younger patients despite the fact that haemodialysis is often poorly tolerated. A home-based treatment also avoids the need for transport which can be expensive and adversely affect the quality of life of a patient. Barriers to PD for older patients include poor vision, frailty, cognitive dysfunction, accommodation issues and a prejudice from renal teams that older patients cannot do PD. In France, where assistance from community nurses has been available for many years, PD is predominantly a treatment for the elderly and older, sicker patients are often preferentially placed on PD. The use of assisted PD is growing in other countries and where this happens, there is a growth in the prevalent PD population. The ability of older patients to use PD as their dialysis modality should not be determined by whether they live in an area where the nephrologist is a PD enthusiast or not. Patients have the right to receive appropriate non-biased information so they can choose the dialysis modality which gives them the best quality of life and suits their and their family's lifestyle.

Chronic kidney disease (CKD) is predominantly a disease of the elderly, with a prevalence of CKD stages 3–5 of 25% in the general population over the age of 70 years and 30% over the age of 80 years compared to11% overall [1]. Although only a small proportion of individuals with CKD will progress to end-stage kidney disease (ESKD), increasing life expectancy means that the number of older people with ESKD is going to escalate in the future. Over the last few years, the UK Renal Registry shows that the median age of patients starting dialysis has been fairly stable at around 65 years over the past few years [2]. The number of 'old elderly' patients, though, is increasing. In the USA, the number of octogenarians and nonagenarians starting dialysis increased from

7,054 persons in 1996 to 13,577 persons in 2003; after allowing for population growth, this represents a 57% increase in the rate of starting dialysis in this age group [3]. In France in 2005, almost 40% of patients starting dialysis were over the age of 75 years [4]. Thus the elderly are the largest and fastest growing group of patients on dialysis, but they are still less likely to be started on peritoneal dialysis (PD). In the United Kingdom, when analyzing modality by age <65 and >65 years, 30% and 17% of incident patients respectively were on PD at day 90 in 2006 [2]. In contrast, in France, where assisted PD using community nurses has been available for many years, PD is predominantly a treatment of the elderly, with 54% of males and 59% of females on PD in January 2006 being over 70 years of age [5]. The elderly are also successfully dialyzed on PD in Hong Kong, which has a PD-first policy: in March 2007, 80% of patients with a median age of 62.3 years were on PD [6].

The elderly on dialysis have unique needs and characteristics and their outcomes vary from that of their younger counterparts. The three year probability of survival for UK patients >65 years old is 42.6% compared to 77.1% for those aged <65 years [2]. Another major difference is likelihood of transplantation; 56.9% of those aged <65 years undergoing renal replacement therapy have a transplant compared to only 20.6% in those 65 years and over [2]. Thus dialysis in this older group needs to be considered as a permanent, lifelong treatment. 'There is more to living than not dying' [7]; quality of life is particularly important for our older patients on dialysis with a limited life expectancy.

Choosing Dialysis Modality for Older Patients

Older patients with ESKD often have considerable comorbidity, not only the vascular disease associated with their renal disease, but also the comorbidity found in many older people, including impaired vision, deafness, poor mobility, arthritis and cognitive problems. They are often socially isolated, live in poor accommodation, have financial problems and are often psychologically depressed due to loss of independence or bereavement. These factors are all problematic for any dialysis modality.

For haemodialysis (HD), vascular disease increases the risk of failure for vascular access. This results in increased reliance on venous access with all the associated risks of infection. Failure of vascular access can also necessitate frequent hospital admissions for unpleasant and painful radiological and surgical procedures. Another problem is cardiac disease which can result in hypotension and arrythmias while on HD. Older patients often therefore feel 'washed out' after a dialysis. Added to this is the need for patients to travel to and from the dialysis unit; many cannot do this independently and require transport provided by the hospital. Not only may some have to travel long distances, but also there are frequently long waits for the transport that is often at antisocial times of day.

PD has the advantage that it is done in the home, thereby avoiding the need for transport. This is an advantage for both the fit and the more frail elderly. For the fit elderly, it means that they can travel, have an active social life and enjoy their retirement. The more frail elderly will also benefit as they will not have the swings of HD and the need for travel to the HD unit will be avoided. The problem is to determine whether such individuals can cope with the rigours of a home treatment. Such decisions are often made by healthcare professionals without full discussion with the patient. Many older patients can be trained to do their own PD, though this may take longer than with younger patients. Family members are often willing to help with all or part of the procedure and, increasingly, use of community nurses enables frail patients to be on PD in their own homes. Providing assistance with PD will therefore expand the proportion of patients able to have PD [8].

Assisted PD– What Is Published?

In France, assisted PD has been standard treatment for older patients for many years. Of 11,744 French PD patients treated in the last decade (1995–2006), 56% were considered unable to perform their own treatment and needing assistance. This was provided by a community nurse in 86% of patients. As expected, the comorbidity of patients on assisted PD was greater than that of patients doing their own dialysis, with median scores on the Charlson comorbidity index of 5 for autonomous patients, 6 for those assisted by their family and 7 for those assisted by a nurse [9]. A detailed analysis by the French REIN registry [9] of 3,512 patients over the age of 75 years starting dialysis between 2002 and 2005 showed that 18% began with PD, with the proportion varying from 3 to 38% depending on region; over half of these patients would have been on assisted PD. Interestingly, compared to UK practice, starting dialysis with PD was significantly associated with older age, congestive heart failure and severe behavioural disorders. The 2-year survival among those who started with PD was 64% and was not different from all of those who started HD. There is little data about quality of life of French patients on assisted PD, but one prospective study of HD and PD patients from Lorraine has demonstrated better quality of life of elderly patients on PD, many of whom would be receiving community assistance [10].

Different models of delivering and funding assisted PD have been developed in various European countries [11]. In France, non-disconnect CAPD with UV-flash is the predominant method used, as this greatly shortens the time needed for the nurse visit – the nurse phones the patient, or a relative, and instructs them to start the drain procedure so when she arrives she just has to remove the old bag and connect the new one, leaving the fluid to drain in and the patient to fold up the bag after her departure. Indeed, assisted CAPD is often the modality of choice for older patients [12]. In other countries, APD is used as the PD modality for assisted patients with 2 visits from the nurse – a longer visit happens in the morning when the patient is disconnected from the machine, the old bags removed and new ones placed on the machine, and then in the evening there is a shorter visit when the patient is connected to the machine. Reported 2-year survival on assisted APD of around 48% makes this a feasible dialysis modality for frail elderly patients [13].

Assisted APD has also been developed in Toronto with up to 2 home visits a day available at some units. The possibility of assistance at home appeared to increase the proportion of patients considered to be eligible for PD [14], though a larger study would be needed to confirm this. Interestingly, of the 22 assisted patients, 15 required chronic support, five graduated to self-care, and two started with self-care but later required assistance.

Overcoming the Barriers to PD

Older patients requiring dialysis should be able to choose their dialysis modality, whether this should be hospital or home-based, and any medical contraindication to a particular modality should be discussed. The information given should be non-biased, and give the pros and cons for both HD and PD relevant to their age and comorbidities. Unfortunately, this is remarkably difficult to do because of the paucity of studies on quality of life and outcomes for older patients on HD and PD. Remarkably, the North Thames Dialysis Study is so far the only prospective study; it showed no difference in survival or quality of life as measured by SF36 between patients over 70 years old on HD or PD [15, 16]. We are currently doing further studies comparing experiential measures of quality of life in older patients on haemodialysis and peritoneal dialysis (BOLDE: Broadening Options for Long-Term Dialysis in the Elderly); the results from this study will be available shortly.

The process of education needs to be tailored for older people. Most education occurs during the predialysis phase when mild cognitive impairment and uraemia make it difficult for patients to understand implications of information and those presenting late never receive this information at all. It may be appropriate to offer choice of modality 2–3 months after starting dialysis, once patients know more about the ups and downs of life on dialysis. As discussed by Oliver [17], a multidisciplinary approach to identify barriers to PD (medical, psychological or social) enables the development of a plan to help overcome them. Using this approach, up to 80% of his patients are regarded as eligible for PD. Approximately half of the eligible patients then choose PD resulting in 40% of the total population starting on PD [14]; this is a much higher rate than in the majority of units.

Conclusion

The ability of older patients to use PD as their dialysis modality should not be determined by whether they live in an area where the nephrologist is a PD enthusiast. Patients have the right to receive appropriate non-biased information so they can choose the dialysis modality which gives them the best quality of life and suits their and their family's lifestyle. Furthermore, the option of PD for the elderly needs to be reconsidered in light of the ever increasing number of older people requiring dialysis, as placing all but a few on HD will be a huge financial burden to any healthcare system. The French experience of community-based PD shows that this can be achieved, and should be replicated elsewhere.

References

1 Coresh J, Astor BC, Greene T, Eknoyan G, Levey AS: Prevalence of chronic kidney disease and decreased kidney function in the adult US population: Third National Health and Nutrition Examination Survey. Am J Kidney Dis 2003;41:1–12.
2 Ansell D, Feehally J, Feest TG, Tomson C, Williams AJ, Warwick G: UK Renal Registry Report 2007. Bristol, UK: UK Renal Registry; 2007.
3 Kurella M, Covinsky KE, Collins AJ, Chertow GM: Octogenarians and nonagenarians starting dialysis in the United States. Ann Intern Med 2007;146:177–183.
4 REIN Registry 2005 Annual Report. Available at: http://www.agence-biomedecine.fr/fr/experts/gref fesorganes-rein.aspx
5 Verger C, Ryckelynck JP, Duman M, Veniez G, Lobbedez T, Boulanger E, et al: French Peritoneal Dialysis Registry (RDPLF): Outline and main results. Kidney Int Suppl 2006;103:S12–S20.
6 Li PKT, Szeto CC: Success of the peritoneal dialysis programme in Hong Kong. Nephrol Dial Transplant 2008;23:1475–1478.
7 Bargman JM: Is there more to living than not dying? A reflection on survival studies in dialysis. Semin Dialysis 2007;20:50–52.
8 Couchoud C, Moranne O, Frimat L, Labeeuw M, Allot V, Stengel B: Associations between comorbidities, treatment choice and outcome in the elderly with end-stage renal disease. Nephrol Dialy Transplant 2007;22:3246–3254.
9 Verger C, Ryckelynck JP, Duman M, Viniez G, Lobbedez T, Boulanger E, Moranne O: French peritoneal dialysis registry (RDPLF): Outline and main results. Kidney Int 2006;70:S12–S20.
10 Frimat L, Durand PY, Loos-Ayav C, Villar E, Panescu V, Briancon S, Kessler M: Impact of first dialysis modality on outcome of patients contraindicated for kidney transplant. Perit Dialy Int 2006:26;231–239.

11 Brown EA, Dratwa M, Povlsen JV: Assisted peritoneal dialysis: an evolving dialysis modality. Nephrol Dial Transplant 2007;22:3091–3092.
12 Lobbedez T, Moldovan R, Lecame M, de Ligny BH, El Haggan W, Ryckelynck JP: Assisted peritoneal dialysis, experience in a French renal department. Perit Dial Int 2006;26:671–676.
13 Povlsen JV, Ivarsen P: Assisted automated peritoneal dialysis (AAPD) for the functionally dependent and elderly patient. Perit Dial Int 2005;25 (suppl 3):S60–S63.
14 Oliver MJ, Quinn RR, Richardson EP, Kiss AJ, Lamping DL, Manns BJ: Home care assistance and the utilization of peritoneal dialysis. Kidney Int 2007;71:673–678.
15 Lamping DL, Constantinovici N, Roderick P, Normand C, Henderson L, Harris S, et al: Clinical outcomes, quality of life, and costs in the North Thames Dialysis Study of elderly people on dialysis: a prospective cohort study. Lancet 2000;356:1543–1550.
16 Harris SAC, Lamping DL, Brown EA, Constantinovici N; for the NTDS Group: Dialysis modality and elderly people: effect on clinical outcomes and quality of life. Perit Dial Int 2002;22:463–470.
17 Oliver MJ, Quinn RR: Is the decline of peritoneal dialysis in the elderly a breakdown in the process of care? Perit Dial Int 2008;28:452–456.

Professor Edwina A. Brown
Imperial College Kidney and Transplant Institute, Hammersmith Hospital
Du Cane Road
London W12 0HS (UK)
Tel. +44 20 8383 5207, Fax +44 20 8383 5169, E-Mail e.a.brown@imperial.ac.uk

Ronco C, Crepaldi C, Cruz DN (eds): Peritoneal Dialysis – From Basic Concepts to Clinical Excellence. Contrib Nephrol. Basel, Karger, 2009, vol 163, pp 270–277

Peritoneal Dialysis in Developing Countries

K.S. Nayak, M.V. Prabhu, K.A. Sinoj, S.V. Subhramanyam, G. Sridhar

Department of Nephrology, Global Hospitals, Hyderabad, India

Abstract

Peritoneal dialysis (PD) is acknowledged worldwide as a well-accepted form of renal replacement therapy (RRT) for end-stage renal disease (ESRD). Ideally, PD should be the preferred modality of RRT for ESRD in developing countries due to its many inherent advantages. Some of these are cost savings (especially if PD fluids are manufactured locally or in a neighboring country), superior rehabilitation and quality of life (QOL), home-based therapy even in rural settings, avoidance of hospital based treatment and the need for expensive machinery, and freedom from serious infections (hepatitis B and C). However, this is not the ground reality, due to certain preconceived notions of the health care givers and governmental agencies in these countries. With an inexplicable stagnation or decline of PD numbers in the developed world, the future of PD will depend on its popularization in Latin America and in Asia especially countries such as China and India, with a combined population of 2.5 billion and the two fastest growing economies worldwide. A holistic approach to tackle the issues in the developing countries, which may vary from region to region, is critical in popularizing PD and establishing PD as the first-choice RRT for ESRD. At our center, we have been pursuing a 'PD first' policy and promoting PD as the therapy of choice for various situations in the management of renal failure. We use certain novel strategies, which we hope can help PD centers in other developing countries working under similar constraints. The success of a PD program depends on a multitude of factors that are interlinked and inseparable. Each program needs to identify its strengths, special circumstances, and deficiencies, and then to strategize accordingly. Ultimately, teamwork is the 'mantra' for a successful outcome, the patient being central to all endeavors. A belief and a passion for PD are the fountainhead and cornerstone on which to build a quality PD program.

The United Nations has developed the Human Development Index, an indicator of the level of human development [1]. Developing countries are countries which have not achieved a significant degree of industrialization relative to their populations. There is criticism of the use of the term 'developing country', which implies inferiority. Cuba, classed as 'developing' has better health outcomes and literacy rates than some States in the USA. This dichotomy also exists in the sphere of renal replacement therapy (RRT) and its costs and funding. Incident rates of reported end-stage renal disease (ESRD) (receiving treatment) are greatest in Taiwan at 418 per million population in 2006, followed by the United States, Mexico and Japan, whereas rates of 100 per million population or lower are reported by Bangladesh, Russia, Pakistan, Iran, Romania, and the Philippines [2].

Why Peritoneal Dialysis over Hemodialysis for the Developing World? Busting the Cost Myth

Peritoneal dialysis (PD) should logically be the preferred modality in developing countries, but surprisingly is not. Many think cost barrier is the single most important factor responsible. This is a myth. Consider the following: The patient on PD gets back to his/her work and thus costs related to travel and loss of pay due to absenteeism are avoided. Also, the QOL and rehabilitation on PD is better [3]. Hidden costs of hemodialysis (HD) are travel expenses, more investigations, more erythropoietin usage, more hospitalizations, and the higher number of variables in the delivery of good quality HD – quality of water, transmission of hepatitis B and C, and maintenance costs of machines. The requirement of expensive equipment for HD should have been an incentive for governments to promote PD instead. Also, if local manufacturing of PD fluids (PDF) is made possible, the cost of the therapy would be about 80% that of HD.

What Ails Peritoneal Dialysis? Why the Inexplicably Low Numbers Around the World?

A worldwide decline in PD numbers especially in the developed economies is of concern. Economics is not the sole factor. Affluent Taiwan has a PD penetrance of 7.6% while equally affluent Hong Kong has 81.3% [2]. Japan has only 3.4% of its patients on PD while South Korea has 21.6% [2]. Important nonmedical factors are physician bias due to a skewed perception of PD as a therapy, lack of structured reimbursement plans for physicians, and social/cultural issues [4]. Regressive measures by industry (cost cutting by reducing funding for training and recruiting clinical coordinators (CC), and servicing patient complaints using untrained call center personnel) and poor postgraduate/fellowship training in PD also contribute. The perception of the government about PD also sometimes lacks long-term vision. PD continues to be nonexistent in many Asian countries, while India and China have a PD penetrance of around 8–12% [5]. This is unfortunate, as the increasing use of PD in developing countries is going to sustain global PD therapy.

Status of Peritoneal Dialysis Programs in the Developing World – Hope for the Future

Asia

Asia is a 'melting pot' for PD practices [6]. Individually driven PD programs with governmental support are thriving, Thailand being an example. A thrust to popularize PD in Thailand advocating a 'PD first' policy with complete financial support from the Government of Thailand, based on cost-effective PD solutions imported from India is paying dividends. In 2008, the first year of implementation, 1,500 patients have been initiated on PD and a further 2,000 patients expect to be initiated in 2009 which is truly remarkable [Dhavee Sirivongs, pers. commun.]. Similar efforts are beginning to bear fruit in Indonesia and Malaysia [7].

Africa

The situation of RRT in Africa is a major challenge for the renal community. PD is plagued by many problems – costs, lack of motivated healthcare teams, and lack of infrastructure. Well-established programs have developed in the more prosperous countries of North Africa (Tunisia, Morocco, Libya, and Algeria) and South Africa. However, the limited healthcare resources in sub-Saharan Africa are well documented. It is heartening that a few PD programs have developed here. The Sudan National Peritoneal Dialysis Program is showing promising results. Public support for growth of PD is sought by a body of social activists, journalists, and businessmen named, 'The Board of Trustees for the Sudan National PD Program' [8].

Latin America

Latin America shows interesting contrasts – Colombia, Guatemala and El Salvador have 30% or more patients on PD. Mexico has the highest numbers of PD patients in the world [9]. On the other hand, Chile, Argentina, Equador and Brazil have significantly lower numbers. The roundtable discussion on the economics of dialysis and chronic kidney disease in Latin America concluded that the choice of RRT modality was linked significantly to the local reimbursement

Table 1. A dozen P's in a pod

1 Patient selection
2 Personnel management
3 Product selection
4 PD prescription and pharmacotherapy of comorbid factors
5 Protein–calorie intake
6 Paraclinical support and parallel services
7 Patient internet connectivity
8 Peritonitis and ESI prevention
9 Preserving the RRF
10 Physiotherapy and rehabilitation
11 Purse management
12 Postgraduate activities

policies [10]. In Colombia, reimbursement allows equal amounts for PD and HD, and the Dialysis Outcomes in Colombia (DOC) study allowed comparative study of outcomes on PD and HD. The clinical results were overall equivalent [11]. There is significant scope for improvement in PD penetration in this region.

Peritoneal Dialysis Program and Strategies at Our Center – A Model for the Developing World

The attributes of a quality PD program can be addressed under 12 key points – all interlinked, each crucial, and all starting with the letter 'P.' I call them the 'Dozen P's in a Pod' (table 1). We believe that we have been reasonably successful in achieving our targets for most of them and our experience would help units working in similar difficult situations.

Patient Selection

Ours is a tertiary care, non-university teaching hospital enrolling 6–8 patients onto PD every month. All patients with ESRD are offered PD as a bridge to transplantation or long-term therapy. Patients are encouraged to start PD while awaiting a cadaveric kidney donation. This avoids scheduling HD and the possibility of contracting hepatitis.

We use automated PD (APD) in selected patients in our ICUs. In patients with suspected acute deterioration of chronic kidney disease, we find better recovery of the acute component in patients put on PD than in those who receive HD [unpubl. data].

Personnel Management

The CC is the key person involved with the patient during the various stages of PD therapy. The CC is basically a paramedical staff member who has been extensively trained in all aspects of PD and is proficient in the local language, culture, and customs. Regular home visits by the CCs are an important part of follow-up care, because the family and the patient realize that continuing support is available [12]. During a home visit, the CC ensures that the exchange techniques are adhered to and compliance is maintained. In short, the CC becomes a link between the patient and the PD center. We have recently begun a CC training course officially recognized by the local University.

Product Selection

Good connectivity reduces spike-related peritonitis in patients on PD. We use the '4GYP System' a novel double-bag disconnect system (J Mitra Industries Ltd., New Delhi, India) with consistently good results, patient satisfaction and affordability.

PD Prescription and Pharmacotherapy of Comorbid Factors

The last word has not been said regarding adequacy and target Kt/V in PD patients [13]. In general, we prescribe anuric patients four exchanges of 2-liter solution bags daily and three exchanges for patients with residual renal function (RRF).Treatment of anemia, renal osteodystrophy, cardiovascular disease, hypertension, dyslipidemias, and diabetes and its complications will decide the patient survival on therapy. We envisage a special role for PD in HIV-related renal failure and cardiac failure.

Protein-Calorie Intake

Ensuring adequate protein intake in the diet of patients in India, with its myriad of dietary preferences, is a challenge. We utilize the services of a dedicated renal nutritionist. The renal nutritionist prepares an individualized plan for the patient's nutritional needs. A cookbook with simple-to-prepare dishes has been created for our patients.

Paraclinical Support and Parallel Services

(a) Microbiology

With continual effort, we have improved our culture detection rates to 64% (2006) from 54% (2005) and 43% (2004). We have incorporated these extra procedures:

- Bactec technology for cultures.
- Agents such Tween 80, Triton-X, and water lysis to lyse neutrophils and extrude internalized bacilli.

(b) Surgical Support

Insertion and repositioning of the PD catheter and correction of mechanical complications is done laparoscopically by a dedicated surgeon. As a matter of policy, we have an arteriovenous fistula created for each PD patient, which is helpful when temporary discontinuation of PD becomes necessary.

Patient Internet Connectivity

We have used the Internet and mobile telephone to aid in treating PD complications. We are able to monitor patients by using a camera phone or digital camera to transmit images to the center. The images may be of the solution bag or of the exit site. Advice is given accordingly. It is not unusual for us to manage patients from a distance of as much as 1,500 km.

Peritonitis and ESI Prevention

Through a concerted effort, we have achieved consistently low peritonitis rates (1 episode every 63.2 patient-months) surpassing earlier good results.

Preserving the RRF

Published literature suggests that PD is superior to HD in preserving RRF [14]. Prevention, early detection and treatment of peritonitis are essential in maintaining RRF.

Physiotherapy and Rehabilitation

Constant psychological support by CCs during home visits, patient support groups, etc. helps build patient morale, and improve QOL.

Purse Management

Making PD more affordable is a constant endeavor of the PD center. Well rehabilitated patients are able to resume work, offsetting the cost of treatment. In select larger patients, we encourage the use of 3 exchanges of 2.5-liter solution bags, meaning that they get 7.5 liters of PD solution for the cost of 6 liters (that is, 2 liters × 3 exchanges).

Postgraduate Activities

A regular teaching program incorporating postgraduate fellows helps everyone keep abreast of the latest developments in the field. This also acts as an impetus to publish our findings. Our unit is a training center for ISPD Asian Chapter fellows. This ensures that we have highly trained staff to take best care of our patients.

Conclusions

The developing world holds the key to the future of PD as a therapy. Though faced with several roadblocks, we believe that the future is not bleak. Sharing and dissemination of knowledge and experience by respected PD practitioners can prove invaluable in sensitizing the medical fraternity, and policy makers to the success and viability of PD. There needs to be a revamp of the medical training curriculum, according PD its rightful place. All avenues of cost minimization need to be explored, without compromising on patient safety. Governments need to be sensitized to the therapy, and helpful trade policies initiated to facilitate export of PDF/other accessories to nonmanufacturing regions. Ways to involve the private sector/NGOs and philanthropists to help promote awareness and initiate PD programs should be explored. The biggest challenge remains increasing patient awareness, and delivering top quality care as patient well-being and satisfaction is paramount.

References

1 United Nations Development Programme's Human Development Report 'Human Development Indices: A statistical update 2008. hdr.undp.org/statistics/
2 USRDS Annual Data Report, 2008. http://www.usrds.org/2008/view/esrd_12.asp
3 Barendse SM, Speight J, Bradley C: The Renal Treatment Satisfaction Questionnaire (RTSQ): a measure of satisfaction with treatment for chronic kidney failure. Am J Kidney Dis 2005;45:572–579.
4 Venkataraman V, Nolph KD: Socioeconomic aspects of peritoneal dialysis in North America: role of non medical factors in the choice of dialysis. Perit Dial Int 1999;19:S419–S422.
5 Li PK, Chow KM: The cost barrier to peritoneal dialysis in the developing world: an Asian perspective. Perit Dial Int 2001;21(suppl 3):S307–S313.
6 Nayak KS: Asia, a melting pot for peritoneal dialysis practices: redefining and expanding peritoneal dialysis horizons in Asia. Perit Dial Int 2004;24:422–423.
7 Hooi LS, Lim TO, Goh A, et al: Economic evaluation of centre haemodialysis and continuous ambulatory peritoneal dialysis in Ministry of Health hospitals, Malaysia. Nephrology 2005;10:25–32.
8 El Elhassan AM, Kaballo B, Fedail H, et al: Peritoneal dialysis in the Sudan. Perit Dial Int 2007 27:503–510.
9 Cusumano AM, Di Gioia C, Hermida O, Lavorato C, on behalf of the Latin American Registry of Dialysis and Renal Transplantation Registry Annual Report 2002. Kidney Int Suppl 2005;97:S46–S52.
10 Pecoits-Filho R, Campos C, Cerdas-Calderon M, et al: Policies and health care financing issues for dialysis in Latin America: extracts from the roundtable discussion on the economics of dialysis and chronic kidney disease. Perit Dial Int 2009;29(suppl 2):222–226.
11 Sanabria M, Muñoz J, Trillos C, et al: Dialysis outcomes in Colombia (DOC) study: a comparison of patient survival on peritoneal dialysis vs hemodialysis in Colombia. Kidney Int Suppl 2008;108:S165–S172.
12 Nayak KS, Sinoj KA, Subhramanyam SV, Mary B, Rao NV: Our experience of home visits in city and rural areas. Perit Dial Int 2007;27(suppl 2):S27–S31.
13 Churchill DN, Taylor DW, Keshaviah PR, and the CANUSA Peritoneal Dialysis Study Group: Adequacy of dialysis and nutrition in continuous peritoneal dialysis: association with clinical outcomes. J Am Soc Nephrol 1996;7:198–207.
14 Misra M, Vonesh E, Van Stone JC, Moore HL, Prowant B, Nolph KD: Effect of cause and time of dropout on the residual GFR: a comparative analysis of the decline of GFR. Kidney Int 2001;59:754–763.

Dr. K.S. Nayak
Head, Department of Nephrology, Global Hospitals
Lakdi Ka Pool
Hyderabad, 500004 (India)
Tel. +91 98480 14555, Fax +91 40 2324 4455, E-Mail drksnayak@gmail.com

Ronco C, Crepaldi C, Cruz DN (eds): Peritoneal Dialysis – From Basic Concepts to Clinical Excellence. Contrib Nephrol. Basel, Karger, 2009, vol 163, pp 278–284

Peritoneal Dialysis for Acute Kidney Injury: Techniques and Dose

Chang Yin Chionh[a,b], *Sachin Soni*[a,c], *Dinna N. Cruz*[a,d], *Claudio Ronco*[a,d]

[a]Department of Nephrology, Dialysis and Transplantation, San Bortolo Hospital, Vicenza, and [d]International Renal Research Institute Vicenza (IRRIV), Vicenza, Italy; [b]Renal Unit, Department of General Medicine, Tan Tock Seng Hospital, Singapore, Singapore; [c]Division of Nephrology, Mediciti Hospitals, Hyderabad, India

Abstract

It has not been clearly shown which modality of dialysis is superior in the management of acute kidney injury (AKI). Most centers in developed countries have adopted extracorporeal blood purification (EBP) strategies, such as continuous or intermittent forms of hemodialysis or hemofiltration, for the supportive management of AKI. On the other hand, the use of peritoneal dialysis (PD) is widespread in developing countries in view of its ease of use, low cost and minimal requirements on infrastructure. The dose of dialysis required for AKI remains controversial, although an augmented dose with a high small solute clearance is advocated until further definitive evidence becomes available. No studies have directly examined the effects of the dose of PD on outcomes in AKI. The targets of dose for PD are inferred from studies conducted with EBP. There are concerns that PD is unable to achieve high clearances required to support a patient with renal failure. However, various techniques have been described which are able to achieve the targets of small solute clearance. These include high volume PD and continuous flow PD. The selection of flexible peritoneal catheters with better catheter function and dialysate flow rates can improve the efficiency of PD. Other aspects of dose should also be examined, including clearance of middle molecular weight toxins as well as adequate fluid removal. With careful selection of techniques to meet the individual demands of the patient, PD is an appropriate modality of dialysis for patients with AKI.

Acute kidney injury (AKI) is a leading cause of mortality amongst critically ill patients. In developed countries, extracorporeal blood purification (EBP) in its various forms is commonly used for AKI. In contrast, PD is regularly employed in developing countries.

In developed countries, this modality of dialysis for AKI should not be forgotten. Disaster planners are increasingly aware of the concept of a renal disaster. A major catastrophe can cause severe damage to the infrastructure. PD is an alternative when reliable power, clean water supply and facilities for water treatment are unavailable. Besides direct trauma, crush injuries are the next most frequent cause of death in disasters. Many victims will develop AKI and the availability of dialysis will be life-saving.

A recent trial by Gabriel et al. [1] suggested that the use of PD was associated with more rapid renal recovery. If the findings of Gabriel et al. [1] are replicable in large-scale studies, PD may return to the forefront in the management of AKI.

There are several reasons why PD usage is declining for AKI. The main concern is the possibility of inadequate clearance. A conventional prescription for chronic PD may be inadequate for patients with AKI. However, various techniques have been described to achieve a higher dose.

Dose in Acute Kidney Injury – How Much Is Enough?

It is important to first determine what the targets of dose are. This is a subject of great controversy. Furthermore, no studies had examined the issue of dose in AKI using PD. As such, the targets have to be inferred from studies based on EBP.

Small Solute Clearance

Patients with AKI are hypercatabolic and require adequate clearance of toxins to avoid complications. Although current consensus guidelines on small solute clearances for chronic dialysis are lower for PD than for HD, it is judicious to aim for clearance targets similar to EBP in AKI.

Studies with EBP have different methods of reporting small solute clearance dose. Indicators of dose such as the fluid replacement rate in hemofiltration cannot be easily translated for application in PD. Therefore, the standardized (std) Kt/V_{urea} was selected as an indicator of dose in this review for the purpose of comparison. The limitations of this measure of dose must, however, be kept in perspective. Urea kinetic modeling (UKM) was developed for the assessment of patients on chronic hemodialysis and is not validated in AKI. Furthermore, the formulae to calculate the volume of distribution of urea consistently underestimates the true value [2]. In spite of this, the application of UKM to AKI persists due to the lack of alternatives.

A study conducted at the Cleveland Clinic Foundation (CCF) [3] found that survival rates were higher in patients who received a Kt/V_{urea} >1.0 per hemodialysis session. Subsequent trials comparing high dose and low dose EBP have yielded conflicting results. In a trial by Schiffl et al. [4], mortality was significantly higher in patients who received a lower dose (std-Kt/V_{urea} 1.85 versus 3.59). On the other hand, a larger study conducted by the Acute Renal Failure Trial Network [5] no significant differences in mortality between the two arms (std-Kt/V_{urea} of 2.1 versus 3.88). Despite limitations of the CCF study, a Kt/V_{urea} >1.0 per session thus persists as a reference point for dose of dialysis in AKI [6]. This is equivalent to a std-Kt/V_{urea} >3.6 per week, assuming 6 sessions a week.

Techniques of Peritoneal Dialysis and Small Solute Clearance

Various techniques of peritoneal dialysis have been described in the literature of chronic PD and these have been adapted for use in AKI. An illustration of various techniques applicable to AKI is provided in figure 1.

Acute Intermittent Peritoneal Dialysis (AIPD)

This is most often utilized for AKI. Frequent and short exchanges with volumes of 1–2 liters and dialysate flows of 2–6 liters/h are performed. Each session lasts 16–20 h and there is usually 2–3 sessions a week. The solute clearance of AIPD is likely inadequate due to its intermittent nature.

Chronic Equilibrated Peritoneal Dialysis (CEPD)

This employs long dwells of 2–6 h with up to 2 liters of dialysate each [7]. It is similar to continuous ambulatory PD and the clearance of small molecules may be also inadequate. However, the clearance of middle molecular weight substances is possibly higher due to the long dwells.

Tidal Peritoneal Dialysis (TPD)

TPD typically involves an initial infusion of 3 liters of dialysate into the peritoneal cavity. A portion of the dialysate, the tidal drain volume (usually 1–1.5 liters), is drained and replaced with fresh dialysate, the tidal fill volume.

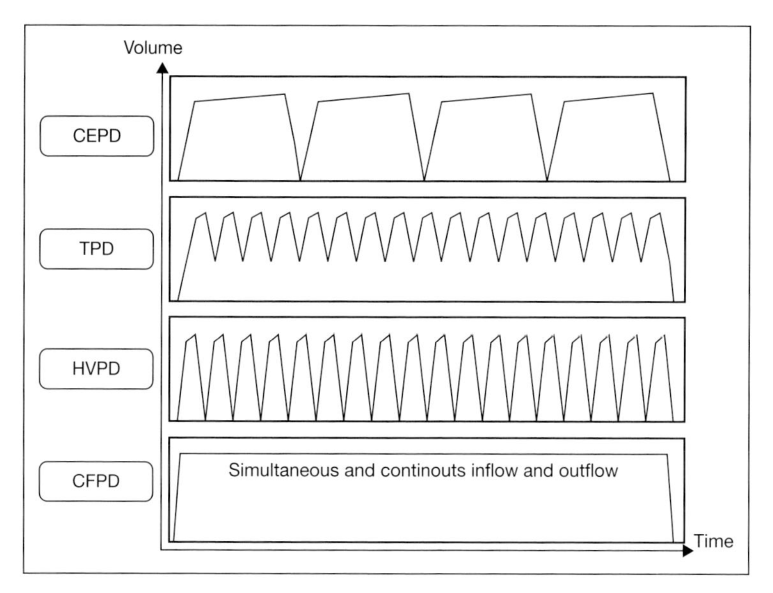

Fig. 1. Illustration of the techniques of peritoneal dialysis used for patients with acute kidney injury.

The reserve volume always remains in the peritoneal cavity throughout the tidal cycle [8].

The rationale behind TPD was to improve dialysis efficiency as time lost during inflow and outflow of the dialysate is minimized. In spite of this, TPD has not consistently achieved a higher small solute clearance when compared to other techniques of PD in chronic dialysis. However, the longer duration of dialysate contact with the peritoneum facilitates clearance of middle molecules which may be useful in AKI.

In a randomized cross-over study comparing CEPD to tidal PD (TPD), the urea clearance of TPD was significantly higher [9]. TPD was able to achieve a mean std-Kt/V_{urea} of 2.43 ± 0.87 [9]. The dialysate fill volume was only 2 liters in this study and a larger fill will achieve a higher urea clearance.

High Volume Peritoneal Dialysis (HVPD)

HVPD is a continuous therapy proposed to achieve high small solute clearances [10]. Very frequent exchanges are undertaken usually with an automated cycler (18–48 exchanges per 24 h, 2 liters per exchange). The total dialysate volume can range from 36 to 70 liters a day [10, 11]. Gabriel et al. [10] reported a delivered std-Kt/V_{urea} of 3.85 ± 0.62 with 36–44 liters of dialysate per day which was comparable to daily hemodialysis.

Continuous Flow Peritoneal Dialysis (CFPD)

The in-flow and out-flow of dialysate occurs simultaneously through two access routes required for CFPD. Dialysate flow occurs at rates close to 300 ml/min and it is possible to achieve a high peritoneal urea clearance (K_{urea}) of 60 ml/min [12]. If the patient with a V_{urea} of 42 liters had a session for 12 h (t = 720 min), the Kt/V_{urea} would be 1.0. The corresponding std-Kt/V_{urea} would be 3.6 with 6 sessions of CFPD a week. Increasing the dialysate flow or the duration of each session can further increase clearance.

Some technical issues must be addressed for CFPD to be adopted widely. A large volume of dialysate is necessary for CFPD unless it is recirculated or regenerated using a combination of filtrative and sorbent technologies. 2 separate catheters are required as originally described, although a dual lumen catheter designed to minimize recirculation is available[13]. Ultrafiltration control remains a problem. Novel techniques such as segmental bioimpedance analysis [14] can be used to measure intra-peritoneal volume and determine the ultrafiltration volume in real-time.

Peritoneal Access

Rigid catheters are often utilized for acute PD. Although easily inserted by the bedside, there is a high risk of peritonitis [15], catheter dysfunction [15] and poor dialysate flow [16]. Flexible catheters are now preferred over rigid catheters and have been shown to be able to accommodate high dialysate flow rates [1, 10].

Other Aspects of Dose

Clearance of Larger Molecules

There is a focus in the clearance of middle molecular weight (MMW) substances in the recent years. The dose of CRRT is accepted to be 35 ml/kg/h in a postdilution continuous hemofiltration based on a previous study [17]. It is not known how this target can be interpreted for use in PD. It is generally assumed that PD has higher clearance of large molecules than hemodialysis

[18]. However, a recent study found that clearance of MMW molecules using high-flux membranes with HD was superior to PD [19]. Besides, the composition of the substances cleared by PD may be very different from what is cleared by EBP. The peritoneal membrane is not a passive hemofilter but a complex biological entity which produces and actively transports substances. Further studies with correlation to clinical outcomes are needed.

Fluid Balance

Adequate fluid clearance is another often overlooked aspect of adequacy. PD is often associated with better hemodynamic stability and also frequently recommended as in the elderly or in patients with congestive heart failure. Adequate fluid removal is therefore easier to achieve. Most studies which utilized PD for AKI did not report difficulties with adequate volume removal.

Conclusions

No data are available on the optimal dose of PD in AKI. The dose of dialysis required for EBP is also controversial. An augmented dose is advocated until more definitive evidence becomes available. Various techniques of PD have been described which are able to achieve a high small solute clearance including HVPD as well as CFPD. Further studies on the use of PD for AKI and its effect on clinical outcomes are necessary.

References

1 Gabriel DP, Caramori JT, Martim LC, et al: High volume peritoneal dialysis vs daily hemodialysis: a randomized, controlled trial in patients with acute kidney injury. Kidney Int Suppl 2008:S87–S93.
2 Himmelfarb J, Evanson J, Hakim RM, et al: Urea volume of distribution exceeds total body water in patients with acute renal failure. Kidney Int 2002;61:317–323.
3 Paganini EP TM, Goormastic M, et al: Establishing a dialysis therapy/patient outcome link in intensive care unit acute dialysis for patients with acute renal failure. Am J Kidney Dis 1996;28(suppl 3).
4 Schiffl H, Lang SM, Fischer R: Daily hemodialysis and the outcome of acute renal failure. N Engl J Med 2002;346:305–310.
5 Palevsky PM, Zhang JH, O'Connor TZ, et al: Intensity of renal support in critically ill patients with acute kidney injury. N Engl J Med 2008;359:7–20.
6 Ricci Z, Ronco C: Dose and efficiency of renal replacement therapy: continuous renal replacement therapy versus intermittent hemodialysis versus slow extended daily dialysis. Crit Care Med 2008;36:S229–S237.
7 Steiner RW: Continuous equilibration peritoneal dialysis in acute renal failure. Perit Dial Int 1989;9:5–7.

8 Agrawal A, Nolph KD: Advantages of tidal peritoneal dialysis. Perit Dial Int 2000;20(suppl 2):S98–S100.
9 Chitalia VC, Almeida AF, Rai H, et al: Is peritoneal dialysis adequate for hypercatabolic acute renal failure in developing countries? Kidney Int 2002;61:747–757.
10 Gabriel DP, Nascimento GV, Caramori JT, et al: High volume peritoneal dialysis for acute renal failure. Perit Dial Int 2007;27:277–282.
11 Trang TT, Phu NH, Vinh H, et al: Acute renal failure in patients with severe falciparum malaria. Clin Infect Dis 1992;15:874–880.
12 Ronco C, Amerling R: Continuous flow peritoneal dialysis: current state-of-the-art and obstacles to further development. Contrib Nephrol. Basel, Karger, 2006, vol 150, pp 310–320.
13 Ronco C, Dell'aquila R, Rodighiero MP, et al: The 'Ronco' catheter for continuous flow peritoneal dialysis. Int J Artif Organs 2006;29:101–112.
14 Zhu F, Hoenich NA, Kaysen G, et al: Measurement of intraperitoneal volume by segmental bioimpedance analysis during peritoneal dialysis. Am J Kidney Dis 2003;42:167–172.
15 Chadha V, Warady BA, Blowey DL, et al: Tenckhoff catheters prove superior to cook catheters in pediatric acute peritoneal dialysis. Am J Kidney Dis 2000;35:1111–1116.
16 Ash SR: Peritoneal dialysis in acute renal failure of adults: the under-utilized modality. Contrib Nephrol. Basel, Karger, 2004, vol 144, pp 239–254.
17 Ronco C, Bellomo R, Homel P, et al: Effects of different doses in continuous veno-venous haemofiltration on outcomes of acute renal failure: a prospective randomised trial. Lancet 2000;356:26–30.
18 Passadakis PS, Oreopoulos DG: Peritoneal dialysis in patients with acute renal failure. Adv Perit Dial 2007;23:7–16.
19 Evenepoel P, Bammens B, Verbeke K, et al: Superior dialytic clearance of beta(2)-microglobulin and p-cresol by high-flux hemodialysis as compared to peritoneal dialysis. Kidney Int 2006;70:794–799.

Dr. Chang Yin Chionh
Renal Unit, Department of General Medicine, Tan Tock Seng Hospital
11 Jalan Tan Tock Seng
Singapore 308433 (Singapore)
Tel. +65 6256 6011, Fax +65 6357 1376, E-Mail kidneyinjury@gmail.com

Ronco C, Crepaldi C, Cruz DN (eds): Peritoneal Dialysis – From Basic Concepts to Clinical Excellence. Contrib Nephrol. Basel, Karger, 2009, vol 163, pp 285–291

Cost Benefits of Peritoneal Dialysis in Specific Groups of Patients

Anabela Rodrigues

Division of Nephrology, Centro Hospitalar do Porto, Hospital de Santo António, Porto, Portugal

Abstract

The cost benefit of peritoneal dialysis (PD) is reflected in generally lower expenditure than in-center hemodialysis while clinical advantages of home therapy are guaranteed. PD offers similar patient survival in comparison with hemodialysis, and opportunities of better adequacy, beyond Kt/V, including continuous removal of uremic toxins, residual renal function protection, nutritional intraperitoneal support, and life satisfaction. Advances on the modality including the use of icodextrine, automated PD, low-glucose degradation products, new solutions and individualized schedules promise to improve the prognosis of risk patients under dialysis such as diabetics and patients with congestive heart failure, the elderly, the fast transporters or the anuric. Patients deserve an individualized prescription taking into account their lifestyle and therapy options in an integrated care plan. PD after renal transplantation failure is an emerging group of PD patients with need for specific protocols, while PD in the acute setting can also be considered in specific situations.

The cost benefits of PD are related to both economical and medical effectiveness. Dialysis costs are becoming a major issue because financial resources are scarce and the number of chronic kidney disease (CKD) dialysis patients is growing. Globally, peritoneal dialysis (PD) solutions account for the major part of PD costs while access, transportation and facilities account more for hemodialysis (HD) expenses. In the majority of the countries, and after an analysis of total therapy expenses, home dialysis is a lower cost modality than in-center dialysis [1] and PD seems to be a cost-effective first-option modality even if transfer to HD is needed later [2]. However, where there is no facility reimbursement for PD, this is less used, while the growing sector of private HD providers, with lower disposable costs, is looking increasingly attractive in countries where PD solutions costs remain a major financial burden.

Considering that similar medical efficacy is offered with either HD or PD, PD being favorable in the earlier years and equivalent in the long-term (same adjusted 5-year dialysis patients survival, USRDS 2008 report), dialysis prescription often does not take into account the patient's point of view. Cost also means the burden of the therapy on patient life-style. Facing the limited survival outcome with either therapies in some group of patients, we should aim at offering quality of life not merely extended life [3]. When treating an individual patient, we should be able to give him an individualized therapy strategy, focusing on predialysis education and closer considerations of patients' preferences.

Major predictors of survival are age, diabetes, cardiovascular disease, and anuria. Clinical outcomes of such riskier group of patients under dialysis must take into account their specificity, with accurate knowledge of the advantages and limitations of each modality of renal replacement therapy. Emerging favorable reports of PD after renal graft failure or in acute settings testimony the feasibility of this therapy.

Diabetics

As in the general uremic population, PD presents clinical advantages as the first-option modality of renal replacement therapy also in diabetic patients, especially if adjusted updated therapy protocols and solutions are prescribed [4, 5]. PD was previously the elective therapy for diabetics due to its better bioavailability of intraperitoneal insulin; however, this strategy has never been implemented much due to the fear of disruption of aseptic procedures with risk of peritonitis, the higher doses needed due to insulin adhesion to the solution bags, hepatic esteatosis, and potential peritoneal membrane changes. On the other hand, PD offers a major advantage in diabetics: preservation of the vascular network, usually frail in these patients. It also allows better hemodynamic tolerance, and avoids anticoagulation and consequent events of hemorrhagic retinopathy.

Exposing diabetic patients to a stress glucose overload seems legitimate, but the facts are reassuring. Patients might need increasing doses of insulin at the beginning of PD or susceptible patient's might develop de novo diabetes or increase in glycated hemoglobin but clinical cost effectiveness is achieved because patients usually benefit from residual renal function preservation, higher renal elimination of advanced glycated proteins, continuous urea nitrogen and middle molecules. Transplantation success is also protected since PD patients show lower rates of delayed graft function after transplantation.

Besides, in spite of the limitations of survival comparative studies in subgroups of dialysis populations [6], we can safely mention that diabetics, at least younger diabetics with lower comorbidity, might benefit more from PD while the majority of more aged diabetics, and those with more comorbidity, can profit from similar PD advantages but will need adjusted therapy to achieve better outcomes [7]. If not adequately managed with glucose-sparing regimens, automated PD and adequate volume control, some patients might present worse results under PD. To sum up, diabetics are a group of patients that will have a gloomier prognosis under dialysis whichever the allocated modality, therefore individual life style and option cannot be neglected.

Cardiovascular Disease and Congestive Heart Failure

Cardiovascular disease is mainly expressed as congestive heart failure, myocardial infarction and arrhythmias: these events occur more often in CKD patients and justify the costly rate of hospitalizations. Much of these patient prognoses are dictated by the predialysis management and once patients begin dialysis the outcomes are universally worse. None of the modalities offer more than some months of survival in patients with such baseline serious morbidity, the point is if modality itself accelerates the atherosclerotic process more or not. There is no evidence of a higher number of major cardiovascular events in one or another modality, after adjustment for baseline condition, although PD is associated with better hemodynamic stability, lower rate of hypotensive episodes and arrhythmias. Control of blood hypertension is less satisfactory once residual renal function is lost but the link of HTA and mortality in dialysis patients is complex and targets of 'adequate' blood pressure are not evidence-based. There have been negative reports about the outcomes of PD in congestive heart failure patients [8]; however, those studies were performed in earlier periods of PD at the time when optimized volume control strategies such as icodextrine and PD plus automated PD could not be prescribed as standard. Negative results from PD might be related to the unmeasured higher grade of cardiovascular morbidity in PD patients [9] or unoptimized treatment schedules in higher risk patients and therefore cannot be generalized. Instead, APD is indeed gaining a role for treating patients with congestive heart failure and has shown improved outcomes [10, 11].

Elderly Patients

Age is an additional major non-modifiable patient risk factor, undoubtedly associated with higher morbidity, hospitalization and lower survival. The point is: once a decision to begin dialysis has been made, which modality can give a patient 'more life than not dying' [3]. And this is also about life style because this group of patients is much more susceptible to life changes.

It will be always onerous to treat these patients and strategies of assisted PD can be assayed [12] although outcomes might depend more on PD skills than on type of modality. Variability of reported results and different regional policies, however, unanimously conclude that PD can be cost effective in elderly patients [13, 14].

It is important to have alternatives and to use complementary strategies such as HD or PD according to the individual and family milieu.

Anuric Patients

This is a risk group of patients but it has been proven before that patients can be managed successfully with APD and icodextrin, fulfilling the recommended small solute removal and fluid control regimen. Adequacy means also integrated care respecting patient-informed options, not to mention vascular capital management. The idea that uraemia might modulate membrane changes with a risk of accelerated process of ultrafiltration failure must be investigated. But PD has now gained more experience and solutions to reduce glucose exposure and improve outcomes, while allowing better patient quality of life. The management of these patients with planned transfer to HD once PD loses its capacity to fulfill the patient's needs can ameliorate global patient survival.

Fast Transporters

There is now greater knowledge about this heterogeneous group of patients, once thought to be unsolvable risk patients. While baseline inherent fast transport can usually be well supported with residual renal function and icodextrin or automated PD, it is now recognized that many patients tend to normalize transport profile without impact on survival [15, 16]. Except if such baseline status is associated with comorbidity, fast transporters are not necessarily more inflamed and indeed APD with icodextrin has effaced its previously reported negative impact. Once more, PD can successfully be maintained respecting the patient's life style.

A different situation is acquired long-term fast transport that can be managed with peritoneal rest and transfer to alternative solutions while icodextrin can prolong PD technical survival. However, transfer to HD must be considered if the fast transport status perpetuates with signs of clinical deterioration or ileum.

Transplant Patients

A recent USRDS report revealed that there has been a 4% increase in patients with graft failure returning to dialysis in the USA. The same occurs in other countries, reflecting an emerging uremic population with specific comorbidity. A challenging renal replacement therapy plan must be induced and PD is feasible after graft failure [17]. We and others have reported similar patient and technique survivals of renal graft failure PD patients in comparison with the first-modality PD group. Although reports are scarce or contradictory, there is the benefit of preservation of residual graft function under a slow tapering immunosuppressant without expression of higher serious infections and peritonitis rate. After withdrawal of antiproliferative drugs but slow reduction of calcineurinics and prednisolone over several months, residual renal function was statistically similar after 1 year of PD treatment, without shortening of peritonitis free survival PD [Bernardo et al: Adv Perit Dial, in press]. So, at least in the short term, PD confers a similar residual renal function protection in renal graft failure patients initiating PD. The psychological benefit of PD as a dialysis modality 'different from the machine' they have dealt with before transplantation deserves mention. Access failure is also a concern in patients with previous HD pursuits. Therefore, as a strategy of vascular network preservation, a course of PD can be essayed, usually with success.

Peritoneal Dialysis in the Acute Setting

A prospective randomized controlled trial was performed to compare the effect of high volume PD with daily hemodialysis on survival in acute kidney injury patients [18]. Metabolic control, mortality rate, and renal function recovery were similar in both groups, whereas high volume PD was associated with a significantly shorter time to the recovery of renal function.

Dialysis clearances closer to high-dose continuous renal extracorporeal replacement therapies were obtained with a 24-hour continuous PD schedule, 2 liters, 35–50 min dwell (36–44 litrers/day) [19].

In spite of the fear of impairment of diaphragm mobilization with reduction of pulmonary compliance and the often prevalent surgical abdominal pathologies that may contraindicate its use, high volume automated PD for acute renal failure is a good alternative to intermittent HD in elective patients or as an alternative in developing countries.

Due to its feasibility, low cost and good hemodynamic tolerance, automated PD is also an alternative to treat congestive heart failure with variable degrees of kidney injury, mainly where other continuous hemodialytic therapies are not applicable [10].

References

1 Just PM, de Charro FT, Tschosik EA, Noe LL, Bhattacharyya SK, Riella MC: Reimbursement and economic factors influencing dialysis modality choice around the world. Nephrol Dial Transplant 2008;23:2365–2373.
2 Shih YC, Guo A, Just PM, Mujais S: Impact of initial dialysis modality and modality switches on Medicare expenditures of end-stage renal disease patients. Kidney Int 2005;68:319–329.
3 Bargman JM: Is there more to living than not dying? A reflection on survival studies in dialysis. Semin Dial 2007;20:50–52.
4 Babazono T, Nakamoto H, Kasai K, Kuriyama S, Sugimoto T, Nakayama M, Hamada C, Furuya R, Hasegawa H, Kasahara M, Moriishi M, Tomo T, Miyazaki M, Sato M, Yorioka N, Kawaguchi Y: Effects of icodextrin on glycemic and lipid profiles in diabetic patients undergoing peritoneal dialysis. Am J Nephrol 2007;27:409–415.
5 Kuriyama S: Peritoneal dialysis in patients with diabetes: are the benefits greater than the disadvantages? Perit Dial Int 2007;27(suppl 2):S190–S195.
6 Davies SJ: Comparing outcomes on peritoneal and hemodialysis: a case study in the interpretation of observational studies. Saudi J Kidney Dis Transpl 2007;18:24–30.
7 Vonesh EF, Snyder JJ, Foley RN, Collins AJ: Mortality studies comparing peritoneal dialysis and hemodialysis: what do they tell us? Kidney Int Suppl 2006:S3–11.
8 Ganesh SK, Hulbert-Shearon T, Port FK, Eagle K, Stack AG: Mortality differences by dialysis modality among incident ESRD patients with and without coronary artery disease. J Am Soc Nephrol 2003;14:415–424.
9 Cavanaugh KL, Merkin SS, Plantinga LC, Fink NE, Sadler JH, Powe NR: Accuracy of patients' reports of comorbid disease and their association with mortality in ESRD. Am J Kidney Dis 2008; 52:118–127.
10 Gabriel DP, Fernandez-Cean J, Balbi AL: Utilization of peritoneal dialysis in the acute setting. Perit Dial Int 2007;27:328–331.
11 Basile C, Chimienti D, Bruno A, Cocola S, Libutti P, Teutonico A, Cazzato F: Efficacy of peritoneal dialysis with icodextrin in the long-term treatment of refractory congestive heart failure. Perit Dial Int 2009;29:116–118.
12 Dratwa M: Costs of home assistance for peritoneal dialysis: results of a European survey. Kidney Int Suppl 2008:S72–S75.
13 Hiramatsu M: How to improve survival in geriatric peritoneal dialysis patients. Perit Dial Int 2007;27(suppl 2):S185–S189.
14 Li PK, Law MC, Chow KM, Leung CB, Kwan BC, Chung KY, Szeto CC: Good patient and technique survival in elderly patients on continuous ambulatory peritoneal dialysis. Perit Dial Int 2007;27(suppl 2):S196–S201.
15 Rodrigues AS, Martins M, Korevaar JC, Silva S, Oliveira JC, Cabrita A, Castro e Melo J, Krediet RT: Evaluation of peritoneal transport and membrane status in peritoneal dialysis: focus on incident fast transporters. Am J Nephrol 2007;27:84–91.

16 Rodrigues AS, Almeida M, Fonseca I, Martins M, Carvalho MJ, Silva F, Correia C, Santos MJ, Cabrita A: Peritoneal fast transport in incident peritoneal dialysis patients is not consistently associated with systemic inflammation. Nephrol Dial Transplant 2005;gfi245.
17 Messa P, Ponticelli C, Berardinelli L: Coming back to dialysis after kidney transplant failure. Nephrol Dial Transplant 2008;23:2738–2742.
18 Gabriel DP, Caramori JT, Martim LC, Barretti P, Balbi AL: High volume peritoneal dialysis vs. daily hemodialysis: a randomized, controlled trial in patients with acute kidney injury. Kidney Int Suppl 2008;S87-S93.
19 Gabriel DP, Nascimento GV, Caramori JT, Martim LC, Barretti P, Balbi AL: High volume peritoneal dialysis for acute renal failure. Perit Dial Int 2007;27:277–282.

Anabela S. Rodrigues, MD, PhD
Division of Nephrology, Centro Hospitalar do Porto, Hospital de Santo António
Largo Professor Abel Salazar
PT–4100 Porto (Portugal)
Tel. +351 222 077521, Fax +351 222 077520, E-Mail ar.cbs@mail.telepac.pt

Ronco C, Crepaldi C, Cruz DN (eds): Peritoneal Dialysis – From Basic Concepts to Clinical Excellence. Contrib Nephrol. Basel, Karger, 2009, vol 163, pp 292–299

Continuous Ambulatory Peritoneal Dialysis and Automated Peritoneal Dialysis: Are There Differences in Outcome?

R. Dell'Aquila, G. Berlingò, M.V. Pellanda, A. Contestabile

Department of Nephrology and Dialysis, San Bassiano Hospital, Bassano del Grappa, Italy

Abstract

The proportion of peritoneal dialysis (PD) patients on automated peritoneal dialysis (APD) has been steadily increasing over the past decade. In the US, the percentage of PD patients on APD has steadily risen from 9% in 1993 to 54% in 2000. In continuous ambulatory peritoneal dialysis (CAPD), PD exchanges are performed manually, while in APD a mechanical device to assist the delivery and drainage of dialysate is employed. In CAPD, the patient or carer must perform at least 4–5 exchanges everyday. Many problems inherent to CAPD such as lack of sustained patient motivation over long periods of time, technique failure and recurrent peritonitis, led to a resurgence of interest in APD. APD has been reported to have several advantages over CAPD including lower incidence of peritonitis, better small solute clearances and reduced incidences of hernias. APD, especially in the form of nocturnal intermittent peritoneal dialysis (NIPD), has also been suggested to offer a number of psychosocial and physical benefits over CAPD mainly on account of fewer connections and being free of fluid in the abdomen during daytime. Such benefits relate to better dialysis acceptability for workers, school students or carers of elderly patients, pain and body image difficulties and reduced intra-abdominal pressures. APD is also considered to be more suitable form of PD in patients who have a rapid rate of solute transfer across their peritoneal membrane (high transporters) because of the ability to perform rapid frequent exchanges with shorter dwell times. It is not still clear if, with APD when compared to CAPD, a more rapid decline in residual renal function is present. Since the direct costs of APD are over 20% greater than CAPD and given this increasing trend towards greater use of APD, the aim of this paper is to understand if there are really differences in terms of quality of life and outcomes in favor of APD when compared to CAPD.

When starting PD, the planning of the initial prescription is sometimes a difficult choice; Finkelstein [1] wrote in 2006 that '...once the decision to start dialysis has been made, the nephrologist needs to decide whether to plan for continuous ambulatory peritoneal dialysis (CAPD) or continuous cycling peritoneal dialysis (CCPD). This decision is usually made after discussion with the patient and patient's family and should be made primarily on quality of life concerns...'. I don't agree with the term 'primarily' since there are many other clinical conditions that can influence the choice of modality, for example the presence of nonreducible hernias, preexisting slipped disc, uterine and/or vesical prolapse in women, membrane characteristics (i.e. low or high transporters), loss of residual renal function (RRF) or ultrafiltration (UF): so quality of life is important but we do not forget clinical issues.

APD has been reported to have several advantages over CAPD including lower incidence of peritonitis [2, 3], better small solute clearances [4] and reduced incidence of hernias [5]. APD has also been suggested to offer a number of psychosocial and physical benefits over CAPD mainly on account of fewer connections and being free of fluid in the abdomen during daytime. Such benefits relate to better dialysis acceptability for young people, workers, school students or carers of elderly patients and reduced intra-abdominal pressures [6, 7].

It is also considered to be a more suitable form of PD in patients who have a rapid rate of solute transfer across their peritoneal membrane (high transporters) because of the possibility of shorter dwell times [8]. It is not still clear if, with APD, when compared to CAPD, a more rapid decline in residual renal function is present [9, 10]. In order to evaluate the possible differences in the outcomes between CAPD and APD we will discuss about several topics as geographical differences in the choice of modality, patient's and technique survival, cardiovascular aspects (water and sodium removal, UF and RRF), peritonitis rates and quality of life.

Discussion

CAPD or (not) APD?
This is the question!

Prevalence and Incidence

Geographical Differences

Utilization of PD fluctuates considerably from country to country [11]. The proportion varies from 0 to above 60% of the total dialysis population and is most often explained by socioeconomic, health care or reimbursement factors [11–14]. Wauters et al. [15] demonstrated that this disparity also exists at the regional or at the center level and may also change over time within the same unit, meaning that other factors must play a role. The so-called Swiss experience shows that while in 1993 PD was used in 18% of the total Swiss dialysis population, this percentage has declined and in 2004 was stabilized at 11%; in the meantime, the proportion of APD had increased from 9 to 43% of all the PD patients. Interestingly, even within the PD treatment modalities, the use of APD versus CAPD seems to be center- and/or region-dependent. While the French-speaking part of Switzerland has largely moved to APD, used in >60% of the PD patients, this trend is less marked and more recent in the German-speaking part. This difference is most probably explained by patient attitudes towards autonomous dialysis and preferences concerning daytime activities.

Yu et al. [16] show a constantly very high percentage of patients in CAPD compared to APD since 1995 for 10 years: in 2005, the percentage of patients on CAPD was 92.3% of all the PD population (3,276 patients). The 2004 Italian Registry of Dialysis and Transplantation report [17] shows a 59% of CAPD and 38% of APD prevalent patients while incident patients in the same year were 56% for CAPD and 34% for APD; also in Italy the choice between the two different modalities seems to be center-dependent; data not published about prevalent patients in two different PD centers (Brescia and Vicenza) report a large difference: Brescia showed in 2003 a 74% of use of APD while Vicenza in 2008 had a 23.5% of prevalent patients in APD.

Patient and Technique Survival

Guo and Mujais [18] in 2003 studied the status of patient and technique survival in three contemporary cohorts of patients on PD in the United States that started peritoneal dialysis in the years 1999, 2000 and 2001 and were followed until February of 2003. The proportion of patients on APD was over 60% in all age groups except for those over 80 years, where slightly more than 40% were still on CAPD, a surprising finding considering the logistic simplification of the therapy afforded by cycler use. The authors demonstrated that the unadjusted Kaplan-Meier estimates showed that CAPD patients tended to have a worse survival than APD patients. In particular the 1-year patient survival was 78.48% for CAPD and 87.24% for APD ($p < 0.001$) while the technique survival was 68.81% for CAPD and 81.30% for APD ($p < 0.001$).

Sodium and Fluid Removal

Ates et al. [19] focused on the importance of fluid and sodium removal on morbidity and mortality in PD patients.

Inadequate fluid and sodium removal as potential risk factors for and mortality have not yet been assessed in PD patients. Recently, the Netherlands Cooperative Study on the Adequacy of Dialysis showed that systolic pressure, as an indicator of fluid overload, is an independent predictor of mortality [20]. However, in that study, total daily fluid removal was not significantly associated with mortality, and the effect of sodium removal on patient survival has not been examined. Adequate sodium and fluid removal as the treatment for hypertension instead of antihypertensive drugs seems to be a more logical approach.

Rodriguez-Carmona et al. [21] studied a group of incident patients treated with CAPD (n = 53) or APD (n = 51) for at least 1 year and found that ultrafiltration and sodium removal rates are consistently lower in incident APD patients than in CAPD patients. Moreover, RRF declines faster during APD than during CAPD therapy, although this difference may be partially counteracted by a detrimental effect of ultrafiltration on RRF.

Ultrafiltration

Brown et al. [22] in the EAPOS Study demonstrated the importance of ultrafiltration in the survival of anuric patients in APD treatment. Fussholler et al. [23] found that APD and CAPD patients showed no significant difference concerning small solute transport data (D/P_{crea} 69.7 ± 1.7 vs. 70.1 ± 1.8%; $MTAC_{crea}$ 11.8 ± 0.8 vs. 11.3 ± 0.7 ml/min). Lymphatic absorption with 149 ± 58 vs. 208 ± 58 ml/4 h and in particular effective ultrafiltration with 291 ± 31 vs. 265 ± 40 ml/4 h did not statistically differ between APD and CAPD patients.

Residual Renal Function (RRF)

In 1994, De Fijter et al. [24], in their prospective, randomized study comparing CAPD and CCPD, revealed a significant decline in RRF over time in both the CAPD and the APD groups. There was no significant difference between the dialysis modalities regarding residual renal function at any time point during follow-up. They compared 11 CAPD patients with 13 CCPD patients that were followed for 24 months and revealed no significant difference in residual renal function.

On the contrary, Hufnagel et al. [10] in 1999 determined and compared prospectively over 1 year two groups of peritoneal dialysis (PD) patients: 18 consecutive new patients starting on APD (12 continuous cyclic peritoneal dialysis (CCPD) patients and 6-nightly intermittent peritoneal dialysis (NIPD) patients)

Table 1. Results of a systematic review of randomized controlled trials comparing CAPD with all forms of APD to assess their comparative clinical effectiveness (from Rabindranath et al. [29])

Outcomes	Differences between CAPD and APD
Mortality	no difference
Infectious complications	no difference
Change of dialysis modality	no difference
Mechanical complications	no difference
PD catheter removal	no difference
Hospital admissions	no difference
Dialysis adequacy meqasures	no difference
Residual renal function	no difference
Quality of life	in favor of APD

and 18 selected patients who had started on CAPD at the same time and were matched for baseline characteristics. Their results were that RRF declined rapidly in APD patients whereas it was well preserved in CAPD patients.

Peritonitis

Yishak et al. [25] reviewed all peritonitis over one decade to compare patient outcomes on APD and CAPD. There were 327 episodes of peritonitis in 198 patients during this period. The rates were 0.57 per patient-year and 0.55 per patient-year on APD and CAPD, respectively. In summary, they stated that the incidence of peritonitis was similar on APD and CAPD. Other studies have found the incidence of peritonitis to be higher in CAPD than in APD [3, 24, 26–28]. In the most recent study, Rodriguez-Carmona et al. [26] reported a rate of 0.66 episodes per dialysis year in CAPD patients compared with 0.36 episodes per dialysis year in APD patients. Both de Fijter et al. [24] and Bro et al. [27] compared APD and CAPD in randomized studies; both found approximately 80% higher rates of peritonitis on CAPD (although the rates were much higher in the de Fijter study).

Quality of Life

The goals for maintenance dialysis treatment are to improve patient survival, reduce patient morbidity, and improve patient quality of life. Bro et al.

[27] performed a randomized prospective study comparing APD and CAPD treatment with respect to quality of life and clinical outcomes. Thirty-four adequately dialyzed patients with high or high-average peritoneal transport characteristics were included in the study. Twenty-five patients completed the study. After randomization, 17 patients were allocated to APD treatment and 17 patients to CAPD treatment for a period of 6 months. Quality-of-life parameters were assessed at baseline and after 6 months by the self-administered short-form SF-36 generic health survey questionnaire supplemented with disease- and treatment-specific questions. Quality-of-life studies showed that significantly more time for work, family, and social activities was available to patients on APD compared to those on CAPD ($p < 0.001$). Although the difference was not significant, there was a tendency for less physical and emotional discomfort caused by dialysis fluid in the APD group. Sleep problems, on the other hand, tended to be more marked in the APD group.

The Cochrane Collaboration Study

In 2007 Rabindranath et al. [29] performed a systematic review of randomized controlled trials (RCTs) comparing CAPD with all forms of APD to assess their comparative clinical effectiveness.

The Cochrane Central Register of Controlled Trials (CENTRAL), MEDLINE, EMBASE and CINAHL, were searched for relevant RCTs. Analysis was by a random effects model and results expressed as relative risk (RR) and weighted mean difference (WMD) with 95% CI. The results of the study are summarized in table 1. The small sample sizes of the included trials would have, however, greatly reduced the power of these trials to detect such differences. Two of the three studies were less than a year in duration. Therefore, these trials are not appropriate for the assessment of long-term clinical outcomes. The results of this review should, therefore, be interpreted with caution owing to the inherent limitations of the included trials.

Conclusions

The use of APD has been expanding rapidly and it is surprising to note that a systematic review identified only three randomized controlled trials with just 139 patients comparing it with CAPD. The small sample sizes of all the studies taken in account led us to interpret all the results with caution. It therefore does appear that the small data from currently available studies are not sufficient enough for us to reach definitive conclusions about the relative clinical effectiveness of CAPD and APD with respect to important clinical outcomes. Larger studies are needed to try to understand if there really are differences between CAPD and APD.

References

1 Finkelstein FO: The initiation of peritoneal dialysis: planning the initial prescription. Contrib Nephrol. Basel, Karger, 2006, 150, pp 42–47.
2 Brunkhorst R, Wrenger E, Krautzig S, Ehlerding G, Mahiout A, Koch K-M: Clinical experience with home automated peritoneal dialysis. Kidney Int 1994;(suppl 48):S25–S30.
3 Holley JL, Bernardini J, Piraino B: Continuous cycling peritoneal dialysis is associated with lower rates of catheter infections than continuous ambulatory peritoneal dialysis. Am J Kidney Dis 1990;16:133–136.
4 Rodriguez AM, Diaz NV, Cubillo LP, et al: Automated peritoneal dialysis: a Spanish multicentre study. Nephrol Dial Transplant 1998;13:2335–2340.
5 Diaz-Buxo AJ, Suki WN: Automated peritoneal dialysis; in Gokal R, Nolph KD (eds): Textbook of Peritoneal Dialysis. Dordrecht, Kluwer Academic, 1994, pp 300–418.
6 Wrenger E, Krautzig S, Brunkhorst R: Adequacy and quality of life with automated peritoneal dialysis. Perit Dial Int 1996;16(suppl 1):S153–S157.
7 Twardowski ZJ, Prowant BF, Nolph KD: High volume, low frequency continuous ambulatory peritoneal dialysis. Kidney Int 1983;23:64–70.
8 European Best Practice Guidelines Working Group in Peritoneal Dialysis: European Best Practice Guidelines for Peritoneal Dialysis: automated peritoneal dialysis. Nephrol Dial Transplant 2005;20(suppl 9): ix21–ix23.
9 Hiroshige K, Yuu K, Soejima M, Takasugi M, Kuroiwa A: Rapid decline of residual renal function in patients on automated peritoneal dialysis. Perit Dial Int 1996;16:307–315.
10 Hufnagel G, Michel C, Queffeulou G, Skhiri H, Damieri H, Mignon F: The influence of automated peritoneal dialysis on the decrease in residual renal function. Nephrol Dial Transplant 1999;14:1224–1228.
11 Gokal R, Horl W, Lameire N: Healthcare systems: an international review. Introduction. Nephrol Dial Transplant 1999;14(suppl 6):1.
12 Nissenson AR, Prichard SS, Cheng IK: Non-medical factors that impact on ESRD modality selection. Kidney Int 1993;40(suppl):S120–S127.
13 Gokal R, Blake PG, Passlick-Deetjen J, Schaub TP, Prichard S, Burkart JM: What is the evidence that peritoneal dialysis is underutilized as an ESRD therapy? Semin Dial 2002;15:149–161.
14 Mignon F, Michel C, Viron B: Why so much disparity of PD in Europe? Nephrol Dial Transplant 1998;13:1114–1117.
15 Wauters JP, Uehlinger D: Non-medical factors influencing peritoneal dialysis utilization: the Swiss experience. Nephrol Dial Transplant 2004;19:1363–1367.
16 Yu AW, Chau KF, Ho YW, Li PK: Development of the 'peritoneal dialysis first' model in Hong Kong. Perit Dial Int 2007;27(suppl 2):S53–S55.
17 Italian Registry of Dialysis and Transplantation 2004 report.
18 Guo A, Mujais S: Patient and technique survival on peritoneal dialysis in the United States: evaluation in large incident cohorts. Kidney Int Suppl 2003;88:S3–S12.
19 Ateş K, Nergizoğlu G, Keven K, Sen A, Kutlay S, Ertürk S, Duman N, Karatan O, Ertuğ AE: Effect of fluid and sodium removal on mortality in peritoneal dialysis patients. Kidney Int 2002;61:1552.
20 Jager KJ, Merkus MP, Dekker FW, et al: Mortality and technique failure in patients starting chronic peritoneal dialysis: results of the Netherlands Cooperative Study on the adequacy of dialysis. Kidney Int 1999;55:1476–1485.

21 Rodriguez-Carmona A, Pérez-Fontán M, Garca-Naveiro R, Villaverde P, Peteiro J: Compared time profiles of ultrafiltration, sodium removal, and renal function in incident CAPD and automated peritoneal dialysis patients. Am J Kidney Dis 2004;44:132–145.
22 Brown EA, Davies SJ, Rutherford P, Meeus F, Borras M, Riegel W, Divino Filho JC, Vonesh E, van Bree M, EAPOS Group: Survival of functionally anuric patients on automated peritoneal dialysis: the European APD Outcome Study. J Am Soc Nephrol 2003;14:2948–2957.
23 Fusshöller A, zur Nieden S, Grabensee B, Plum J: Peritoneal fluid and solute transport: influence of treatment time, peritoneal dialysis modality, and peritonitis incidence. J Am Soc Nephrol 2002;13:1055–1060.
24 De Fijter CWH, Oe PL, Nauta JJP: Clinical efficacy and morbidity associated with continuous cyclic compared with continuous ambulatory peritoneal dialysis. Ann Intern Med 1994;120:264–271.
25 Yishak A, Bernardini J, Fried L, Piraino B: The outcome of peritonitis in patients on automated peritoneal dialysis. Adv Perit Dial. 2001;17:205–208.
26 Rodriguez–Carmona A, Perez–Fontán M, Falcon TG, et al: A comparative analysis on the incidence of peritonitis and exit-site infection in CAPD and APD. Perit Dial Int 1999;19:253–258.
27 Bro S, Bjorner JB, Tofte–Jensen P, et al: A prospective, randomized multicenter study comparing APD and CAPD treatment. Perit Dial Int 1999;19:526–533.
28 Locatelli AJ, Marcos GM, Gomez MG, et al: Comparing peritonitis in continuous ambulatory peritoneal dialysis patients versus automated peritoneal dialysis patients. Adv Perit Dial 1999;15:193–196.
29 Rabindranath KS, Adams J, Ali TZ, Daly C, Vale L, MacLeod AM: Automated vs. continuous ambulatory peritoneal dialysis: a systematic review of randomized controlled trials. Nephrol Dial Transplant 2007;22:2991–2998.

Roberto Dell'Aquila, MD
Department of Nephrology & Dialysis, St Bassiano Hospital
IT–36100 Bassano del Grappa (Italy)
Tel. +39 424 888487, Fax +39 424 889493, E-Mail Roberto.DellAquila@aslbassano.it

Ronco C, Crepaldi C, Cruz DN (eds): Peritoneal Dialysis – From Basic Concepts to Clinical Excellence. Contrib Nephrol. Basel, Karger, 2009, vol 163, pp 300–305

The Wearable Artificial Kidney: Is Peritoneal Dialysis the Solution?

Claudio Ronco

Department of Nephrology, Ospedale San Bortolo, Vicenza, Italy

Abstract

The evolution of technology in hemodialysis has a new challenge in the development of miniaturization, transportability and wearability of devices applicable for renal replacement therapy. Although we are not there yet, a new series of papers have recently been published showing promising results on the application of a wearable artificial kidney. Some of them use extracorporeal blood cleansing as a method of blood purification while others use peritoneal dialysis as a treatment modality. Many of the challenges imposed by an extracorporeal wearable artificial kidney can be overcome by the use of the PD technique in a wearable system. A PD-based wearable artificial kidney has been demonstrated to be feasible and potentially applicable in chronic patients. We should make an effort to make a quantum leap in technology making the wearable artificial kidney a reality rather than a dream.

For many years the standard care for uremia has been based on dialysis therapy. Since the early eighties however, after continuous ambulatory peritoneal dialysis (PD) had been described [1], some interest has grown on the possibility to develop a truly wearable or portable dialysis system [2]. Some reports of 'wearable' hemodialysis devices appeared in the literature showing, however, limitations in terms of excessive size and weight, low performance and most of all, uncertain safety [3, 4]. Recently, new interest in the field has been spurred by a series of papers based on a truly wearable technology [5–7]. The idea of a miniaturized device working 24/7 is not new, but only now, with the advent of nanotechnology and miniaturization, adequate supplies can provide the technological platform for a wearable artificial kidney (WAK) aiming at new targets of efficiency and safety.

Rationale for a Wearable Artificial Kidney

There is a multidimensional rationale for a development of a WAK system. The outcomes of chronic renal replacement therapy remain dismal regarding the quality of life, morbidity and mortality. A growing body of literature indicates that more frequent and prolonged dialysis treatment is associated with strikingly improved outcomes [8]. As patients are shifted from the typical three-dialysis treatments per week regime to one of daily dialysis, significant improvement of the quality of life was reported (i.e. liberalization of the diet, fluid restrictions) alongside substantial reductions in medication consumption, complications, psychological symptoms, admissions and hospitalization [9]. The reported advantages of daily dialysis are improved volume control, elimination of the need for phosphate binders, no sodium retention, improvement in appetite and nutrition, less hypertension, decreased need for blood pressure drugs, no hyperkalemia, decreased expected morbidity and mortality, no hyperphosphatemia from bone disease, lesser degree of anemia, no metabolic acidosis cardiovascular disease and stroke, improved serum albumin. Thus, it seems logical to approach the treatment of uremia with a therapy that mimics the characteristics of the human kidney and working continuously and every day. Although continuous ambulatory PD is accomplishing such task with success, this therapy is only applied to 10% or less of the population and further solutions should be pursued. A WAK system may provide the type of therapy considered clinically adequate, assuming some technical problems are solved and some difficulties are overcome. Today, we have technologies that were not even imaginable a few years ago. We should take advantage of them to make a quantum leap in dialysis and to make important discoveries in the field. The development of WAK require some advances in terms of biomaterials, miniaturization and vascular access to the circulation. The last probably represents the most important aspect considering the high morbidity connected with the use of catheters and percutaneous vascular accesses. A new paradigm should be pursued to drawn blood from the circulation and to return it to the patient. New nonthrombogenic biomaterials can be helpful to maintain the circuit patent with minimal or no anticoagulation for several hours or days. Finally, new materials and production processes may contribute to miniaturization of devices and to reduction of their weight. Such research will make possible important discoveries and advances useful also for the current dialysis technology. With the recent technological breakthroughs, the major dilemma that arises is whether society wishes to maintain the status quo and bear the present morbidity, mortality and cost of treating CKD patients, or move forward to create new, innovative and cost-effective dialysis devices with the potential for alleviating the plight and misery of uremic patients. Such innovations may provide daily or even continuous dialysis, without imposing an unbearable burden on already scarce financial healthcare resources. In the US alone, the number of CKD patients has been growing steadily and currently approaching 400,000. The total cost of treating these patients tops USD 30 billion a year. Furthermore, the cost of CKD to society during the current decade is estimated at USD 1 trillion [10]. Even so, the mortality in CKD patients remains unacceptably high, reaching that of metastatic carcinoma of the breast, the colon or the prostate.

Thus, there is a growing need for a practical around-the-clock solution that will afford to ESRD patients the ability of receiving significantly increased dialysis dosages, while increasing efficiency and reducing the overall cost and the utilization of manpower.

Wearable Artificial Kidney Challenges

There are several challenges to overcome for a rapid development of WAK and for its wide application: the device must be truly wearable and, therefore, light, independent of the electrical outlet, and cost effective. The amount of dialysate has to be minimal with an effective regeneration process done by a cheap and safe sorbent device. The ergonomic design is a must, offering a combination of a friendly used interface [8] and a miniaturized, easy-to-wear device. Some of the recent experiments have demonstrated the feasibility of the concept and the potential for innovations in the near future. Many improvements are still needed, but if we keep doing things as we have done so far, we will end up with the same outcomes and results. A paradigm shift may be required and if we explore parallel fields we might advocate for the artificial kidney what has been done for computers, pacemakers and telephones. It is up to us to make this dream come true now, next year or never.

Peritoneal Dialysis-Based Wearable Artificial Kidney

PD is the most common form of home-based renal replacement therapy. In spite of significant improvements in technique survival and reduction of infectious complications, PD still remains under utilized in most countries. Several barriers to the expansion of the PD program have been identified as possible causes of PD underutilization. Among them, the commitment of patients and the time spent in performing the technique, together with the limitations imposed to the normal life, represent the most common factors pushing patients away from the choice of PD. A wearable PD system would probably represent a new option to increase the use of PD and to expand the PD program worldwide.

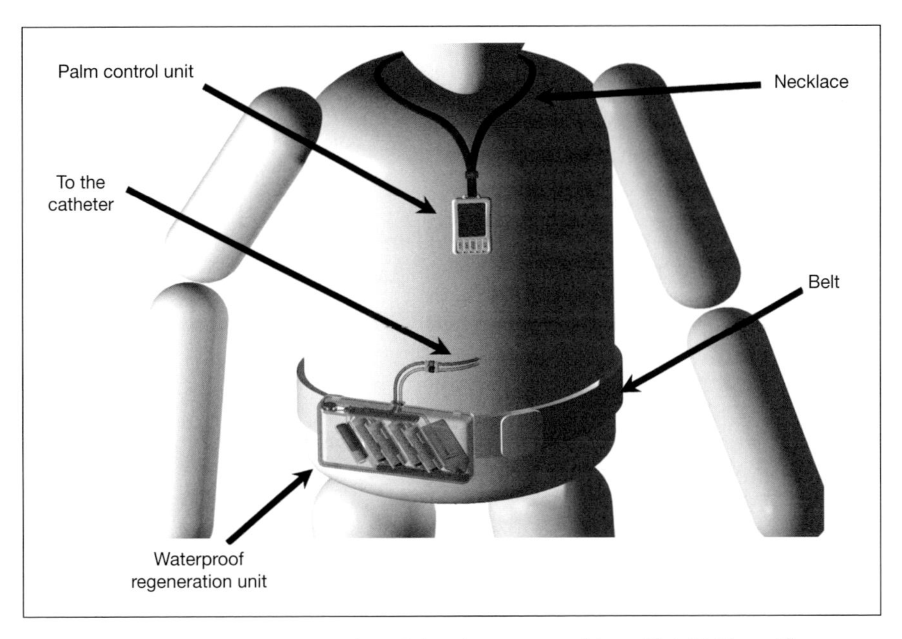

Fig. 1. Schematic representation of the Vicenza wearable artificial kidney. The system is based on a continuous flow PD technique in which the effluent dialysate is regenerated by a sorbent system built in a special wearable and disposable belt.

Looking from another perspective, many of the challenges imposed by an extracorporeal WAK can be overcome by the use of the PD technique in a wearable system.

An innovative wearable system called the Vicenza wearable artificial kidney for PD (ViWAK PD) has been recently presented [11]. The system is conceived to perform continuous PD requiring some maneuvers only in the morning and in the evening, and leaving the patient free during the day and the night. The concept of the ViWAk PD is based on the fact that a friendly user interface and a miniaturized system reducing the number of maneuvers and liters of solution required to perform PD may contribute to a better acceptance by the patient and a wider application of this form of renal replacement therapy (fig. 1).

The system is based on a combination of a long overnight dwell exchange and a continuous flow PD during the day performed with a special catheter and a special minicycler utilizing a mixture of sorbents for regenerating the PD solution.

A peritoneal-based automated wearable artificial kidney (AWAK) has also been presented by Lee et al. [12]. The system requires no extracorporeal circulation and is therefore 'bloodless'. Because AWAK is designed to continuously regenerate and reuse the spent dialysate in perpetuity, it is also 'waterless.' A sorbent-based assembly regenerates both the aqueous and the protein components (AqC and PrC) of the spent dialysate, producing a novel, autologous protein-containing dialysate. The regenerated AqC has the same composition as the commercially available peritoneal dialysate, but contains bicarbonate instead of lactate and has a more physiological pH. The regenerated PrC is recycled back into the peritoneal cavity, thereby ameliorating or eliminating protein loss. Depending on the steady-state protein concentrations that can be achieved (under the condition of continuous dialysate regeneration and recycling), the PrC also has the potential of both augmenting ultrafiltration and mediating the removal of protein-bound toxins. Additional sorbents can be incorporated into AWAK for the removal of middle molecular weight uremic toxins. At a regeneration rate of 4 liters/h, AWAK provides a dialysate flow of 96 liters/day (8–12 times the current rate).

Conclusions

Wearable solutions to perform round-the-clock dialysis and ultrafiltration provide steady-state metabolic-biochemical and fluid balance regulation, thereby eliminating 'shocks' of abrupt changes in these parameters that characterize the current dialytic modalities.

Several problems remain to be solved but PD-based WAK systems may represent a breakthrough in this field. Dialysis-on-the-go, made possible by PD-based WAKs such as viWAK and AWAK, represents a proof of concept of 'wearability' and automation, frees end-stage renal failure patients from the servitude that is demanded by the current dialytic regimens, and will probably represent a stimulus for further developments. These experiences may result in a step forward in the research of wearable devices and may contribute to a paradigm shift for the treatment of uremia.

References

1 Popovich RP, Moncrief JW, Dechard JF, Bomar JB, Pyle WK: The definition of a novel portable/wearable equilibrium dialysis technique (abstract). Trans Am Soc Artif Intern Organs 1976;5:64.

2 Shaldon S, Beau MC, Deschodt G, Lysaght MJ, Ramperez P, Mion C: Continuous ambulatory haemofiltration. Trans Am Soc Artif Organs 1980;26:210–212.

3 Murisasco A, Reynier JP, Ragon A, Boobes Y, Baz M, Durand C, Bertocchio P, Agenet C, El Medí M: Continuous arterio-venous haemofiltration in a wearable device to treat end stage renal disease. Trans Am Soc Artif Organs 1986;32:567–571.

4 Shettigar UR, Kablitz C, Stephen R, Kolff WJ: A portable hemodialysis/hemofiltration system independent of dialysate and infusion fluid. Artif Organs 1983;7:254–256.

5 Gura V, Beizai M, Ezon C, Polaschegg HD: 'Continuous renal replacement therapy for end-stage renal disease: the wearable artificial kidney (WAK)'; in Ronco C, Brendolan A, Levin NW (eds): Cardiovascular Disorders in Hemodialysis. Contrib Nephrol. Basel, Karger, 2005, vol 149, pp 325–333.
6 Gura V, Ronco C, Nalesso F, Brendolan A, Beizai M, Ezon C, Rambod E, Davenport A: A wearable haemofilter: first human study of slow continuous ambulatory ultrafiltration. Kid Int 2008; 73:497–502.
7 Davenport A, Gura V, Ronco C, Beizai M, Ezon C, Rambod E: A wearable haemodialysis device for patients with end-stage renal failure: a pilot study. Lancet 2007;370:2005–2010.
8 Blagg CT, Ing TS, Berry D, Kjellestrand CM: The history and rationale of daily and nightly hemodialysis. Contrib Nephrol. Basel, Karger, 2004, vol 145, pp 1–9.
9 Nesrallah GE, Chan CT, Buoncristiani U: Cardiovascular risk modification with quotidian hemodialysis. Contrib Nephrol. Basel, Karger, 2004, vol 145, pp 55–62.
10 Lysaght MJ: Maintenance dialysis population dynamics: current trends and long-term implications. J Am Soc Nephrol 2002;13(suppl 1):S37–S40.
11 Ronco C, Fecondini L: The Vicenza wearable artificial kidney for peritoneal dialysis (ViWAK PD). Blood Purif 2007;25:383–388.
12 Lee DB, Roberts M: A peritoneal-based automated wearable artificial kidney. Clin Exp Nephrol 2008;12:171–180.

Claudio Ronco, MD
Department of Nephrology
Ospedale San Bortolo
IT–361000 Vicenza (Italy)
Tel. +39 0 444 753 650; Fax +39 0 444 753 949, E-Mail cronco@goldnet.it

Ronco C, Crepaldi C, Cruz DN (eds): Peritoneal Dialysis – From Basic Concepts to Clinical Excellence. Contrib Nephrol. Basel, Karger, 2009, vol 163, pp 306–310

What Will Be the Role of Industry in the Future?

Peter Rutherford

Medical Director (Renal Europe), Baxter Healthcare SA, Zurich, Switzerland, and Visiting Professor and Fellow, Glyndwr University, Wrexham, UK

Abstract

Over the last 30 years peritoneal dialysis has provided effective therapy for many patients living with chronic kidney disease. Developments are required for this therapy in the future to meet the needs of an increasingly elderly population with complex needs. The population, patient and professional issues relating to the role of industry in dialysis therapy are discussed. A collaborative approach is required to address these issues and meet the needs of patients in the future.

Since peritoneal dialysis became more widely available over the last 30 years it has been shown to be an effective therapy for many thousands of patients with end stage renal failure. Key developments have made the therapy more safe (improved connector technology reducing peritonitis), effective (non-glucose solutions such as icodextrin and amino acid peritoneal dialysis (PD) fluids) and accessible to more patients (patient-focussed PD cycler to allow APD at home). At the same time, there have been great advances in the scientific understanding of peritoneal membrane biology, the importance of avoiding high glucose concentrations and the potential benefits of biocompatible fluids. Combined with the observation that patient training and retraining is of vital importance, clinical outcomes with modern PD therapy can be excellent including in elderly patients and patients with comorbidity, e.g. diabetes [1].

Nevertheless significant issues and questions for PD therapy remain:

(1) How can home therapy with PD expand in service delivery environments where there is high haemodialysis (HD) capacity and where private provision of dialysis seem not to favour home-based therapies?

(2) Are developments in PD fluids possible to deal with the long-term damaging effects of glucose exposure?
(3) PD access, exit site infection and peritonitis are still common reasons for PD failure – what can be done to improve in these areas of practice?

Looking ahead over the next 10 years in PD, it is interesting to examine the issues across the health arena that will face both industry and physicians providing therapies for people living with chronic kidney disease (CKD). These challenges are diverse and will require collaboration to work through for the benefits of the patient.

Population Issues

There is a growing recognition of the impact of changes in population demographics on the incidence of long-term conditions, including CKD. Improved survival from other comorbid illnesses is leading to an estimated tripling in the number of persons aged 80 and over by 2051 [2] alongside a rise in the median age of the European population. This will undoubtedly lead to an increased incidence of CKD stage 5 and requirement for dialysis therapy, but in a population with comorbidity and frailty who are often dependent on others for care. This will be a challenge for dialysis companies – how to provide effective therapies for such a population – bringing simplicity into an area of complexity. To allow home therapy with PD there will be a need for support systems, perhaps remote e-health solutions as well as access to support from staff. In addition, the global aspect of the rise in incidence in CKD needs to be remembered. The highest incidence of long-term conditions including CKD is to be found in the developing world where there is the additional challenge of lack of dialysis provision [3]. This is another challenge with different needs and requirements of PD therapy where cost is a major driver. In this respect, partnerships will be required between physicians, industry and governments to determine the best way to provide clinical and cost-effective dialysis therapy for a large number of patients in challenging environments.

Patient Issues

The average age of patients commencing chronic dialysis continues to rise and this produces challenges for physicians. The proportion of patients with more than one long-term condition is between 30 and 65% and this rises with age [4]. This comorbidity leads to clinical risk with complex therapies applied by a variety of physicians. In one study [5], a hypothetical patient with 4 common long-term conditions was required to take 12 different medications (at 5 different times of the day), have 5 dietary manipulations, 10 patient tasks, 18 clinician tasks daily and was at risk of 20 potential treatment interactions. Add dialysis to this complexity and the risks rise significantly. So industry will need to develop effective therapies that simplify regimes for patients and acknowledge co-morbidity. At the same time, people are becoming more aware of choices around therapy options and this growth of 'consumerism' in health care will continue. Patients will want to be more aware of choices, will desire a more coordinated approach to their care and will be more aware of targets and goals of therapy. A recent study [6] across 8 countries including several in Europe highlighted areas where people are concerned about their current management. Physicians and industry need to strive to 'join up' care across disease boundaries.

Provider Issues

Providers of care for dialysis patients balance and address safety, effectiveness and cost of therapy. There will be more attention on the real costs of therapy for the patient receiving dialysis therapy comparing hospital to home dialysis. Home therapies continue to be cost effective [7] compared to thrice weekly centre HD and this will be of interest to providers at a local level and at a macro level (government level). Industry will need to continue development of effective therapies but with total therapy cost in mind during the development and implementation phase.

Professional Issues

A set of challenges are being described which physicians, providers and industry need to consider now. With the changes in birth rates, employment choices, migration patterns and the need to care for older relatives, the number of nurses in Europe will fall. A Royal College of Nursing report from the UK estimated that there would be approximately 20% fewer nurses by 2015. How can the increasing patient need for dialysis be met then by therapies that are nurse intensive (e.g. hospital HD)? Industries will need to plan dialysis therapies which reduce nurse number dependency and of course home therapies and in particular PD carry obvious advantages.

Physician Issues

The presentations in the Vicenza PD course have highlighted current clinical practice, current scientific knowledge and pointed the way to new developments in PD therapy. There is an increasing and correct demand for evidence-based therapies. Developments in systems (connectology, cyclers, etc.), solutions (long and short dwell) and support (e.g. online technology, health) will need to be and indeed should be supported by clinical trials of adequate design, performed in accordance with all regulatory requirements and published in an ethical manner. Industry needs to develop interventions after listening to physicians over patient needs and clinical benefits. Clinical trials will need active commitment from industry to fund but then from physicians to randomize patients into clinical trials. Clinical trials in dialysis patients are few in number compared to in other specialties and there have been some clinical trials in PD that have failed to complete. More active partnership working is required to bridge the data gaps and provide evidence for new therapies. Another major challenge now and in the future is something that the Vicenza PD course attempts to help – the reducing educational and training exposure to PD in Nephrology education. This knowledge and competency gap influences physician confidence and comfort in PD therapy and as the number of PD patients in Europe remains static or declines, the impact of the problem on the future of PD therapy will only grow. But this area is a challenge for industry involvement. A recent editorial in JAMA [8] described some of the problems with industry involvement in research and education and one of its recommendations is that 'continuing medical education courses should not condone or tolerate for-profit companies having any input into the content of educational materials or providing funding or sponsorship for medical education programs'. This is a challenge that needs to be addressed in an ethical and transparent manner within individual countries but will be an increasing challenge in the future.

PD has been shown to be a clinically effective therapy across all patients groups from children to the elderly and across developed and developing nations. Industry need to continue development and trials of interventions to meet the needs of patients being guided by physicians and working in partnership with them. In the current global financial situation there are challenges for both industry and dialysis providers to ensure that dialysis is affordable. PD has and will have an important role in therapy options for people living with CKD. A recent editorial [9] in *Peritoneal Dialysis International* is correct in remarking: 'Sterile arguments about the superiority of one technique over another are redundant. What is required is an understanding that a patient's renal replacement requirements evolve over time.' PD in the future, developed with industry, implemented by physicians and meeting patient needs, is a key therapy over a patient's life with CKD.

References

1 Yeates K, Vonesh EF, Fenton SS: A new analysis comparing survival of patients receiving hemodialysis versus peritoneal dialysis in Canada: 1991–2000. Perit Dial Int 2008;28(suppl 4):S5.
2 Data from European Communities – 'Long term population projections at national level' March 2006. www.epp.eurostat.cec.eu.int/
3 White SL, Chadban SJ, Jan S, et al: How can we achieve global equity in provision of renal replacement therapy? Bull WHO 2008;86:229–237.
4 Wolff JL, Starfield B, Anderson G: Prevalence, expenditure and complications of multiple chronic conditions in the elderly. Arch Intern Med 2002;162:2269–2276.
5 Boyd CM, Darer J, Boult C, et al: Clinical practice guidelines and quality of care for older patients with multiple comorbid diseases. Implications for pay for performance. JAMA 2005;294:716–724.
6 Schoen C, Osborn R, How SKH, et al: In chronic condition; experiences of patients with complex health care needs, in eight countries, 2008. Health Affairs, November 2008.
7 Just PM, Riella MC, Tschosik EA, et al: Economic evaluations of dialysis treatment modalities. Health Policy 2008;86:163–180.
8 deAngelis CD, Fontanarosa PB: Impugning the integrity of medical science; the adverse effects of industry influence. JAMA 2008;299:1833–1835.
9 Fluck R: Transitions in care: what is the role of peritoneal dialysis? Perit Dial Int 2009;28:591–595.

Dr. Peter Rutherford
Medical Director, Renal-Europe, Middle East and Africa
Baxter Healthcare SA
PO Box
CH–8010 Zurich (Switzerland)
Tel. +41(0) 44 8786 395, Fax +41(0) 44 8786 450, E-Mail peter_rutherford@baxter.com

Ronco C, Crepaldi C, Cruz DN (eds): Peritoneal Dialysis – From Basic Concepts to Clinical Excellence. Contrib Nephrol. Basel, Karger, 2009, vol 163, pp 311–319

Role of the International Society for Peritoneal Dialysis (ISPD) in the Future of Peritoneal Dialysis

Wai Kei Lo

Department of Medicine, Tung Wah Hospital, Hong Kong, SAR, China

Abstract

The International Society for Peritoneal Dialysis (ISPD) was established in 1984. Throughout the years, the ISPD has been playing a pivotal role in the development of peritoneal dialysis (PD) through organizing congresses, publishing the Peritoneal Dialysis International formation of treatment and training guidelines, and supporting international studies. In recent years, it has enhanced its educational programs through organizing PD courses in developing countries, online education videos and a function 'Questions about PD' on its website. Several regional chapters – Asian, North American and Latin American – have been formed to target the special needs of different regions. To move forward, apart from enhancing the current activities, good use of cyber technology for out-reaching and educational purposes, and collaboration with other international or national societies particularly in the area of national policy making are envisaged.

The Birth of CAPD and the ISPD

Like many important inventions in human history, pioneer works are often neglected, not given the appropriate attention or even being rejected. In 1976, the abstract titled 'Equilibrium Peritoneal Dialysis' by Moncrief and Popovich from Austin, Tex., USA submitted to the American Society for Artificial Internal Organs was not accepted for presentation [1], yet it had produced ripples that were augmented as it spread across the world. It drew the attention of Dr. Karl Nolph of the University of Missouri to implement this new form of peritoneal dialysis (PD) in Missouri with a new name 'continuous ambulatory peritoneal dialysis' (CAPD) [2, 3]. Dr. Jack Rubin, a former fellow of Dr. Dimitrios Oreopoulos from Toronto Western Hospital of Canada went to Missouri and brought the message of CAPD to Dr. Oreopoulos in 1977, Dr. Oreopoulos converted all his intermittent PD patients to CAPD with peritoneal dialyzate in plastic bags, founding the prototype of CAPD today [2, 4]. The works of Dr. Nolph and Dr. Oreopoulos soon excited many nephrologists around the world with the establishment of CAPD programs here and there. CAPD soon gained support from the National Institute of Health and the Medicare of the USA, and a National CAPD Registry was formed in 1981, which much preceded the establishment of the United States Renal Data System (USRDS) in 1988.

The interest in CAPD spreading across the world was reflected by the uprising number of conferences specifically for PD in the next few years. In 1978, Dr. Alejandro Trevino-Becerra organized the First International Peritoneal Dialysis Symposium in Chapal, Mexico. The Second Symposium was later held in Berlin in 1981. On the other side of the globe, the First International Symposium on CAPD was held in Paris in 1979, and the Second Symposium in Austin, home town of Moncrief and Popovich. In the southern hemisphere, an International Symposium on Chronic Peritoneal Dialysis was held in Buenos Aires, Argentina, and a Pan Pacific Symposium on Peritoneal Dialysis was hosted in Melbourne, Australia, in 1980 [2]. Within the USA, Dr. Nolph on behalf of the University of Missouri organized the First National CAPD Conference at Kansas City, attracting over 350 participants including some from 11 other countries [2]. Its success has led to a yearly conference by the University of Missouri and subsequently it had developed into the Annual Dialysis Conference with around 2,500 participants every year. In 1982, Dr. Giuseppe LaGreca from Vincenza, Italy, organized The First International Course on Peritoneal Dialysis in his home town. It was thereafter repeated once every 3 years, alternating with hemodialysis and intensive care nephrology. It remains to be the consistently most attended PD course in the world.

The idea of forming of an international PD society was conceived by the forefathers of CAPD in Vincenza in 1982. This dream finally came true at the Third International Symposium on Peritoneal Dialysis in Washington, D.C., USA June 17–20, 1984, which was attended by 980 participants from 30 countries, the International Society for Peritoneal Dialysis was formed with John Maher being its founding President. It began with a small membership including Karl Nolph (USA), Dimitrios Oreopoulos (Canada), Jim Winchester (USA), Jack Moncrief (USA), R.P. Popovich (USA, bioengineer), Giuseppe LaGreca (Italy), Jonas Bergstrom (Sweden), Ram Gokal (UK), Fred Boen (The Netherlands), Alejandro Trevnio-Becerra (Mexico), Alberto Locatelli (Argentina), Barbara Prowant (USA, nurse) and many others important leaders in PD that I won't go into details [5]. The *Peritoneal Dialysis Bulletin*, which was started and edited by Dimitrios Oreopoulos as a work of Toronto Western Hospital since 1980, then became the official journal of the Society and was later transformed into

Peritoneal Dialysis International (PDI) in 1988. It should be emphasized that the name of this newly formed society is International Society 'for' instead of 'of' Peritoneal Dialysis. The choice of the preposition 'for' probably reflects the enthusiasm of the founders of the society towards PD.

The ISPD and PD: from 1984 to 2008

Since the establishment of the ISPD in 1984, the most noticeable activities of the ISPD are no doubt its congresses and its official journal.

After the third meeting in Washington D.C., USA, congresses of the ISPD were then hosted in Venice, Kyoto, Thessaloniki, Stockholm, Seoul, Montreal, Amsterdam, Hong Kong and Istanbul. The next congress will be hosted in Mexico City in 2010. Among these 11 congresses, three were in the Americas, three in Asia, four in Europe and one in a city that bridges Europe and Asia. Since the turn of the millennium, the participant number had grown beyond 2,000 and come close to 2,500. The ISPD congress is now the PD conference with the highest attendance (though the Annual Dialysis Conference has attendance slightly more than the ISPD congresses, it has as many hemodialysis sessions as with that of PD). Undoubtedly, it is the most important platform for presenting research findings, updating knowledge and exchange of ideas in all aspects related to PD. After 32 years from the first international symposium on PD in Mexico in 1978, hosting the next Congress in Mexico City in 2010 carries a quite a symbolic meaning of going back to its birth place, not to mention that Mexico is the country with highest penetration of PD.

One big step in bringing PD to different localities of the ISPD is the establishment of regional chapters. The ISPD is to my knowledge the only nephrology society that has regional chapters. The idea of a chapter emerged after the Congress in Seoul in 1998. It was the year right after the publishing of the DOQI guideline on adequacy of PD which recommended a target weekly Kt/V of 2.0 in PD patients. During the Seoul Congress where works from many Asian nephrologists in the area of adequacy of dialysis were presented, an Asian nephrologist came to me saying that western nephrologists were not taking serious consideration of the Asian findings and situation, and that Asian nephrologists should not follow the Western world and he asked me to consider organizing an Asian PD organization of our own. While echoing his comments, I thought that establishing a separate society from the ISPD was not very constructive. I then proposed the idea of having an Asian Chapter to the then ISPD President Ram Gokal. This idea was embraced enthusiastically by Ram Gokal and the council of the ISPD. I was delegated by the council to organize the first Asian Chapter Meeting (ACM) in 2002 in Hong Kong. The contents of the ACM were tailored to the needs in Asia and it was very well received. The close to 1,000 attendees from different Asian countries were excited by the achieved development and success of PD. Further, ACM had been held once every 2 years thereafter in Hyderabad, India and Hiroshima, Japan, respectively. This year, it will be held in Beijing, China. Before the first ACM, PD had already started growing in some countries in Asia. After the establishment of the Asian Chapter, the bloom was phenomenal, particularly in China and India. Most recently, the government of Thailand announced supporting a PD-first policy starting from January 2008 with, consequently, a very rapid growth in number of new PD patients in Thailand. This can be regarded as one of the remarkable achievements of the ACM.

Following the success of the Asian Chapter, the North American Chapter (NAC) Meeting was established with its first meeting in Chicago in 2005 and then in Miami in 2007 and it will go to Vancouver later this year. The NAC not only organizes meetings, it has formed a structure with committees that target education, research, awards and policy advocacy. The Latin American Chapter (LAC) was formally formed during the Miami meeting in 2007 and is going to hold its first independent chapter meeting (but the second LAC meeting) in Iguazzu at the junction of Brazil, Argentina and Paraguay in mid-2009. These chapter meetings, though not big meetings, were designed to target the specific issues of concern in each region that a big Congress can rarely do. In line with the development in Latin America, some ISPD guidelines have been translated into Spanish.

During the early years of CAPD development, new knowledge on various aspects of PD was coming up every year. There were so many aspects of PD awaiting scientists and clinicians to explore and study, yet little of these works were published by the mainstream general nephrology journals. The *Peritoneal Dialysis Bulletin* was the most suitable platform for those small works to be published and read. Despite its small size, it was the 'must-read' nephrology journal for anyone who practices PD in those days. Though a small bulletin, it contains one-sixth of all PD publications in the year 1985 and no other journal had published more than 20 PD articles a year except for the *Peritoneal Dialysis Bulletin* [6].

With the transition into *Peritoneal Dialysis International* (PDI) in 1988 and adoption of the format of mainstream medical journal and peer review system, the impact factor of the official journal of the ISPD gradually climbed. It has soon established itself as one of the most important nephrology journals with impact factor ranking among the top, with its peak achievement of impact factor 2.8, ranking number 4 among all nephrology journals in the year 1998 [7]. In the late 1990s when people were getting more concerned on publishing their works in journals with higher impact factors, the top PD articles tended to go to other journals. This produced a negative impact on the PDI. As the gap in the impact factor from the top journals widens, fewer high-quality manuscripts were submitted to PDI, further widening the gap. Then came a dilemma to the editorial board of the PDI: to push up the impact factor by rejecting more manuscripts or to maintain its conventional value – a platform for education, stimulation of research and publishing valuable works from young investigators? Facing this challenge, the new editorial board from 2004 under Dr. Peter Blake reformed the journal by adding high quality review articles and new sections like PD in the developing world, and keeping the level of publications with modest adjustment of the acceptance criteria. The fruit was seen in 2006 where the impact factor had risen significantly to 2.38. The PDI still continues to be the *must-read* journal for anyone who practices PD.

Behind the front-stage, the ISPD puts much emphasis on helping the development of PD in the developing countries and supporting scientific research. To the developing countries, the ISPD is providing fellowship scholarship to undergo clinical or research training in high standard PD centers aboard. Each year, there are around 6–10 such fellows. In addition, the Asian Chapter is offering similar fellows to Asian nephrologists to receive training in other Asian centers, with a philosophy that the culture of Asian training centers are closer to their original culture than the western centers, and thus easier for the fellows to bring the knowledge and practice back to their own countries.

In 1986, our founding president John Maher wrote: 'Nevertheless even a congress and a publication are not enough: the Society wants to promote scientific excellence' [6]. In line with this vision, the ISPD began awarding outstanding abstracts and young investigators, and started funding international studies of high scientific values. The ISPD has been supporting international PD research studies that involve more than one country. The applications are scrutinized by the International Study Committee of the ISPD on basis of their scientific value and the necessity for a multinational study. The studies that were funded by the ISPD in the past 10 years included IDEAL (Initiating Dialysis Early and Late) study, GLOBAL Fluid Study, International EPS DNA Bank: Genetic Markers of EPS in PD, and International Pediatric Peritoneal Dialysis Network These studies had a board dimension involving many countries and a long study period. Results that are of great impact to the PD community are to be anticipated.

ISPD Guidelines

The 1990s was the era of guidelines. To the nephrology community, we are so well familiar with K/DOQI guidelines, European Best Practice Guidelines, etc. In addition, many national societies also publish their own guidelines in many different aspects of nephrology. The DOQI (Dialysis Outcome Quality Initiatives) guideline was first published in 1997 by the National Kidney Foundation of the USA as a response to well recognized poorer outcome of dialysis patients (both

PD and hemodialysis) in the USA than many other countries, thinking that it could be a result of inadequate dialysis achieved in many patients. The DOQI guideline later covers more aspect of dialysis and earlier stages of chronic kidney disease and the name was modified to KDOQI (Kidney Disease Outcome Quality Initiative) guideline. The appearance of DOQI in 1997 had stimulated the birth of European Best Practice Guidelines and many other national guidelines afterwards.

But long before that, the ISPD was beginning to lay down guidelines. The first ISPD guideline was published in 1989 in the field of peritonitis [8]. The ISPD formed an Ad Hoc Committee under the lead of William Keane to lay down guidelines on the most important complication of PD in those days. This guideline was formulated on evidence from the published literature, though at the time the current-day evidence medicine practice format was not yet formulated. Catching up with updated research findings and evidence, the guideline was revised every few years by the ISPD in 1993, 1996, 2000 and 2005 [9–12]. It is the most authoritative document that shapes the treatment of peritonitis in the whole world. In addition to the peritonitis guideline, guidelines published by the ISPD included recommendations for training requirements of nephrology trainees and nurses (1994), PD catheters and exit site practices (1999, 2005), lipid disorder management (1999), ultrafiltration management (2000), PD curriculum for trainee (2000), encapsulating peritoneal sclerosis (2000), PD patient training (2006) and solute clearance and ultrafiltration (2006) [13–21]. They provide important guidance to PD. What makes the ISPD guidelines stands out from other guidelines is that they are meant to be applicable to situations around the world instead of meeting the need of a certain country or region. Thus, they carry a nature of universality which other guidelines may not. In fact, the ISPD is the only 'international' society under nephrology that constantly laid down guidelines to help clinical management.

Moving Towards the Future

Now we have entered the era of the internet. So the ISPD has to adapt to this era. Catching up the pace of the internet era, the PDI has installed on-line submission, on-line review, and searchable archives on the website. The ISPD education committee has produced teaching videos posted on its website. The latest example is the 'Teaching Nurses to Teach – Peritoneal Dialysis Training' education video jointly produced by the ISPD and the University of Pittsburgh. To further extend the educational function to day-to-day PD management in every corner of the world, a new function 'Questions About PD' was installed since July 2008. People can submit their queries to the web and someone from the ISPD would answer that question. Just in 8 months, over 100 questions have been received, mostly from nonmembers from many different countries. This indicates that firstly there is a strong need of guidance for day-to-day practice in many different places and secondly, quite a lot of nonmembers would visit our website. Thus, enhancing our services on-line would be able to reach more people from different corners of the world and meet their needs.

In the past, PD courses are often organized by PD industries or national societies with the endorsement of the ISPD. This is a very economic and effective way in organizing educational or training courses. However, with such practice, the ISPD is playing a relative passive role in selecting the target countries in promoting PD education. Since 2007, the ISPD has become more active in soliciting partnerships in organizing PD courses in countries or regions that had a great need for such courses. By now, these PD courses had taken place in Khartoum, Sudan (2007), Nairobi, Kenya (2008), Cairo, Egypt (2008) and Nigeria (2009) with the African Association of Nephrology and the Egyptian Society for PD, and in Rio de Janeiro of Brazil (2007) with the World Congress of Nephrology. These courses were well attended with around 300 local attendees. In some sense, the ISPD resembles the work of COMGAN of the International Society of Nephrology. I believe that this should be a future direction of the ISPD. Looking forward to working with COMGAN or other bodies is something the ISPD will do as well.

Needless to say, the ISPD should continue to promote the scientific development of PD. This requires further enhancing the role of the PDI, supporting scientific research and training through various means, and updating management guidelines. However, to promote PD development, it is not just pure science, research and clinical practice. Socioeconomical factors and health policy are important barriers to PD development in many countries. This is certainly an area that the ISPD should address. Policy committees have now been developed in certain societies like the American Society of Nephrology. However, the political situation differs tremendously from country to country and it is extremely difficult for the ISPD which is still a relatively small society with just over 1,100 members, to have the sufficient manpower to work in this aspect in different countries. But at least, we can target some countries with a high need for this. In this aspect, our regional chapters should be given the support to carry these specific tasks as the NAC is planning to do [22, 23]. Working with national societies to overcome these barriers is something that we should explore, as we have seen in the successful example in Thailand mentioned above.

Epilogue

We learn by reading history. Lessons from the past may vanish if historical documents are just archived but not read. Restudying the vision of our forefathers of PD would brighten our hearts and eyes towards the future of PD. The future of PD development shall rest on several major areas – scientific advancement, clinical practice advancement, education and training, overcoming the socioeconomical and political barrier, and reaching out to needy localities. Surely, the ISPD should continue to provide good scientific platforms for presentations and publications, enriching educational programs and supporting PD training, reaching out through the internet and regional chapters and courses, and possibly engaging in more active policy analysis and lobbying. There are surely more paths to be explored. To accomplish our mission, we need to strengthen our society by increasing its size and financial capacity, we need to collaborate with other nephrology communities like the International Society of Nephrology, and national societies as Maher [6] wrote in 1986: 'We anticipate that our societies will be mutually supportive, not competitive'. Let the spirit of our forefathers keep on lighting up the road ahead.

Appendix

The ISPD Mission Statement in 2006

The ISPD promotes and advances knowledge and practice of the therapy of PD. The long-term objective of the ISPD is to improve the well-being and care of patients with kidney failure. It achieves this through its education programs and guidelines, by disseminating information through its publications, national and regional meetings, and by promoting research into basic and clinical aspects of PD designed to maximize healthcare benefit.

References

1 Popovich RP, Moncrief JW, Decherd JF, Bomar JB, Pyle WK: The definition of a novel portable/wearable equilibrium dialysis technique (abstract). Trans Am Soc Artif Intern Organs 1976;5:64.
2 Nolph KD: 1975 to 1984 – an important decade for peritoneal dialysis: memories with personal anecdotes. Perit Dial Int 2002;22:608–613.
3 Popovich RP, Moncrief JW, Nolph KD, Ghods AJ, Twardowski ZJ, Pyle WK: Continuous ambulatory peritoneal dialysis. Ann Intern Med 1978;88:449–456.
4 Oreopoulos DG, Robson M, Izatt S, Clayton S, deVeber GA: A simple and safe technique for continuous ambulatory peritoneal dialysis (CAPD).Trans Am Soc Artif Intern Organs 1978;24:484–489.
5 Maher JF: Peritoneal dialysis acquires a society; the bulletin gains a sponsor. Perit Dial Int. 1984;4:119–120.
6 Maher JF: The International Society for Peritoneal Dialysis: Its current status. Perit Dial Int. 1986;6:166–167.
7 Oreopoulos DG: Peritoneal Dialysis International: Its past, present, and future. Pert Dial Int 2006;26:540–546.
8 Keane WF, Everett ED, Fine RN, Golper TA, Vas S, Peterson PK, Gokal R, Matzke GR: Continuous ambulatory peritoneal dialysis (CAPD) peritonitis treatment recommendations: 1989 update. The Ad Hoc Advisory Committee on Peritonitis Management. Perit Dial Int 1989;9:247–256.

9 Keane WF, Everett ED, Golper TA, Gokal R, Halstenson C, Kawaguchi Y, Riella M, Vas S, Verbrugh HA: Peritoneal dialysis-related peritonitis treatment recommendations. 1993 update. The Ad Hoc Advisory Committee on Peritonitis Management. International Society for Peritoneal Dialysis. Perit Dial Int 1993;13:14–28.
10 Keane WF, Alexander SR, Bailie GR, Boeschoten E, Gokal R, Golper TA, Holmes CJ, Huang CC, Kawaguchi Y, Piraino B, Riella R, Schaefer F, Vas S: Peritoneal dialysis-related peritonitis treatment recommendations: 1996 update. Perit Dial Int 1996;16:557–573.
11 Keane WF, Bailie GR, Boeschoten E, Gokal R, Golper TA, Holmes CJ, Kawaguchi Y, Piraino B, Riella M, Vas S: International Society for Peritoneal Dialysis Adult peritoneal dialysis-related peritonitis treatment recommendations: 2000 update. Perit Dial Int: 2000;20:396–411.
12 Piraino B, Bailie GR, Bernardini J, Boeschoten E, Gupta A, Holmes C, Kuijper EJ, Li PK, Lye WC, Mujais S, Paterson DL, Fontan MP, Ramos A, Schaefer F, Uttley L: ISPD Ad Hoc Advisory Committee. Peritoneal dialysis-related infections recommendations: 2005 update. Perit Dial Int 2005;25:107–131.
13 Recommendations of the International Society for Peritoneal Dialysis for training requirements of nephrology trainees and nurses. Perit Dial Int 1994;14:117–120.
14 Gokal R, Alexander S, Ash S, Chen TW, Danielson A, Holmes C, Joffe P, Moncrief J, Nichols K, Piraino B, Prowant B, Slingeneyer A, Stegmayr B, Twardowski Z, Vas S: Peritoneal catheters and exit-site practices toward optimum peritoneal access: 1998 update. Official report from the International Society for Peritoneal Dialysis. Perit Dial Int 1998;18:11–33.
15 Flanigan M, Gokal R: Peritoneal catheters and exit-site practices toward optimum peritoneal access: a review of current developments. Perit Dial Int 2005;25:132–139.
16 Fried L, Hutchison A, Stegmayr B, Prichard S, Bargman JM: Recommendations for the treatment of lipid disorders in patients on peritoneal dialysis. ISPD guidelines/recommendations. International Society for Peritoneal Dialysis. Perit Dial Int 1999;19:7–16.
17 Blake PG, Breborowicz A, Han DS, Joffe P, Korbet SM, Warady BA, International Society for Peritoneal Dialysis Standards and Education Subcommittee: Recommended peritoneal dialysis curriculum for nephrology trainees. The International Society for Peritoneal Dialysis (ISPD) Standards and Education Subcommittee. Perit Dial Int 2000;20:497–502.
18 Mujais S, Nolph K, Gokal R, Blake P, Burkart J, Coles G, Kawaguchi Y, Kawanishi H, Korbet S, Krediet R, Lindholm B, Oreopoulos D, Rippe B, Selgas R: Evaluation and management of ultrafiltration problems in peritoneal dialysis. International Society for Peritoneal Dialysis Ad Hoc Committee on Ultrafiltration Management in Peritoneal Dialysis. Perit Dial Int 2000;20:S5–21S.
19 Kawaguchi Y, Kawanishi H, Mujais S, Topley N, Oreopoulos DG: Encapsulating peritoneal sclerosis: definition, etiology, diagnosis, and treatment. International Society for Peritoneal Dialysis Ad Hoc Committee on Ultrafiltration Management in Peritoneal Dialysis. Perit Dial Int 2000;20:S43–S55.
20 Bernardini J, Price V, Figueiredo A: Peritoneal dialysis patient training, 2006. Perit Dial Int 2006;26:625–632.
21 Lo WK, Bargman J, Burkart J, Krediet RT, Pollock C, Kawanishi H, Blake P: The International Society for Peritoneal Dialysis (ISPD) guideline on targets for solute and fluid removal in adult patients on chronic peritoneal dialysis. Perit Dial Int 2006;26:520–522.
22 Prichard S: Why the North American chapter of the ISPD? Perit Dial Int 2005;25:12–13.
23 Mehrotra R, Burkart J: Education, research, peritoneal dialysis, and the North American chapter of the International Society for Peritoneal Dialysis. Perit Dial Int 2005;25:4–15.

Wai Kei Lo
Department of Medicine, Tung Wah Hospital
12 Po Yan Street
Hong Kong, SAR (China)
Fax +852 28587340, E-Mail wkloc@hkucc.hku.hk

Author Index

Subject Index